K

ENVIRONMENTAL PROBLEMS
IN EASTERN EUROPE

Environmental problems and transboundary pollution are now of international concern. Recent political changes in Eastern Europe have revealed how extensive has been the environmental damage of four decades of socialism. This legacy is one of the region's chief obstacles in the transition to a market economy.

Environmental Problems in Eastern Europe presents case studies of Albania, Bulgaria, Czechoslovakia, Hungary, Poland, Romania and (former) Yugoslavia. Each chapter describes and analyzes the major causes and consequences of air, water, soil and vegetation pollution. The book focuses in particular on the effects on the peoples of Eastern Europe and on the environmental quality of life.

F.W. Carter is Head of Department of Social Sciences at the School of Slavonic and East European Studies, London University.
D. Turnock is Reader in Geography at the University of Leicester.

ROUTLEDGE *NATURAL ENVIRONMENT: PROBLEMS AND MANAGEMENT SERIES*
Edited by Chris Park
Department of Geography, University of Lancaster

Offering a contemporary treatment environment topics, this series adopts an interdisciplinary international approach, and is an important source of information for the academic, the practitioner, and the student of environmental affairs.

ENVIRONMENTAL PROBLEMS IN EASTERN EUROPE

Edited by F. W. Carter and D. Turnock

London and New York

First published 1993
by Routledge
11 New Fetter Lane, London EC4P 4EE

Simultaneously published in the USA and Canada
by Routledge
a division of Routledge, Chapman and Hall, Inc.
29 West 35th Street, New York, NY 10001

© 1993, 1996 F. W. Carter and D. Turnock

Typeset in 10/12 pt September by
Leaper & Gard Ltd, Bristol
Printed and bound in Great Britain by
T.J. Press (Padstow) Ltd, Padstow, Cornwall

British Library Cataloguing in Publication Data

A catalogue reference for this title is available from the British Library

ISBN 0–415–13757–8

Library of Congress Cataloging in Publication Data

Environmental problems in Eastern Europe / edited by F.W. Carter and
David Turnock
p.   cm. – (Routledge natural environment–problems and
management series)
Includes bibliographical references and index.
1. Pollution–Europe, Eastern–History–20th century.
2. Transboundary pollution–Europe, Eastern–History–20th century.
3. Environmental health–Europe, Eastern–History–20th century.
4. Pollution–Europe, Eastern–Case studies.   I. Carter, Francis W.
II. Turnock, David.   III. Series.
TD186.E58   1993   92-35163
363.73'2'0947–dc20   CIP

CONTENTS

v

FIGURES

TABLES

CONTRIBUTORS

Francis W. Carter is Head of the Department of Social Sciences at the School of Slavonic and East European Studies, London University. He has written extensively on East European problems in both the contemporary and historical perspective, especially in the urban context. Present interest is focused on contemporary changes in the area, with particular reference to energy problems, environmental degradation and urban conservation.

Derek R. Hall is Head of the Geography Unit at Sunderland University. He has been interested for many years in the socialist states of both Europe and the Third World but has specialized on the geography of Albania since the mid-1970s. Despite travel difficulties he has visited the country on several occasions and published papers on all aspects of human geography, especially tourism and transport. He is currently studying the potential for tourism in various parts of Eastern Europe in association with international organizations such as the World Conservation Union.

Don Hinrichsen is an author and consultant (for environment, development and conservation issues) based in London. He is engaged by several UN agencies. He was previously Editor-in-Chief of the *World Resources Report* published in Washington; and editor of *AMBIO* (the international journal of the human environment) published by the Royal Swedish Academy of Sciences in Stockholm. He has also served as foreign correspondent for radio, newspapers and magazines in Athens, Berlin and Copenhagen; and has written a number of books: most recently *Our Common Seas: Coasts in Crisis* and *WWF Atlas of the Environment*.

Barbara Jancar-Webster is Professor of Political Science at the State University of New York at Brockport. She is the author of *Environmental Management in the Soviet Union and Yugoslavia* which received the International Studies Association's 'Sprout Award' in 1990 for the best book in the field of comparative/international studies published in that year. She has studied and travelled widely in east-central Europe and the former Soviet Union and has published over fifty articles on environmental problems, women's issues and other matters relating to the former Communist countries. In 1992 she was SUNY–Moscow University exchange scholar in the Department of Environmental Economics of the MGU's Faculty of Economics.

Istvan Láng is Secretary-General of the Hungarian Academy of Sciences based in Budapest, having previously been Deputy Secretary-General of the Academy's Biological Section and a scientific researcher at the Hungarian Institute for Soil Science and Agricultural Chemistry. He has contributed to international organizations as a member of the World Commission on Environment and Development (Brundtland Commission); at present he is a member of the Executive Board of the International Council of Scientific Unions and of the Editorial Advisory Board of the World Resources Institute in Washington.

David Turnock is Reader in Geography at Leicester University. He has been researching on the geography of Eastern Europe since the 1960s and has published widely on a range of economic and historical themes. His particular interest is Romania where he is currently engaged in an ESRC-funded research project dealing with the consequences of the transition to a market economy in the Carpathian regions of the country.

PREFACE

This book was planned during the 1988 Anglo-Bulgarian Seminar which was held at the University of London's School of Slavonic and East European Studies. We felt that environmental problems in Eastern Europe were gaining increasing recognition both inside and outside the region. Therefore it seemed an opportune moment to undertake a review of the situation in each country and to offer a comprehensive view to complement the *ad hoc* surveys already available. We were pleased when Routledge took up the idea with their series on 'Environmental Studies' in mind, and wish to record our thanks to Tristan Palmer of Routledge and the series editor Chris Park (Lancaster University) for their support and encouragement. The book was planned at a time when few commentators anticipated the early demise of state socialism and so we are pleased to find that we can cater for the great increase in interest in East European problems in general since the revolutions of 1989. But we must stress that the research does not take detailed account of current trends, being primarily a 'base line' study of the environmental situation which is just one of a complex of issues complicating the transition to a market economy in each country of Eastern Europe.

We are grateful to our four contributors who have been free to present their assessments in their own way so as to take full advantage of their experience and expertise. The differences in structure and illustration reflect the lack of any rigid plan – an approach which we considered appropriate in view of the differences in source material between the various countries. We regret that there is no detailed coverage of the former German Democratic Republic: although the unification of Germany has in a sense detached this component from the old bloc of socialist countries it was our intention that coverage should be given and it is unfortunate that material could not be secured in time to meet the requirements of the publication schedule. We also regret that the Yugoslavia chapter was prepared before the process of disintegration in that country had escalated to the point of secession by several constituent republics, therefore we are unable to recognize the independence of Croatia and Slovenia through provision of chapters on these two countries aside from the discussions dealing with the former Yugoslav state as a whole. Similar sentiments must be issued with regard to Bosnia-Hercegovina and Macedonia whose quest for recognition by the world community was gaining momentum when these notes were written.

Francis W. Carter and David Turnock
March 1992

PREFACE TO THE 1996 EDITION

As editors, we are very pleased that the book has been sufficiently well received for a paperback edition to be justified. Unfortunately, the tight production schedule has made it impossible to revise the chapters which first appeared in the 1993 edition. We have been able to add two new chapters: one to review the general situation which has unfolded during the transition years and another to present a case study of environmental action, both indicating the range of approaches that are now being adopted. This compromise, however, is particularly unfortunate in that it again fails to recognize events in the Former Yugoslavia where the collapse of the old federation was evident even before the first edition appeared. Also it fails to acknowledge properly the 'Velvet Divorce' and breakup of Czechoslovakia which occurred in January 1993. All we can do is hope that continued interest in the book will be sufficient to allow a more comprehensive revision to be made in due course, so that the present political realities can be fully taken into account.

Francis W. Carter and David Turnock
August 1996

1

INTRODUCTION

F. W. Carter and D. Turnock

This book was planned in 1988 at a time of increasing public awareness about the serious environmental problems of Eastern Europe. More recently, European environment ministers have urged governments of the former Eastern bloc countries to establish a critical timetable for new standards to bring local principles into line with those in the rest of Europe. As we progress towards a more unified continent, there is a need for a comprehensive overview of the growing environmental menace in Europe resulting from the requirement of air pollution control, water and waste-water treatment, reclamation of contaminated land and hazardous waste treatment and disposal. In east-central Europe much of this situation can be laid at the feet of four decades of central planning by Communist parties, which failed to adhere to declared priorities for sound environmental management; instead, there was serious ecological damage of the sort previously attributed only to Western capitalist regimes.

Admittedly, some pollution black spots already existed in this area prior to the Second World War but they were mainly confined to the limits of the Budapest–Łódź–Leipzig industrial triangle. The early years of post-war development saw the initiation of projects under socialism which were relatively harmless environmentally; however, there was a powerful urge to rapidly develop their economies and eliminate areas of material underdevelopment. There ensued a proliferation of large-scale programmes of mining and heavy manufacturing which increased the risks of environmental damage by alarming proportions. Adoption of the Stalinist model of economic development led to emphasis on heavy industry, including cement, chemicals and metallurgy, along with the drive to exploit all available energy resources through domestic thermal power-stations, which all helped to take their toll on the environment. Moreover, the Marxist concept of regarding natural resources as 'free goods' has only added to the problem by encouraging waste.

Over several decades naturalists have issued warnings about the impact of high-level environmental damage, while doctors have increasingly stressed the dangers of polluted air and water on the health of the population. Economists similarly appreciate the heavy costs afflicted by pollution in agriculture, forestry and tourism, whilst sociologists are conscious of the spin-off on the quality of life

1

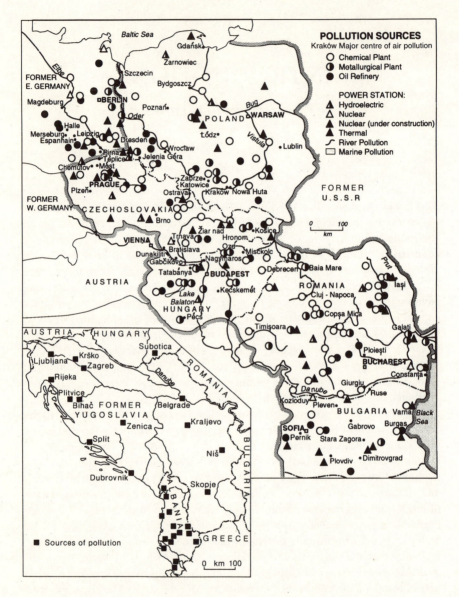

Figure 1.1 Eastern Europe: major pollution centres

Source: Based on J. Thompson (1991) 'East Europe's dark dawn', *National Geographic* 179 (6), p. 44.

created by a blighted environment. Unfortunately, these socialist governments saw the financial expenditure on reducing pollution levels as unacceptably too high and replaced them with less effective palliative measures. In fact, these governments tended to protect themselves from the realities of environmental degradation by giving a low priority to pollution research and monitoring. This has resulted in a lack of reliable data which is so necessary for the preparation of a realistic strategy – now so urgent in these countries – and has led to a confusion about contemporary strategies and aims, and lack of accord both at domestic and international policy levels, to effectively deal with the regional degradation of the environment. In fact, in the past, publication of pollution material was deliberately deterred, whilst factory employees were dissuaded from commenting on the weaknesses of certain industrial processes.

The combination of large natural deposits of sulphurous brown coal/lignite, a wasteful and unyielding ideological and economic system, and the adherence to an outmoded and inadequate heavy industrial infrastructure, all contributed to an easily identifiable problem of environmental degradation sometimes reaching severe proportions. Fortunately, during the 1970s and 1980s there was increasing awareness placed on pollution control, although the accompanying legislative support was often found wanting; nevertheless, this experience has been useful in demonstrating what was needed in the immediate future if problems were to be solved. It is now realized that more research is needed if pollution forecasts are to improve. It is absolutely imperative that government legislation lays down clear standards which can be applied over a wide range of conditions, and these laws must be implemented by state courts.

There are indications that these lessons have already been learnt, because even before the revolutionary upheavals of 1989 new legislation was being introduced in many east-central European countries. Even so, there are still fears that Western companies may try to take advantage of the more relaxed legislation in east-central Europe, encouraging the deliberate selling and use of technologies which are at present prohibited in the exporting country. There is obviously a need for legislation to protect these former Communist countries from unclean Western industries migrating eastward. However, without the liberalization of trade it would be difficult for countries to exert pressure on the monopoly companies which in the past were mainly responsible for high pollution emission rates, particularly in the so-called 'black triangle' covering Czechoslovakia, Poland, and the former East Germany.

The scale of transboundary pollution has also come under closer scrutiny. East-central Europe has not only signed trade agreements with Western companies to dispose of waste material, it has also been obliged to import air and water pollutants from Western Europe as well as those generated by immediate neighbours, as was seen in the case of Poland with transboundary deposits from the eastern parts of the former East Germany adding to its own pollution problems. It is hoped that the European Commission's important changes in attitude to environmental research through the Environment Research

3

Programme initiated in June 1991 by the Council of Ministers, will open all parts of the programme to the countries of east-central Europe, whose environmental problems are of concern to the whole continent. By means of a clause in the council's decision, these countries may be included in specific projects, as appropriate, without the need for separate agreements. Special arrangements will also be made to finance them.

Environmental damage can be seen in all parts of east-central Europe, but commentators are agreed that there are quite massive variations in the levels of air, water and soil/vegetation pollution which constitute the heart of the problem. It should therefore be a straightforward task for the geographer to highlight the uneven distribution. However, it is extremely difficult to produce a set of maps showing the extent of the variations according to a range of criteria. As this book will show, much published research deals with local problems and a simple collation of this material cannot provide a satisfactory overall view. Official statistical yearbooks vary a great deal in the attention given to pollution and while some countries like Hungary and Poland are relatively forthcoming, other states such as Albania and Romania have virtually nothing to say on the subject.

However, the emission of atmospheric pollutants correlates closely with thermal power stations and units of heavy industry, where sulphur dioxide is a key criterion, although all large cities are significant contributors, along with nitrogenous oxides, because of the prominence of motor traffic and domestic heating systems. Atmospheric pollution largely arises from the types of energy resources which have been available and the ways they have been utilized. For example, in the former East Germany and in Czechoslovakia, these states have been bequeathed by nature with an abundant supply of brown coal/lignite. As a result of the Communist states' propulsion towards self-sufficiency, together with the relative dearth of alternative resources of energy, one should not be amazed at the obvious heavy exploitation they have received, particularly as the basis for all kinds of industrial development.

Typical of such a situation is the former state of East Germany. Whilst the unification of Germany has made a special chapter on this country less appropriate, its impact should not be ignored. The centre of the problem in that country lay with the use of lignite, a soft, brown, low-quality coal found and used there in abundance. It provided over two-thirds of the state's domestic energy needs and was used largely as a raw material for the chemical industry and as a fuel in thermal electric power-stations. By 1978, hard coal reserves were virtually worked out, resulting in greater use of the only other fuel in profusion, namely lignite, with its large burn-off rate, mostly in the form of acrid fly ash and noxious sulphur dioxide.

The inevitable result was an ever-present cloud of dust and soot which enveloped the country (see Figure 1.1). For example, Leipzig was covered in 400,000 tonnes of SO_2 annually, life expectancy was six years less than the national average, and four-fifths of the children aged below 7 years developed chronic bronchitis or heart problems. The dubious title of 'most polluted city in Europe' was claimed by Bitterfeld, with its coal mines, thermal power-station and chemical plant. The River Mulde directly received its untreated waste (e.g. dioxin), while the nearby River Elbe had mercury levels 250 times above the EC limits. Clearly, such problems were attributable to a lack of sophisticated technology and hard currency – factors which may now change as a result of unification.

The production and use of energy in east-central Europe has had detrimental effects not only on air pollution but also either directly or indirectly upon surface water quality. In the case of pollutants discharged into rivers, the main offenders are again large industrial complexes, but another adversary in this complex picture may be alluded to in the form of untreated municipal sewage discharge from the large urban settlements. Sewage treatment is in an abysmal state in most east-central European countries, and has proved one of the surprises to emerge from this review of environmental problems in the region.

On the surface, many rural areas may be seen as relatively pure and peaceful: peasant communities still follow traditional farming practices and maintain a colourful range of distinctive cultural activities. It is certainly true that the worst environmental problems evident in the countryside arise from polluted air drifting in from the conurbations and causing extensive damage to crops and woodlands. From the results of a United Nations Forest Damage Survey in Europe in 1989 it appears that in some regions, especially east-central Europe, damage has increased. In forests at higher elevations, and in those more than sixty years old, the woodlands have suffered much heavier defoliation than in younger stands and at lower elevations. This is particularly so in the mountainous regions of former East Germany, Czechoslovakia and Poland, where trees over several thousands of hectares are either dead or seriously damaged.

Soil degradation may be found throughout east-central Europe; widespread damage to the land has resulted from the so-called 'chemical war' which the farmers have been waging against the environment through the application of excessive fertilizer dressings – the latest in a series of dislocations brought about by agricultural expansion. Soil pollution has resulted not only from pesticide poisoning but also from heavy metals and acid deposition. Opencast mining has led to the destruction of topsoil and arable land, where on occasions the considerable amount of mining and quarrying located in rural areas has been supplemented by local thermal power-stations dependent on lignite working often located well distant from urban centres. The countryside around such places led to a difficult milieu for inhabitation because of the substantial fall-out of fly ash from thermal power-stations and heavy traffic levels on some country roads, as well as the disfigurement of the landscape through quarries, waste tips and processing units. The dumping of municipal and industrial waste adds to this problem along with wind and water erosion as a straight result of agricultural mismanagement or malpractice. Finally, in some cases, heavy tourist pressure and the inadequate planning of beauty spot management can result in damage to farmland. Within the villages there may also be cases of visual pollution through the intrusiveness of modern infrastructure such as electricity cables and telephone wires thoughtlessly located, together with political propaganda. Moreover, the imposition of modern architectural styles sometimes debases respected local traditional building forms to the detriment of the whole neighbourhood. So the rural areas are by no means immune from environmental damage in its various guises.

However, it has always been difficult to go beyond these basic generalizations and, to date, no book has reviewed the situation in detail in every country of the region. The present work seeks to remedy this deficiency. Since generalization is so difficult the editors decided that it would be desirable for experts on the various countries of east-central Europe to provide a detailed profile in each case. It was also recognized that each country had distinct characteristics with regard to its resources, its record of environmental protection, its published data on environmental problems, and in the scope for public opinion to make itself felt. Therefore it seemed appropriate to avoid imposing a rigid framework and instead to invite each contributor to present each national profile in the most effective way, subject only to general advice on the need to refer to all significant aspects of pollution, to assess the literary sources, and to examine relevant legislation, public pressure and current political trends.

The chapters have reached back into the pre-socialist period although the prime focus rests with the post-Second World War era of central planning and totalitarian government. This approach has been a realistic one under the circumstances because although the unforeseen revolutions of 1989 have provided a stimulus for the expression of public opinion throughout east-central Europe, there are still wide variations with regard to other criteria and generalization remains difficult. Even so, a final chapter provides an overall view and discusses a number of broad themes. These may be regarded not only as a package of conclusions but also as a stocktaking exercise, undertaken in the light of the national surveys, which may provide a base for further work in the near future. For without doubt the east-central European pollution scenario will continue to run its course in the years ahead and greater comparability – across both Eastern and Western Europe – should arise from the growth of international co-operation. Environmental problems are an element of the international ambience of states and nations, implying deeper and ever-increasing interdependency at all levels of co-operation; surely the possible harmonization of Europe post-1992 will clearly provide a stimulus for investment in pollution control as member states aspire to conform to future environmental laws and their stipulations. Looking even further ahead, the needs of environmental control may have to adapt to newly created and as yet unknown environmental stimuli, but at least the basis will have been laid for what is to come in the twenty-first century.

2

ALBANIA

D.R. Hall

INTRODUCTION

Preamble

As the final draft of this chapter was being rounded off in December 1991, the plight of Albania appeared to be at an all-time low. Agricultural production had fallen by 55 per cent in the first half of the year, with food distribution systems virtually at a standstill. Industrial production had been reduced by 33 per cent, exports by 80 per cent (Carrier 1991), and thousands of Albanians were desperate to flee their country.

History of low level of development

Albania's level of economic development, the lowest in Europe, had been exacerbated in recent years by the dogmatic manner in which autarky had been pursued by the country's Stalinist leadership. Article 28 of the 1976 Constitution, drawn up as the rift with China was widening, stipulated that no credits or loans could be taken from foreign governments or organizations.

> The granting of concessions to, and the creation of, foreign economic and financial companies and other institutions or ones formed jointly with bourgeois and revisionist capitalist monopolies and states, as well as obtaining credits from them, are prohibited ...
>
> (PSR of Albania 1977: 17)

Foreign aid, joint ventures and other foreign interests in the Albanian economy were therefore summarily precluded, with the result that much Albanian trade was based either on barter or through the maintenance of an import–export equilibrium.

The Albanian economy had stagnated in the 1980s (Sandström and Sjöberg 1991), and by the time processes of economic and political change were gathering momentum in the early 1990s, the Albanian population of 3.3 million represented a very weak consumer society: but a low consumption level coupled with

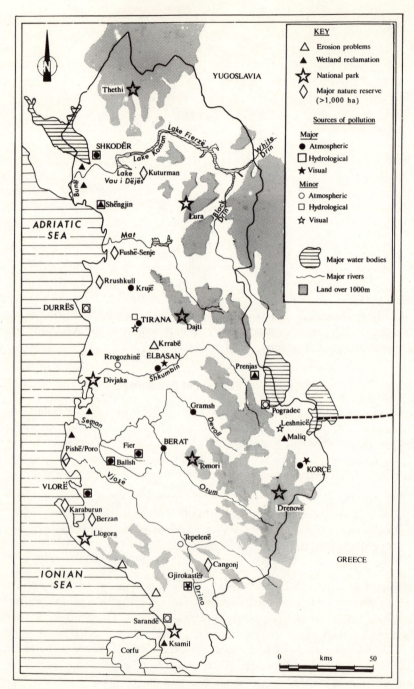

Figure 2.1 Pollution in Albania

an emphasis upon import substitution and recycling resulted in relatively little material waste. The individual ownership of motor vehicles, for example, remained forbidden from the 1950s until 1991, by which time they still numbered less than 10,000 (Atkinson *et al.* 1991). Idiographic examples of substitution within an economy of shortage include the use of recycled strips of pneumatic tyre as bindings in hot houses, potentially lethal multiple sections of glass to replace broken windows in buses, and the use of rubber tyres for fuel in small bitumen plants.

Article 28 was formally abolished in July 1990, and a new draft constitution, to bring Albania politically and economically more in line with the rest of Europe, was published in the daily *Zeri i Popullit* in April 1991 for debate in the newly pluralized Albanian parliament. By the end of the year, however, a formal document had yet to be legalized (EIU 1991a).

Resource base

Although only the size of Wales or the state of Maryland, Albania is notable for the beauty and variety of its natural landscapes. Varied relief and climatic conditions, significant water and mineral reserves, a diversity of fauna and flora, and often stunning Ionian and Adriatic coasts characterize one of the hitherto lesser-known and least exploited corners of Europe.

The country is particularly rich in energy sources and minerals. While oil and natural gas, bitumen, some brown coal and hydro-electricity resources are found in the south, most minerals and water-power potential are located in the northern half of the country. Extracted metals include manganese, chrome, bauxite, copper and nickel, and non-metallic minerals are phosphorus, calcium, clay, gypsum and quartz. In the late 1980s annual output amounted to just over two million tonnes of coal and just over one million tonnes each of chromite, ferrous-nickel ore and copper ore (Bërxholi 1990b). The northern rivers have been harnessed to an increasing extent in recent years, with hydro-power providing some 80 per cent of the country's electricity needs; the 285-km-long River Drin provides 50 per cent of the total (see Figure 2.1). Brown coal is burnt in a number of power-stations, with deleterious atmospheric and health consequences, as, for example, at Korçë. However, consideration of long-term energy provision has stimulated studies into the potential for low-cost solar, wave, wind, geothermal and photosynthetic means of energy generation (Karadimov 1989).

Although the country's post-war birth-rate has declined from over 35 per thousand in the mid-1960s to a 1988 figure of 25.3, there has remained a very explicit pro-natalist policy, with access to contraceptive methods being exceptional until very recently. As a result of a virtual halving of the death-rate since 1960, the country's natural increase of 2 per cent per annum (2.01 per cent average 1979–89) – about five times the European average – has been placing severe pressures on the country's ability to feed itself. At 111 inhabitants per square kilometre in 1989 (compared to just 28 in 1923), Albania has the highest population density in the Balkans (Vejsiu 1990). The low average age of population (27 years), with 43 per cent of the total under 20 years old, has seen a large body of young people coming to the forefront of demands for change within the country.

Stalinist policies

> Rapid industrialization, in order to liquidate the technical and economic backwardness, was considered as the main link for successful accomplishment of the major tasks of socialist construction ... it was decided that by the end of 1955 the volume of industrial production would be more than 3 times greater than in 1950 and 12 times greater than before the war.
>
> (Omari and Pollo 1988: 133–4)

The combined emphasis upon Stalinist heavy industrialization, with belching smokestacks viewed as symbols of socialist modernization to be extolled and emulated (for example, see Hall 1987a: 52–4), and dogmatic notions of 'self-reliance' (Gurashi and Ziri 1982), reached its zenith in the late-1970s with the construction of the Elbasan metallurgical complex, the country's largest industrial plant. Begun by the Chinese, this project was abandoned by them when Sino–Albanian political relations were broken off in 1978. Apparently denied access to the original plans and to pollution control technology, the Albanians subsequently laboured over a number of years to complete the project. The plant directly employes a 10,000-strong work-force, and has some 300 other enterprises dependent upon it. Including the substantial atmospheric pollution damage extending across the valley in which Elbasan lies, it has been estimated that the plant costs the Albanian exchequer the equivalent of $40 million a year (Atkinson *et al.* 1991): the health effects on its work-force can only be guessed at. Renovation to bring it up to an acceptable standard would cost $100 million (BBC 1991d). The future fate of the Elbasan plant will act to symbolize changing attitudes to the Stalinist past and to its environmental legacy.

Official policies for protection/conservation

Early examples of Albanian environmental legislation include those concerning control of hunting (November 1951) and forestry protection (October 1963) (IUCN 1967, IUCN-EEP 1990, Borisov *et al.* 1985). National parks were created under the auspices of these laws. Until the early 1970s, however, most ecological effort was concentrated on the protection of rare animals and plants.

Today, protected areas are classified into three categories (Atkinson, *et al.* 1991, Bogliani 1987, IUCN-EEP 1990):

1 National parks have the dual role of providing areas for public access, recre-

ation and education, and of protecting the landscape. Within them, a total protection regime is imposed, with no permanent human occupation or exploitation. Hunting and ancient grazing rights are prohibited, although villagers from neighbouring settlements are permitted to gather dead wood. Tourism is encouraged, with limited motor access, and each park has a forest lodge for accommodation. There are currently six internationally recognized national parks (Dajti, Divjaka, Llogora, Lura, Thethi, Tomori (see Figure 2.1), extending in size up to 4,500 hectares (see Table 2.1). In addition to those parks listed in Table 2.1, Albanian sources also refer to the 1,350 ha Bosdovec national park at Drenovë (Bërxholi and Qiriazi 1986: 38, Hoda 1991), noted for its black pine and beech forests.

2 'Integral' reserves are areas strictly protected for nature conservation: public entry is prohibited. A total protection regime is again imposed.

3 'Orientated' reserves allow traditional human activities to be pursued, and have the dual role of providing areas for education and of protection of the landscape and wildlife. Some human activities such as fishing are permitted.

Additionally, some 12,200 ha are designated as game reserves (Cobani 1989). Protected animals include the brown bear, lynx and chamois, while bird colonies on the islands of Kuna (in Lezhë district) and Ada (Shkodër district), as well as the pelican colony in Divjaka, are also protected. Overall, some 350 types of bird are found within the country. A number of eliminated species have been re-introduced: for example, 15,000–20,000 pheasants are now bred annually in an 8-ha breeding station near Lezhë (Petrela 1990). Artificial winter feeding is organized for roe deer, chamois and wild boar (Bërxholi and Qiriazi 1986: 39–41). However, frogs are exported in large numbers to France for culinary purposes.

Following the upheaval brought about by three changes of government during 1991, the administrative framework for enforcing and overseeing environmental protection has been in flux. A Central Environmental Protection Commission, established in 1979, operated from within the Council of Ministers, and included representatives of ministries and research institutes. Its major role was to monitor the work undertaken by ministries, local authorities, institutions and farms. Later reorganization saw the establishment of a Ministry of Public Economy and Protection of the Environment, which, in May 1991, was actually demoted, for reasons of streamlining governmental structure, to committee status, although its chairman was to be of ministerial level (BBC 1991g). A watchdog role has also been supposedly undertaken by the Ministry of Health's Department of Hygiene and Environmental Protection, particularly in relation to the workplace environment. A number of ministries have had their own sections for environmental protection, although a major responsibility for nature conservation has fallen within the purview of the Ministry of Forest and Water Economy. Hitherto, each administrative district's executive committee had one person responsible for environmental matters, although rarely were they trained for the purpose.

Table 2.1 Protected areas in Albania

Name	Conservation management category*	Size (ha)†	Year designated	Characteristics
1 National parks				
Dajti	II	–/4,000	1966	Beech and Bosnian pine, mixed woodland flora with Mediterranean maquis
Divjaka	II	1,194/4,000	1966	Aleppo and Stone pine forest, salt marsh and dune flora on the Adriatic coast
Llogora	II	1,040/3,500	1966	Black pine forests
Lura	II	1,300/4,000	1966	Bosnian, Macedonian, Scots and black pine, beech and spruce; alpine meadows
Thethi	II	2,700/4,500	1966	Pine forest
Tomori	II	–/3,000	1956	Beech forests and Bosnian pine; alpine meadows
2 Major nature reserves (1,000 ha or larger)				
Berzan	IV	1,000		
Cangonj	IV	3,000		
Fushë-Senje	IV	4,200		
Karaburun	IV	12,000		
Kuturman	IV	4,000		
Pishë-Poro	IV	5,500		
Rrushkull	IV	1,800		
3 Specially protected areas				
Divjaka	SPA	1,000	1956	Pelican reserve within the coastal national park

Additionally, there are a further eleven category IV protected areas of less than 1,000 ha and six category VIII areas.

Sources: Atkinson *et al.* 1991: 11–14; Basha 1986: 22; Bërxholi and Qiriazi 1986: 37–8, 1990; Çobani 1989: 49; IUCN 1990: 23; IUCN-EEP 1990; Hoda 1991; PADU 1990; Polunin 1980: 108–11; author's field observations, 1991.
Notes: *For definitions of the conservation management categories see IUCN 1990: 9–14.
†Sizes are taken from Bërxholi and Qiriazi ('national parks') and IUCN/IUCN-EEP respectively. Çobani refers to 'about 8,400 ha of national parks', Bërxholi and Qiriazi cite 18,000 ha.

Several scientific institutions have directly and indirectly contributed to environmental protection work:

1   the Academy of Sciences, established in 1972, derived from the Committee for Science and other university institutes located in Tirana, comprising three sections: Social Sciences (humanities), Natural Sciences, and Technical Sciences. The Institute of Hydrometeorology, Seismological Centre, Centre for Geographical Study, and Laboratory for Hydrological Research, comprise the latter. A national system of 220 meteorological and 210 hydro-logical stations was established in 1949, and membership of the World Meteorological Organization has been held since 1975. A system of seismo-logical stations, now numbering thirteen, was inaugurated in 1968, and contributes to UNESCO Balkan projects. Zoological sciences have been weakly represented as the structure of the Academy has tended to favour the humanities, although the Centre of Biological Research within the Natural Sciences section has played an important role. In 1978 the Academy held a conference on environmental protection, in 1980 a symposium on the 1979 earthquake, and in 1985 a colloquium on geographical studies (Lloshi 1990: 14–15);

2   the Natural Sciences Faculty of the State University, established in 1957 (Anon 1985b);

3   the Natural History Museum, which belongs to the University;

4   the 14-ha Botanical Gardens, which are jointly supported by the University and Tirana city authorities;

5   the Institute of Scientific Research in Hygiene and Epidemiology in Tirana, whose work includes analysis of the physical and chemical properties of drinking water, protection from environmental pollution, the study of noise and vibration, dietary nutrition and occupational diseases (Cikuli 1984: 77).

During 1991 higher education was reorganized: the Luigj Gurakuqi University in Shkodër (a centre of pre-war intellectual life) was reopened, as was the Tirana Agricultural University. The former 'Enver Hoxha' State University in Tirana was divided into two institutions: Tirana University and the Polytechnic University. In all, there are now nine higher education institutions (Zanga 1991b).

Although still an essentially rural society (Sjöberg 1991), with some 65 per cent of the population living in the countryside, Albania's erstwhile penchant for scarring the landscape with hastily constructed polluting factories and poorly finished apartment blocks has been only marginally offset by positive attitudes towards urban conservation and protection. Many religious buildings were destroyed or rebuilt for non-religious functions during and after the 'cultural revolution' of 1967, and many old town centres have been destroyed through comprehensive redevelopment based on structure plans of the 1970s (Hall 1987a: 51). Selective urban conservation, however, has been pursued through the designation of protective zones in much of the old residential districts of

Berat and Gjirokastër ('museum cities') (Anon 1973a; Riza 1971, 1978, 1981; Strazimiri 1987), and in the historic parts of Durrës and Krujë (Karaiskaj and Baçe 1975, Riza 1975, Toci 1971), while restoration has been carried out on an individual site or street basis in a number of other urban centres such as Korçë, Shkodër, Tirana, Vlorë and Elbasan (Anon 1973b, 1977; Bërxholi 1985: 27; Miho 1987; Muka 1978; Thomo 1988). Major archaeological sites, notably at Apollonia and Butrint, have also been under the protection of the state (Adhami 1988, Baçe and Condi 1987, Zarshati 1982), although this did not prevent the theft of 15 third century BC sculptures from the former site during November 1991 (McDowall 1991d).

Urban green and open spaces have also been emphasized in recent development and redevelopment programmes. The southern city of Vlorë, for example, had over 50 ha of such space by the mid-1980s, with an aim to provide 10 square metres of greenery for each of the city's 65,000 inhabitants by 1990 (Bërxholi 1985: 41). Such green spaces provide important lungs for humans and wildlife alike. In the open spaces of central Tirana, for example, nightingales and scops owls can be regularly heard.

International co-operation/agreements

Until the death of the country's post-war Stalinist leader Enver Hoxha in April 1985, the country tended not to participate in any international gatherings or treaties, arguing that each was dominated by at least one of the superpowers. However, participation by viral disease experts at an international congress in Cairo during April 1985, soon indicated a change of direction and a new official desire to participate in scientific gatherings (Anon 1985a). Indeed, subsequently Tirana has acted as host to a number of international conventions, such as the 1987 inaugural meeting of the Balkan seismology working group, and in May 1990 the first gathering of Balkan energy ministers, which agreed to establish working groups for environmental protection.

Most notably for environmental matters, Albania was a party to the February 1988 Belgrade foreign ministers' communiqué on tourism, transport and protection of the environment, and there was significant Albanian participation at the Balkan Scientific Conference on environmental protection in the area, held in Varna, Bulgaria in September of that year (Çobani 1988, Dingu 1988, Dollani 1988, Frashëri 1988, Gjiknuri 1988, Hoxha 1988, Mara 1988, Puka 1988, Tartari 1988). In July 1989 the country signed the World Heritage Convention, although no Albanian sites were listed in a subsequent United Nations inventory (IUCN 1990: 227). This was followed, in 1990, by Albania becoming the last Mediterranean country to be a party to the UNEP Regional Seas Programme (Barcelona Convention), proposing the inclusion of six coastal sites under the convention. However, no Albanian biosphere reserves have been notified under the UNESCO Man and the Biosphere (MAB) Programme, nor is the country yet a signatory to the 1971 Convention on Wetlands of International Importance

(Ramsar Convention). The last Communist government was said to have been favourably disposed towards preserving the country's remaining wetlands (Atkinson *et al.* 1991), although concern had been expressed over three projects for soil improvement and land reclamation which overlapped or adjoined coastal wetlands considered to be of international importance (IUCN-EEP 1990: 7, Kusse and Winkels 1990). In October 1991 Albania gained access to the World Bank and the IMF, and in the following month an environment and energy mission of the European Bank for Reconstruction and Development (EBRD) visited the country following its admission to that body (Dyvik 1991, Murphy 1991).

Changing attitudes towards the environment

Article 20 of the 1976(–1990) Constitution pointed to environmental protection as being a duty of the state, all institutions and citizens. Yet in practice, these statements carried little force given the state's 'Stalinist' approach to heavy industry. However, programmes against waste, continuing low levels of consumerism, and a policy of recycling materials within the context of an economy of shortage had also long characterized the Albanian approach, if not, primarily, for environmental reasons.

A broader dimension to Albanian environmental concern nevertheless emerged after the 1989 upheavals in Eastern Europe (Hall 1990b, 1990c, 1990e), marking the outward sign of a further official change of approach. Some pronouncements, reflecting divisions within the ruling elite and unsupported by legislative process, began to express criticism of previous practice. For example, early in 1990 the Minister of Health pointed to 'impetuous' development processes as the cause of substantial deforestation and atmospheric pollution. Later, in May, in one of the then most explicit official comments on the country's environmental situation, *Zeri i Popullit* carried an article by the chairman of the Foreign Economic Co-operation Committee, arguing that the coal industry, mining 2.4 m tonnes annually, was both uneconomic and environmentally harmful, and that it was unacceptable and irresponsible for the country's leadership to permit the industry to continue in its present manner. This was just the tip of a large iceberg. Ironically, however, just as in agriculture, the poor production level of the mining industry had recently stimulated increased state subsidies – whereby prices paid would be a function of production costs – which, while attempting to boost production, could only be a recipe for further state-sanctioned environmental degradation as more difficult and sometimes fragile environments were more intensively worked.

Following the participation of some 30,000 athletes in the May 1990 Tirana marathon, the following month saw a rather hollow mass environmental protection campaign instigated under the title of 'Week of the Mediterranean Environment', with lectures held in the capital on anti-pollution measures. People in the rest of the country were mobilized for 'cleaning and restoration' activities (BBC 1990). However, the hastening of political and economic change from the end of 1990 (AMC 1991), and the drafting of a new constitution provided the impetus for a much wider public debate of environmental concerns than was possible hitherto, although this was overtaken by the near-anarchic condition of the country during the second half of 1991 (see, for example, McDowall 1991a, 1991b, 1991c).

Following the legalization of opposition political parties in December 1990, an Albanian Ecology Party was established, with Namik Hoti as its chairman. In the run-up to the country's first post-war multi-party elections, held on 31 March 1991, its three manifesto pledges were environmental improvement, national unity, and peace. The party failed to win any seats in the new parliament, and its specific environmental policies remained unclear at the time of writing. Unlike the major opposition Democratic Party, however, which appeared to alienate the collectivized rural peasantry by declaring a wish to privatize the land, the Ecology Party opposed rural land sales. In the political arena, the Ecology Party was later, in 1991, joined by the Green Party, while a Society for the Preservation and Protection of the Environment was established largely by members of the scientific establishment (Murphy 1991).

THE PRE-SOCIALIST PERIOD

While Albania was far from being the totally undeveloped malarial swamp that a number of post-war histories would have had the world believe, environmental pressures from economic exploitation and development had been relatively low, although reclamation of some of the mosquito-infested lowland wetlands had begun.

Industry was restricted in scale, scope and location. With no large industrial establishments, processing and manufacturing was little more than a cottage industry. Eighty-eight per cent of all industrial enterprises employed less than fifteen workers, and by 1938 only 4.5 per cent of national income was derived from industry (Prifti 1978: 52). It was mostly limited to food processing, with nine diesel-powered modern flour mills, two olive oil refineries (Elbasan, Vlorë), a fish-preserving plant at Shkodër, a brewery (at Korçë, built by the Italians) and four raki distilleries. Tobacco processing was also undertaken and cigarette manufacture took place at Durrës, Shkodër, Elbasan, Berat and Vlorë, each of which also contained a soap factory. Building materials were represented by brickworks at Elbasan, Korçë and Tirana, and cement plants at Shkodër and Elbasan, the latter begun in 1940. In all cases, manufacturing was limited to domestic requirements and was concentrated in a handful of towns. Power was usually provided by internal-combustion diesel engines, except for a small hydroelectric plant serving Korçë. Two Italian companies provided electric lighting for Tirana and Durrës.

Oil was first raised in 1918, and by the end of the 1930s exploration and development of some 400 wells was largely under Italian auspices. The crude was piped to the coast near Vlorë for direct export to Italy. The capacity of the pipe-line was said to be 5,000 barrels a day in 1937, and special wharves were constructed to handle two million barrels annually. The first shipment took place in 1935. Just as there was no oil refining capacity, so too Albania possessed no facilities for the processing of the country's copper and chromium resources. Lignite was worked to a limited extent for domestic fuel and for brick-making and lime-burning.

Tourism was limited (Hall 1991): at Durrës, latterly the focus of both domestic and international tourism, there were just fifteen cottages with 73 rooms and one hotel with 28 rooms in 1945 (Bërxholi 1986: 27).

A public health service was first created in 1922, although it was hampered by a lack of funds and personnel. At that time tuberculosis was very common, with a particularly high incidence of adolescent pulmonary tuberculosis among the Islamic female population. High mortality levels in the Gjirokastër, Shkodër and Krujë districts were notable. Syphilis was widespread in rural areas. Intestinal parasites were common, and undernourishment was everywhere apparent. Cancer and cardiovascular diseases were reported to be rare. Anthrax was the major epizootic affecting humans, being widespread in the coastal plain. Most notably, malaria was endemic in all areas below 1,200 metres, and hyperendemic in parts of the coastal plain and in the river valleys. The country's census of May 1930 enumerated a total population of just over one million (1,003,000). In December 1933 the Director-General of Health reported 250,000 cases of malaria (McKinley 1935) – one quarter of the total population. Under the auspices of the Health Committee of the League of Nations, surveys were under-taken in 1923 and 1924 to prepare plans for the eradication or control of malaria (Haigh 1925). While the inaccessibility of many of the country's villages hampered the spread of infectious diseases, this situation was gradually being broken down during the inter-war period, with increasing seasonal and more regular migrations and movement between upland and lowland regions. Methods of recruitment for the army also contributed to this process (Mason *et al.* 1945: 119–22). Average expectation of life at birth in 1938 was just 38.3 years (KPS 1989: 37).

Some draining of marshlands was actually undertaken by the Rockefeller Institute, and this was said to have reduced the incidence of malaria. After their invasion in 1939, the Italians had grandiose plans for reclamation schemes, involving much of the coastal plain, Lake Maliq and the Vjosë valley. Of the country's total land area in 1938 only 11 per cent was given over to arable farming or fruit growing: largely confined to upper courses of rivers, upland plateaux and basins. Thirty-one per cent was pasture, 36 per cent woodland, 11 per cent was potentially productive but uncultivated, and the remaining 11 per cent barren and unproductive. It was recognized, however, that the malarial lowlands were potentially very fertile. Indeed, an Italian company obtained a concession of 3,000 ha at Sukth, near Durrës. Italian settlers drained and irrigated the land for arable cultivation to produce a model farm (which was to

17

become a model state farm for visitors after the war). This enclave flourished virtually outside of the Albanian domestic economy. Albanian sources claim that after the Italian invasion of 1939 plans were laid for 'tens of thousands' of settlers to establish themselves on the coastal plains. Overall, Albania was to have absorbed at least a million settlers from southern Italy (Pollo and Puto 1981: 225). During 1933–4, when the Albanian government had entered into secret negotiations with France in an effort to circumvent economic subservience to Italy, the French government had demanded exclusive rights to exploit the Albanian subsoil and sought control of much of the coastal plain for French settlers (Pollo and Puto 1981: 207–8).

Upland erosional problems were exacerbated by the considerable herding of sheep and goats. In 1938 the value of wool exports was the highest for any single item of trade; cheese was also significant. Goats were bred for mohair and milk. By the late 1930s it was recognized that many forest areas, containing all of the best European hardwoods as well as conifers, had been devastated by reckless cutting for fuel or charcoal burning. Walnut had almost disappeared. In the late 1930s about 390,000 cubic feet were annually exported, mostly to Italy. Preservation and planting had not yet been considered (Mason *et al.* 1945: 246–7).

DATA SOURCES AND PROBLEMS

As previously noted, until the late 1980s, acknowledgement of environmental problems was subordinated to economic growth and increased food production. This, together with Albania's well-known hitherto reticence in publishing accessible statistical data (for example, see Sjöberg and Sandström 1989), saw very little published material, and even less public debate, on environmental 'problems', although uncharacteristically, the last three years of Communism saw the availability of published annual statistics (KPS 1988, 1989, 1990). But little of the data in these compilations is of immediate relevance for the purposes of this text. No data on any form of pollution is provided, and no released demographic data, for example, have ever alluded to birth defects, occurrence of premature births or other possibly environmentally induced abnormalities. Similarly, observations on health and incidence of disease have hitherto only referred optimistically to living conditions, recreational opportunities and mass immunization programmes, citing the eradication of malaria and syphilis and the 'appreciable reduction' of tuberculosis (from 296 new cases per 100,000 population in 1951 to 35 in 1981). By the early 1980s, the most commonly encountered diseases, in descending order, were said to be tracheo-bronchitis, angina, influenza, localized pneumonia, and diseases of the heart and blood vessels (Cikuli 1984: 33, 73).

However, a wide range of environmentally related literature based on a not insubstantial amount of research has been available for some time, although often important publications of the country's Academy of Sciences have had a very limited circulation. The Institute of Hydrometeorology, for example, has produced works on Albania's climate and hydrology, including a climatic atlas, and has published monthly and annual bulletins since 1956, and a 'studies' series since 1960 (*Studime meteorologjike dhe hidrologjike*). The Centre for Biological Research produces an annual journal (*Vjetari punimeve shkencore të QKB*). The Seismological Centre has published works on seismic zone definition, an appraisal of the 1979 earthquake, translated into English (Dede 1983), and a catalogue of the country's earthquakes, as well as its own monthly and annual bulletins (in Albanian and English). Since 1986 the Laboratory of Hydraulic Research has produced the occasional *Kërkime dhe studime hidraulike* (Hydraulic research and studies). Soil studies and the production of pedological maps of the country have been undertaken since 1954 by the Institute of Pedological Studies and its predecessors (Skarço 1984: 66–7). Other environmentally related research publications include the Geography Institute's series on geographical studies (Kabo *et al.* 1985, 1987, 1988), the Institute of Folk Culture's French language journals *Culture Populaire Albanaise* (published since 1982) and *Ethnographie Albanaise* (since 1979), and the Archaeology Institute's periodical *Iliria* (from 1971). An occasional relevant paper may also appear in the Social Science Institute's *Studia Albanica*, which first appeared in 1964. The State University's important Natural Sciences faculty has published its *Bulletin i Shkencave të Natyrei* (Journal of Natural Science) since the 1950s. The Higher Agricultural Institute (Tirana, Korçë) regularly produces monographs and other publications (Temo 1985: 55–6), as do, jointly, the University's Faculty of Medicine and the Directorate of Health Education, including a bulletin of hygiene and epidemiology (Cikuli 1984: 88). The Institute for Monuments of Culture, overseen by the State Committee for Culture and Art produces the twice yearly *Monumentet*, containing archaeological and architectural studies of the country's cultural monuments.

In the early 1980s, a Centre for Scientific and Technical Information and Documentation was established within the Academy of Sciences. It produces a number of bulletins, including *Njohuri, të dhëna a fakte shkencore dhe teknike* (Scientific and technical information, data and facts).

More pedagogical studies include a range of texts on the geography of Albania (for example, Bërxholi *et al.* 1988, Ostreni 1972, Xherahu and Baruti 1975), together with a popularized foreign language version (Bërxholi and Qiriazi 1986), a monumental work on Albanian flora encompassing more than 2,500 line drawings (Demiri 1983), studies of plants with medical and other applications (for example, Ahmataj and Ceku 1984, Nano 1987, Piperi and Kajno 1990), and atlases (for example, Anon 1971, Bërxholi 1990a), as well as the recently released 1:200,000 physical map of the country (Bërxholi and Qiriazi 1990).

Some regional monographs have been produced in popular form in foreign languages (Bërxholi 1985, 1986; Basha 1986), while a number of districts have had their own *Almanak*, a wide-ranging occasional compilation of papers on local, including environmental, matters (for example, Kraja 1976; Toçka 1980,

1981a, 1981b, 1982, 1985), albeit centrally edited and published.

Original sources which will contain increasing levels of public debate on environmental matters, are the former ruling Party of Labour (now Socialist Party's) daily newspaper *Zeri i Popullit*, and, since January 1991, the Democratic Party's rival *Rilindja Demokratike*, and increasing numbers of new publications (for example, see Zanga 1991a).

As yet unpublished sources include the papers presented by Albanian partici-pants of the September 1988 conference Environmental Protection in the Balkans, held in Varna, Bulgaria (Çobani 1988, Dingu 1988, Dollani 1988, Frashëri 1988, Gjiknuri 1988, Hoxha 1988, Mara 1988, Puka 1988, Tartari 1988), and those of recent missions to the country (for example, Kusse and Winkels 1990).

Generally, however, the Albanian literature on environmental concerns has been fragmented and uncoordinated, ranging over specialist periodicals such as *Probleme Ekonomikë* and *Bujqesia Shqipërisë* (Albanian Agriculture) and the familiar glossy and hitherto bland periodicals produced for foreign consumption, such as *New Albania* (see, for example, Pano 1990) and *Albania Today* (see, for example, Çobani, 1989). Optimistic works published on the material culture and man-made environment, especially in the areas of archaeology and architecture (see, for example, Adami 1983, Meksi 1983, Thomo 1988), or on man's trans-formation of the environment, such as through agricultural development (see, for example, Anagnosti 1985, Anon 1982, Baçi 1981, Skarço 1984) have tended to overshadow more 'scientific' works on the natural environment *per se*, although this balance has been changing in recent years. For example, the Academy of Science's Centre for Biological Research, the University's Natural Sciences Faculty, and the Higher Agricultural Institution have been collaborating to produce a five-volume work on the country's flora (Lloshi 1990: 41–2).

ATMOSPHERIC POLLUTION

Urban and industrial sources

Until recently, official pronouncements had argued that constant attention coupled with the clear separation between industrial and residential zones guaranteed a minimum of atmospheric problems in Albania's major settlements. This was 'confirmed' by analyses of air quality in twelve centres (Karadimov 1989). However, even before the country's recent opening, it was being admitted that pollution from various sources was increasing in all urban centres. For example, work on the climatic factors influencing the content and transport of atmospheric pollutants urged that levels of soot and sulphur dioxide concen-trations receive particular attention (Çobani 1988). Carbon dioxide and nitrogen oxide are also problematic.

Since 1988 there has been a requirement for new industrial developments to be constructed at least 4 or 5 kilometres from residential areas. But pre-existing industrial concentrations within the built-up areas of such major settlements as Tirana, Shkodër and Korçë pose major atmospheric, visual, noise and olfactory pollution problems. A handful of examples have been documented of polluting industries being removed from near residential areas (Karadimov 1989), including the relocation of an asphalt and concrete plant in Tirana. There is little evidence that atmospheric pollution technology has been widely employed thus far: Ellenburg and Damm (1989), for example, argue that industrial emission filters are unknown. Western assistance should begin to address this deficiency with, for example, the installation of flue gas desulphurization plant.

Major atmospheric polluters (see Figure 2.1) include: (a) the Elbasan metallurgical complex, which is a major source of untreated pollutants, directly impacting on the town and surrounding forest land; (b) cement works at Elbasan, Krujë, Vlorë, Tirana and elsewhere: 746,000 tonnes of cement were produced in 1988 compared to 16,000 tonnes in 1950 (KPS 1989: 75); (c) oil and lignite-burning power-stations such as those in Tirana and Korçë, employing locally mined coal: that mined near Korçë has a notably high ash and sulphur content. Post-war lignite production has grown dramatically from 41,000 tonnes in 1950 to 2.18 million tonnes in 1988 (KPS 1989: 75); (d) petrochemical complexes at Ballsh and Fier; (e) battery factories at Berat and Gramsh; (f) the copper wire factory at Shkodër, with 11,600 tonnes production in 1988 compared to 3,100 in 1970 (KPS 1989: 75); (g) PVC and tannin plants at Vlorë; (h) small yet locally heavily polluting bitumen plants, such as those in Rrogozhinë, near Sarandë, and in Tepelenë.

A number of major settlements are situated in basins and valleys, or are located adjacent to high ground, such that pollutants tend to become trapped. This feature is often viciously exacerbated by temperature inversions, as experienced in the country's highest city Korçë (at 800 metres) on calm evenings, when the centre and major residential areas of the town may become literally covered with the sulphurous outpourings of the lignite-burning power-station. No data recording the extent of such pollution are yet publicly available, although Professor Vasil Trojani at the State University has been using methods of direct access to individual plants in order to gain appropriate material for a future publication (Trojani 1991).

While the level of motorized vehicles within the country is currently low, the use of low grade, sulphurous fuel, coupled with poor maintenance of often long time-expired Chinese and Soviet bloc machines does produce relatively high concentrations of vehicle emissions within a number of towns, again as yet unmeasured. In rural areas and some urban back-streets, poor road maintenance or lack of hard core produces high levels of atmospheric dust, especially through the long dry summer.

The domestic burning of wood and low grade coal adds to urban pollution problems, particularly in winter and under conditions of temperature inversions. While natural gas is piped to a number of urban homes (Hall 1990a), wood-burning ovens are produced in quantity within the country and are domestically ubiquitous.

Pollution within working environments

Visits and other idiographic evidence have indicated that health and safety regulations at the workplace have often been flouted. For example, in the former Mao Ce Dun textile mills in Berat, the largest in the country, repeated visits have revealed lax use of face masks despite high and very visible levels of airborne fibres; extractor facilities here appear relatively ineffective, judging from accumulated deposits on machinery, window sills and other surfaces. Such problems have been barely addressed, with female workers appearing to take the brunt of the health hazards.

Social security provision has implicitly recognized the nature of working conditions. Miners, metallurgical workers, and airmen are eligible for pensions at 50 after 20 years' service. Workers in oil, cement, glass and rubber industries (together with teachers) are eligible at 55 after 25 years' service. All other workers are eligible for a pension at 60 after 25 years' service. However, in all employment categories women are entitled to pensions five years earlier than men and with five years' less service (Cikuli 1984: 43). Invalidity pensions range from 40 to 85 per cent of the normal wage.

WATER POLLUTION

Albania is particularly rich in water resources: average annual precipitation in the upland areas is the highest in Europe at around 1,500 mm, and the average annual water flow is 42,000,000,000 cubic metres (Frashëri 1988). The sources of Albanian rivers are invariably above 1,000 metres. Rivers flow fast over relatively short courses to the sea and provide substantial hydro-electric potential. Conversely, until large storage lakes were developed in conjunction with the HEP schemes (Koman, Fierzë), navigation of most Albanian waterways was virtually impossible.

By the early 1980s only 56 per cent of the country's villages were supplied with drinking water (Shkodra and Ganiu 1984: 36). In 1986, the Ninth Congress of the Party of Labour resolved to ensure that by 1990 sufficient quantities of pure water would be available for all the country's settlements. Subsequently, particular attention was centred on the construction of aqueducts to supply towns and villages (Frashëri 1988). Little detailed literature is yet available to indicate how far substantial spillages in the oilfields, run-off of pesticides, and leaching of livestock and human manure have polluted water tables and wells. There are, however, some considerable localized problems (Trojani 1991).

Poor domestic and industrial plumbing and maintenance of pipes, aqueducts, irrigation canals, village taps and standpipes permits bacteria and other harmful additions to enter domestic water supplies. Critically, in view of recent near drought conditions, it also results in an enormous inadvertent waste of water.

Marine pollution

The impact of untreated urban domestic and industrial waste, offshore discharges, and pollution originating in Italy, Yugoslavia and Greece, have begun to take their toll on Albania's marine environment. The polluted nature of water adjacent to the ports of Shëngjin, Durrës, Vlorë and Sarandë, where waste and spillages from vessels are locally problematic, is increasingly impacting upon nearby tourist resorts. This is exacerbated by the discharge of polluted rivers reaching the Adriatic coast, notably the Bunë on the Yugoslav border, the Mat south of Lezhë, and the Shkumbin, Seman, and Vjosë between Durrës and Vlorë. It has been reported that pollution is adversely affecting fisheries and doubtless other fauna and flora (PADU 1990). Some 260 kinds of fish can be found in Albanian waters. Sardines are caught in the Bay of Vlorë, and at Durrës and Shëngjin; grey mullet is caught in coastal lagoons and other varieties fished include white perch, barbel, cod and tuna (Bërxholi and Qiriazi 1986: 43). However, Albanians have never been notable fish eaters, and the need to secure the country's maritime and lake borders has constrained the development of a fishing industry.

Tourism along Albania's varied Ionian and Adriatic coasts has been concentrated in and around Durrës Beach, which has accommodated up to 60,000, mostly domestic, holidaymakers per month in summer (Bërxholi 1986: 30), and a handful of other centres such as Vlorë and Sarandë (Albturist n.d., 1989; Hall 1984b, 1984c, 1990d, 1990f, 1991). With a number of Albanians arguing that tourism was the country's only short-term salvation, plans to render tourism a major element of the Albanian economy (BBC 1991d, 1991g; Shilegu 1991) saw joint-venture contracts signed in early 1991 for substantial developments along the coast and inland, including a marina at Sarandë, a tourist village in Durrës and at four other Adriatic resorts, and an 18-storey 260-room Sheraton hotel in Tirana, all to be completed by 1995 (EIU 1991b: 44). New hotels in Durrës, Divjaka, Vlorë and Berat as well as the refurbishment of existing ones in Durrës and Shkodër, were to be completed by Yugoslav, Italian, French, Swiss and German interests under contracts worth half a billion dollars. Although by the end of the year little progress had been made on such projects, increasing pressures, particularly along the relatively unspoilt Ionian coast, cannot but bring environmental deterioration in their wake. Sewage, solid waste and other refuse disposal, degradation of habitats through recreational pressures, substantially increased traffic and the need for improved transport infrastructures and other utilities will all take their toll, imposing adverse impacts upon the very marine and landward environments which can provide the basis for such tourism expansion.

In the extreme south of the country, careless local recreational activity coupled with flotsam from the many Greek and Italian vessels passing through the Corfu Straits has already contributed to a substantial littering of the coast with waste material at such locations as Ksamil. However, the country has shown a willingness in recent years to undertake more active participation in international approaches to combating pollution problems in the Mediterranean basin (Pano 1990). New political and economic conditions within the country, involving Western assistance, should permit the application of waste treatment and other anti-pollution technologies.

In early July 1991, the country's first ever 'National symposium on the coastal space' was held in Tirana, with Italian attendance (Anon 1991b). Later that month, alongside Italian, Greek, Yugoslav and EC representatives, Albania became a signatory to the 'Declaration on the Adriatic Sea' in Ancona. An initiative was launched to clean up the northern stretch of the Adriatic and to keep the Albanian coast pollution-free, establishing guidelines for 'environmentally-friendly' tourism development (Anon 1991a).

River pollution

In the coastal plains, the more important rivers are visibly polluted by industrial wastes. The level of toxicity from metals is not made known, but with little waste treatment apparent, untreated discharges pose a threat to both aquatic life and human health. Run-off from increasing use of chemical fertilizers is also a contributory factor. The lower reaches of many streams and rivers in the central part of the country may already be organically dead. Even in upland areas careless mining activity has damaged water bodies, such as the Shkumbin running red downstream from the ferro-nickel workings of the Librazhd and Prenjas areas. It has been acknowledged for some time that waste from paper mills (such as near Lezhë) and plants associated with non-ferrous metals has damaged aquatic flora and fauna along a number of rivers and estuarine areas, and calls have been made to find more suitable ways of protecting ecological balance in such areas (Tartari 1988).

Lake pollution

Albania's lakes are recognized as wetlands of international importance. They have acted as breeding and feeding areas for white and Dalmatian pelicans with eleven pairs of the latter being located in 1985 (and five birds at Divjaka in 1989), the pygmy cormorant, a large colony of which is thought to be at Lake Shkodër, various herons (Karadimov 1989), and several other waterfowl species. The precarious position of some of these is a matter of concern: wetland modifications are thought to have caused recent decline. Lake Ohrid has particular interest due to a rich endemic and relict fish and aquatic fauna (Carp 1980): for example, twenty-four varieties of carp are unique to the lake (Bërxholi 1986: 30). The country's four major natural lakes are shared with Yugoslavia (Shkodër, 368 sq. km; Ohrid, 362.6 sq. km), Greece (Little Prespa, 23.5 sq. km) or both (Prespa, 285 sq. km). While this raises questions of trans-boundary pollution, particularly with increasing urban and industrial development in the vicinity of

Lake Ohrid, it also provides an opportunity for possible regional co-operation within Albania's new-found outward-going approach. This will be particularly necessary following an Albanian–Yugoslav agreement to build a road around the lake to link Ohrid, Pogradec and Struga, and to further develop hotels and shops in the Albanian town (BBC 1991f). Much more pollution from visitor use may be expected to result.

The smaller lakes of the northern mountains, such as Lura, appear to have largely retained their character, although domestic tourism pressures are increasing. In the extreme south-west of the country, the larger Lake Butrint (16.3 sq. km) has been developed in recent years for mussel cultivation. However, diversion of some of the lake's freshwater sources for irrigation purposes has increased the lake's salinity to the detriment of its fauna.

SOILS, FLORA AND FAUNA

Albania can be topographically divided into a sub-alpine and alpine zone, situated north of the River Drin, with peaks attaining heights of over 2,000 m; a mountainous zone of alternating deciduous montane forests of beech and cultivated basins, between the Drin and Osum valleys; a broad Mediterranean and transitional deciduous forest zone with scattered patches of Mediterranean pine south of the Osum valley featuring limestone mountains and valleys; and a coastal plain of phrygana, maquis and cultivation, dissected by numerous rivers (Grimmett and Jones 1989). Two-thirds of the country contains sedimentary deposits and one-third igneous rock. The coastal zone consists mainly of alluvial material: of 250,000 ha, 220,000 ha have been reclaimed, mostly from mosquito-infested marshland. Of the remainder, 18,000 ha are to be developed for agriculture and intensive fisheries, and 12,000 ha will be utilized as nature reserves (Kusse and Winkels 1990).

Of exploitable land, almost half is covered by forest and bush, some 20 per cent by natural pasture and 30 per cent by cultivation. Such an ecosystem relationship favours floral and faunal diversity: the country contains some 3,500 species of flora, with several hundred endemic and sub-endemic species. These are well documented within the country (see, for example, Demiri 1983), and are again attracting increasing attention in the West (see, for example, North 1990; Almond 1988a, 1988b; Polunin 1980; see also Turrill 1929). Some 865 specimens are claimed to have curative properties, and Albania is Europe's leading exporter of medicinal herbs, supplying around 20,000 tonnes worth $20 million.

Farming practices

Albania has maintained the lowest level of urbanization in Europe – just 35.5 per cent of the population were recorded as living in towns in the 1989 census: an increase of 2 per cent over 1979. This reflects an explicit rural-led development programme and the maintenance of a higher natural increase of the rural population (Hall 1987a; Sjöberg 1989, 1991). Since 1946 cultivable land has increased by 240 per cent. However, between 1950 and 1986 the population increased by 250 per cent with the amount of arable land per person decreasing by 10 per cent. This process accelerated during the 1980s, and has exerted pressure on the authorities to rethink their pro-natalist approach. Arable land currently comprises 700,000 ha, 450,000 ha (64 per cent) of which is on sloping ground. Expansion of arable acreage has hitherto entailed an improvement in desalination methods (although salination remains a problem), reclamation from coastal and lakeside lagoons, and an extension of cultivation in upland areas, often through extensive terracing (Hall 1987a, Kusse and Winkels 1990). Collectivized agriculture absorbed a rapidly growing work-force during the 1970s and early 1980s, but with privatization and uncertainty, the 70,000 new recruits which enter the national job market every year can no longer be accommodated in the agrarian sector. Recent rural to urban migration and waves of emigration partly reflect this problem.

The Communists' socio-economic policy had positively discriminated in favour of rural against urban and upland rather than lowland areas (Hall 1987a). But with an increased urgency to stimulate food production, and the establishment of a new Agricultural Bank to sanction and monitor the viability and execution of rural projects, a fourfold spatial designation was introduced early in 1990 for funding agriculture, increasing state subsidies for marginal, inefficient, farming areas. The zones – lowland, hilly, mountainous, and the north-east – would receive successively higher prices for their produce by virtue of their higher production costs (EEN 1990: 2). But this was likely to increase inefficiency and the likelihood of detrimental ecological consequences in marginal and fragile environments. The policy is now in ruins.

In the waning months of Communism, in stark contrast to previous policy, during 1990 a programme of encouraging the development of private plots was begun, particularly aimed at stimulating vegetable production. For the first time peasants would be allowed to rear cattle on their own plots, and co-operatives were asked to transfer some of their stock to their members for this purpose: in hilly and mountainous areas 2,000 sq. m to each member in addition to their existing private plots, and in lowland areas peasants were to be given up to 2,000 sq. m including their plots (EIU 1990: 36). While clearly intended to raise food productivity and incomes, prolonged drought tended to undermine the programme. None the less, the domesticated animal population increased dramatically although a significant growth in the number of sheep and particularly goats was also evident, which, through indiscriminate grazing, was likely to exert adverse impacts on a number of habitats and fragile environments, creating conditions for renewed erosion in upland areas.

The subsequent coalition government's land and private property laws of July and August 1991 dissolved all co-operatives but forbade the purchase or sale of land. In the consequent rush to regain lands held before the war, former rural landowning families were unilaterally claiming and fencing off territory, even destroying buildings such as schools (Zanga 1991b) and installations such as oil wells (Kristo 1991) in the process.

The use of Chinese-designed crawler tractors has reduced the problem of soil compaction on the alluvial arable lands of the reclaimed coastal plain. Seed bed preparation and drilling is usually poor: heavy equipment gives an uneven surface, such that seed drills are said to fail to seed up to 25 per cent of the ground (Atkinson *et al.* 1991). Recent German aid in this area has been ameliorating the problem. Many tasks are still labour intensive, however, undertaken by human hand and animal power, often using improvised implements and vehicles. Reversion to smaller private holdings will only exacerbate this situation in the short term. However, as the country's economy opens and personal mobility is enhanced, whole areas of upland agriculture, particularly that which enjoyed substantial investment of financial and labour resources in the 1960s and 1970s, could be abandoned as rural to urban and upland to lowland migration gains pace. While a shedding of surplus rural labour might appear to present an ideal opportunity for agricultural mechanization, difficult topography coupled with capital shortages will severely constrain such an application.

The debris of non-biodegradable polythene being used for hot houses, vehicle coverings and other purposes is having an increasingly pervasive presence in the Albanian countryside. Destruction through high winds and lack of maintenance has seen such polythene covers ripped away from hot house supports and scattered over increasing distances, with such adverse effects as the clogging of irrigation channels, the littering of fields and beaches, and the potential choking of livestock. This is apparent in several areas of the coastal plain, in the vicinity of Lake Ohrid, and in a number of upland basins.

Use of chemical pesticides and fertilizers

Problems resulting from the increased use of chemical pesticides and fertilizers in agriculture have recently brought a recognition of the need for control (Hoxha 1988). Insecticides have been used on a large scale: about 800 tonnes in 1960, 5,900 in 1970, and 10,200 in 1980. Application of chemical fertilizers has grown considerably in the post-war period (see Table 2.2), rising overall from just 1,000 tonnes in 1950 to 96,500 in 1988, representing an increase from 2.6 to 132.2 kg per hectare of tilled land (KPS 1989: 89, Papajorgji 1989: 43).

Impacts on woodland

Forests in Albania cover about 1 million hectares. Some evidence of acid rain exists, particularly in the country's northern forests (Atkinson *et al.* 1991). International collaboration would appear to be indispensable in this as in other aspects of environmental protection (Bouvier and Kempf 1987). Of more significance is deforestation, which, while historically rooted, continues today in an

Table 2.2 Growth of use of chemical fertilizers in Albania, 1950–88 (in thousand
tonnes)

Type of chemical fertilizer	1950	1960	1970	1980	1988
Nitrogenous	0.1	4.6	22.5	68.8	69.3
Phosphate	0.3	2.2	20.7	21.0	25.0
Potash	0.6	1.2	0.9	3.9	2.2
Total fertilizers used	1.0	8.0	44.1	93.7	96.5
Kg of 'active substance' per hectare of tilled land	2.6	17.0	75.0	133.5	132.2

Source: KPS 1989: 89.

official and unofficial basis, not least to meet the large demands of domestic
woodburning ovens.

Erosion problems

Partly to meet increasing food demands for a rapidly growing population, vast
areas of hillside were dogmatically terraced in the 1960s and 1970s (Anagnosti
1985, Skarço 1984). The characteristic slogan of the 1966–70 fourth five-year
plan was 'let us take to the hills and mountains and make them as beautiful and
fertile as the plains' (Skarço 1984: 63–4).

Over 100,000 ha of virgin land were broken in for cultivation during that plan
period and a further 92,000 ha between 1971 and 1975. However, many of the
hillsides should never have been terraced, the top soil being too shallow and the
scrub too fragile for working. In many cases, bedrock has appeared. Terracing
may have been more effective had the terraces been properly built up with stones
and then planted immediately. Fruit trees have been planted often in areas too
high and dry. Olive planting has had better results, but ground vegetation is too
sparse to prevent soil erosion, rapidly leading to gullies between the trees
(Atkinson *et al.* 1991). A number of problems resulted from the deployment of
mass mobilization, particularly the involvement of untrained 'volunteer' urban
youth in the terracing work, a policy which has now ended.

Afforestation affected almost 200,000 ha between 1951 and 1988, with a
further 467,000 ha of forestry improvement during that period (KPS 1989: 90).
This programme helped to stabilize some areas of hitherto fragile conditions. The
planting up of susceptible areas with stabilizing trees has usually involved *Pinus
pinea* and *P. radiata*. The districts of Pukë, Shkodër, Librazhd, Kukës and Fier
were the main focus of this activity. In the Vlorë district, for example, by the mid-
1980s 4,500 ha of forest had been improved and 4,000 ha of denuded land reafforested (Bërxholi 1985: 31–2). Fruit and olive tree planting has latterly
involved individual rather than contour terracing, and this policy could be seen
being implemented in such districts as Sarandë and Pogradec.

28

The impacts of irrigation and reclamation

Since 1940 major programmes of wetland drainage have been undertaken as part of the drive against malaria, with most marshes being converted to agricultural lands irrigated through a series of reservoirs. Rivers, especially in the coastal belt, have been regulated and canalized. Between 1946 and 1983 some 9 per cent of the country was drained, and, between 1966 and 1970 at the height of the programme, 20 per cent of all capital investment was devoted to major hydrological projects. Inland, drainage of the Maliq lake and marshes, started in 1946, had resulted in a loss of some 60,000 ha of wetlands by 1974 and the improvement of a further 170,000 ha for agricultural production. The area is now Albania's main sugar-beet source.

With rainfall unevenly distributed throughout the year – Albanian agriculture often suffers from a shortage of water between April and September – water improvement projects have been closely associated with irrigation, which by 1988 had been extended to some 58.2 per cent of all cultivable land (compared to 47.4 per cent in 1970 and 10.5 per cent in 1950), representing an irrigation capacity of 416,800 ha, the figure for 1970 being 283,800 ha and for 1950 39,300 ha (KPS 1989: 89, Papajorgji 1989: 43). Irrigated flood plains provide important habitats for wetland birds and other fauna (Grimmett and Jones 1989).

Other impacts of human activity on wildlife

The Albanian fauna is rich and varied, with both Mediterranean and European characteristics. Officially, hunting is severely restricted but is likely to gain greater significance as a means of accruing hard currency. In practice, it appears to have been widespread and locally safeguarded. With increasing human and economic pressures, deforestation may threaten the habitats of such rare animals as the brown bear, of which there was an estimated 500–800 in the northern mountains in 1987 (Atkinson *et al.* 1991).

OTHER ENVIRONMENTAL PROBLEMS

Noise pollution

By European standards, Albania is a relatively quiet country. A low level of development in a still essentially rural society, with limited numbers of motorized vehicles and consumer durables, contributes to this superficial impression. Within an economy of shortage, however, poor plumbing and maintenance of equipment and machinery of all kinds incrementally adds unnecessary noise to domestic and working environments.

Idiographic evidence of noise pollution within workplaces reveals employees of both sexes working excessively noisy machinery in textile mills, engineering workshops and other industrial enterprises, rarely using ear protection equipment. As yet no published data are available in this field.

Until 1990 there was no international overflying of Albanian air space, apart from the handful of weekly civil flights into the rural Rinas airport north of Tirana, and the occasional training flights of the Albanian air force's ageing Chinese-built MiGs. Now both international overflying and an increase of flights into Rinas are taking place. Not since the 1930s have domestic civil flights operated (for example, see Mason *et al.* 1945), other than air rescue operations and crop spraying from biplanes, but with increasing tourism pressures these may well reappear, using landing strips at some half-dozen centres.

From 1991, the private ownership of motor vehicles, encouragement of inward investment, and a substantial hastening of the modernization process – including an ardent desire of the Albanian youth to acquire personal stereos – have been making inroads into the erstwhile outward tranquillity.

Visual pollution

Ever since the period of worsening relations with China in the mid-1970s, national defensive measures have included two forms of installation. Air raid shelters have been built into urban residential and industrial areas. These are often distinguished by heavy metal doors above small elevations at ground level, although they tend to be unobtrusive. Defensive concrete 'pill-boxes', however, numbering at least several thousand, have been scattered throughout the country, visually scarring the landscape and encroaching on valuable agricultural land. They are particularly numerous in 'strategic' areas such as near borders, ports and along the key Durrës–Tirana trunk road. The Democratic Party pledged to dismantle them at the earliest opportunity (*Newsweek*, 8 April 1991: 52).

In both urban and rural areas, such is the housing shortage that families are often moved into apartments and houses before the external decor is finished. In particular, completion of the outer facing of urban apartments has often considerably lagged behind their occupation, and in many cases low-rise brick-built flats have never been faced. This often reflects a shortage of materials and distribution problems (Hall 1990a), and presents a very shoddy outward impression of such residential units.

Insensitively located rural apartment blocks, usually associated with the administrative centres of state farms, can often be found in otherwise harmonious landscapes, as in the coastal village of Lukovë, where three four-storey units dominate a ridge some 200 metres above the Ionian Sea. Slightly less prominent are the eleven four-storey blocks of the nearby coastal village of Piqeras.

As a postscript to the lifetime personality cult, some particularly hideous large statues of Enver Hoxha were erected after his death in 1985, most notably in the centre of Tirana (toppled and broken up by rioting crowds in February 1991) and on a dominant bluff in his home town of Gjirokastër (protected by special armed units until later removed). The relatively new full-size statue in Korçë was removed during the troubles of early 1991, only to be temporarily replaced by a marble bust, far too small for the surviving plinth, and a four-strong armed guard. During December 1990 the remaining statues and busts of Stalin were removed, including that in central Tirana facing Lenin, for whom the same fate befell the following year, across the Boulevard of the Nation's Martyrs. Large posters and other political exhortations have been largely abandoned since the Communists' loss of power in June 1991. In rural areas they had symbolized an interesting continuity by taking the place of religious roadside shrines on mountain passes.

A greater visual blight has been imposed by officially sponsored hillside 'graffiti', often marked out in bare limestone, extolling the virtues of the party and 'Comrade Enver'. It appears that these confections were usually put together by members of the armed forces (for example, see Gardiner 1976: 61–2). Although by no means unique to Albania, one of the most grotesque examples has been an enormous 'ENVER' prominently splayed across five ridges facing the town of Berat and extending to a height of some 30 metres. Others include huge examples of 'PARTI ENVER' on the mountainside facing Gjirokastër and over-looking the village of Borsh on the Ionian coast, and a pervasive 'POPULL PARTI ENVER' spread over three ridges by the village of Lin next to Lake Ohrid. Many such exclamations have been removed, damaged or allowed to be over-grown. Indeed in April 1991, following anti-government rioting, in the town of Kavajë the grafitto 'ENVER HITLER' was observed painted on to a wall. In the same month, however, the hero town of Përmet still sported a range of the old ideological tapestry, ranging from '1908' (Hoxha's birth year) etched into the surrounding hillside, to a large floral display of 'PARTI ENVER' embellishing the town centre.

Other aspects of visual pollution include:

1 industrial areas with disused parts, obsolete equipment and unusable machinery left lying around, as in the Elbasan metallurgical complex;
2 hideously patched up industrial plant, such as at the Laç superphosphate works;
3 badly constructed and designed industrial buildings, such as the iron-nickel mine at Prenjas and copper smelters at Fushë Arrëz;
4 poorly repaired or unsurfaced roads and open areas in towns and cities which turn virtually into lakes after heavy rain, such as the centre of Fushë Krujë, with the open bus station becoming a desultory quagmire;
5 ugly, prefabricated concrete irrigation canals, such as those near Kavajë, and poorly constructed railway viaducts, as between Elbasan and Prenjas. Indeed, much of the railway system was constructed by 'volunteer youth' (Hall 1984a, 1985, 1987b), and is now being overhauled to render it physically capable of supporting a modern transport system;
6 as well as polluting nearby rivers, the ferro-nickel mining activity of the

Prenjas–Pishkash area has left red trails along most of the roads of the area, as well as open cast scars (Dede 1986).

Olfactory pollution

A strong smell of sulphur emanates from many parts of the country's oil fields, but is particularly notable on the Ballsh–Tepelenë road, where poorly maintained oil storage facilities are located. It has been known to make travellers passing through the area physically sick. Similarly, the sulphurous compounds resulting from the burning of lignite are especially noticeable by their smell in a number of towns.

In public buildings toilet facilities are often inadequately supervised, and in a number of rural settlements open drainage poses problems, particularly in summer.

Health and safety factors

As noted earlier, idiographic evidence points to health and safety legislation in factories and other employment centres being flouted in relation to protection from airborne pollution, noise, potentially dangerous open machinery, wet floor surfaces and inadequate lifting and other mechanical aids, ventilation and lighting. Outdoors, poorly constructed scaffolding, a lack of safety harnessing, and an absence of hard hats appear commonplace on construction sites. The improvisation necessary to keep agricultural machinery and equipment working in the absence of appropriate replacement parts may also pose safety hazards. As yet no data on health and safety at work have been made available.

Many safety considerations are flouted by transport undertakings. Because of a shortage of glass (and recent vandalism), some buses and railway carriages have been operating with cracked, broken or missing windows, and both modes are usually very seriously overloaded with passengers. The latter often also applies to the many road freight vehicles which carry large numbers of 'hitch-hikers' in open and often potentially lethal conditions. Road lighting is at best poor at night; most bicycles and many motorized vehicles also fail to be illuminated after dark. The substantial use of slow unlit animal transport is an additional road safety hazard at all times of the day, but particularly at night. Due to the great age and nature of upkeep of road vehicles, buses and trucks regularly spew out clouds of thick diesel fumes.

Radioactivity

The monitoring of natural radioactivity has been possible since a laboratory was established in 1969 under the auspices of the State University. Systematic monitoring of the atmosphere began in 1971, following the establishment of the Institute of Nuclear Physics within the Academy of Sciences. Research undertaken after the Chernobyl accident into air, water and foodstuff samples found radioactivity levels of up to 1,500 times the official 'norm'. Measurements of iodine-131 taken in samples of milk, for example, detected concentrations of up to 30,000 Bq/kg. Monitoring is to be extended to enable the compilation of a national map of natural radioactivity (Dollani 1988).

THE INTRODUCTION OF WESTERN TECHNOLOGY AND EXPERTISE

The changing political and economic context

The Albanian leadership's initial reaction to the 1989 events in the rest of Eastern Europe was to tinker with the economic, political and judicial system in the hope of forestalling major unrest within the country. During the first half of 1990, a range of political, cultural and economic measures were outlined, aimed at improving grass-roots democracy, reducing bureaucracy, permitting religion, reversing the relatively isolationist foreign policy, and emphasizing a commitment to the provision of foodstuffs.

Most significantly, mid-year saw a formal acknowledgement of the need for foreign investment and mechanisms for the establishment of joint ventures on Albanian soil. Small-scale industrial and agricultural privatization was also initiated. Prior to this, diplomatic relations had been finally established with West Germany in 1987, and a subsequent protocol covering co-operation in mining, energy, and the chemical and food industries saw trade between the two countries dramatically improving during 1988, with Albanian exports increasing by over 80 per cent (raw materials by 160 per cent) and imports by nearly a quarter. Under a further agreement signed in March 1989, Bonn was to provide Albania with DM10 million in 1989 for financial co-operation, and a similar total in 1990 for technical co-operation, covering mining, iron and steel, oil production and livestock breeding.

This newly found relationship was to act as something of an economic Trojan Horse for a country whose stance of relative self-reliance had restricted its access to modern technology and thus its capacity to more effectively exploit and process its natural riches. In particular, oil, chrome and copper continued to pose productivity problems: during 1990 oil production was reduced by more than half to just 1.5 million tonnes, largely due to the obsolescence of the Soviet and Chinese technology being used, producing a recovery factor of only 12 per cent (Anon 1991b). The West German injection gave the Albanians a taste for Western aid and technological support, and during 1990, with further explicit acknowledgement of the need for foreign assistance and the passing of joint venture legislation, strong interest in the country's oil and gas activities, particularly in the offshore potential, came from a number of Western multinationals. An American report published in March 1991 estimated Albanian onshore oil reserves to be 2 billion barrels, worth $40 billion at current prices (EEN 1991a,

1991b). Indeed, while in New York, the then Albanian party leader Ramiz Alia spent some time with the now late head of Occidental Petroleum, Armand Hammer. By the end of 1991 five multinational companies were engaged in offshore exploration in co-operation with the Albanian government (BBC 1991e).

Virtually all plant which dates either from the pre-1961 Soviet period or from China up to the mid-1970s is obsolete, being usually copies of earlier, often already outdated Western technology. Estimates of the sum required to modernize the country's textile, steel, chrome, oil and other mining industries and transport infrastructure over the next decade have been put at $10 billion (EEM, 11 January 1991: 2), representing $3,000 for every Albanian man, woman and child.

As the smallest, least developed and hitherto most introspective of the countries of Eastern Europe, the impact of any foreign aid, assistance and technology is likely to be proportionately greater in Albania than elsewhere in the region. After fifteen years of self-proclaimed autarky, however, there is a critical need for foreign aid and support for environmental improvement. This should be secured through a number of means:

1   directly, through the funding of environmental enhancement and training programmes:

    • from the European Bank for Reconstruction and Development (EBRD), which, formally instituted in April 1991, has the promotion of improved environmental standards as one of its two major priorities for Eastern Europe. Initially, schemes for the Baltic, Danube and Silesia were to be instituted (for example, see Donovan 1991), but within a month of its inauguration, a high-level Albanian delegation, led by the country's then prime minister Fatos Nano, visited the bank to negotiate an aid package. In November 1991 an EBRD energy and environment mission visited Albania;

    • from the European Community via an equivalent of the programmes extended to other East European countries. Italy has taken a notable lead. In the autumn of 1991 the Rome government approved soft credits worth $55 million over three years to finance infrastructure improvements and machinery imports, particularly for food processing equipment;

2   indirectly, through foreign involvement in hitherto environmentally adverse economic development, such as mining and oil production. For example, included in Albano–Italian co-operation ventures is a plan to reconstruct the Kalimash chrome enrichment plants (EIU 1991b: 45).

On the other hand, as one of the country's newly established economic development priorities (BBC 1991d, 1991g; Shilegu 1991), foreign-assisted international tourism development will bring a range of hitherto barely encountered environmental problems. The total number of foreign tourists in 1989 was

20,000 and in 1990 30,000 (EIU 1991b: 44). Yet as part of a two-year $250 million contract for the development of tourism, transport and light industry, five tourist villages, with a total of no less than 10,000 beds, are planned near the southern and relatively isolated Ionian coastal town of Sarandë, itself with only a population of 20,000. This is involving Iliria Tourist, an 80–20 per cent joint venture between Iliria Holdings, a largely expatriate Albanian-run organization with its headquarters in Switzerland, and Albturist, the state tourism organization.

Further, all joint ventures are currently free from planning obligations, rendering local environments and economies, however free or regulated, increasingly vulnerable to the depredations of international decisions. In September 1991 the country's first multi-party cabinet for half a century established a Foreign Investment Agency and pegged the exchange rate of the lek to the ECU (Carrier 1991). The relatively cheap labour now offered by Albania could encourage significant inward investment once stability has returned to the country (and indeed to the region). Indeed, in the second half of 1991 the government was planning the establishment of tax-free zones, mooted for the port of Vlorë and the industrial centres of Shkodër and Kavajë. Foreigners who invested in 'growth export industries' would be exempt from taxes for up to five years (Anon 1991a).

Agricultural reform and aid

Early in the process of economic change, several agricultural missions, most notably of German, Italian and Dutch specialists, visited Albania to provide advice on a range of agrarian problems. In 1989, for example, Fiatagri became involved in pilot projects for mechanizing the sowing and harvesting of 300 ha each of maize and rice, employing machinery previously untried within the country. The Italian company has also provided laser technology to aid soil levelling. Albanian reports of these projects have been optimistic, quoting a 50 per cent increase in maize yields with a 50 per cent reduction in costs after the first year (Anon 1990), and a doubled yield with costs reduced by 58 per cent, and in the case of rice by 75 per cent, at the end of the two-year trial period (BBC 1991c). In the autumn of 1991 a joint agribusiness venture with the French Ducros company, based in Durrës, gave a monopoly on processing all cultivated oleaginous crops and indigenous flora (BBC 1991a). This company would be able to operate tax-free for at least three years (Anon 1991a).

A FAO mission in December 1989 pointed to the need to strengthen the knowledge of specialists in the fields of saline soils improvement and land reclamation, through seminars, feasibility studies and pilot projects, and recommended extending the laboratory of the Institute of Pedology and the completion of a national soil survey at a scale of 1:10,000 (Kusse and Winkels 1990).

CONCLUSIONS

An economy of shortage prevails in Albania. Innovation, improvisation, import substitution, recyling and low levels of consumerism and material waste are key elements of that context. Many traditional skills have been retained thus far. The nature of the transformation of that situation and its speed and consequences remain to be seen.

Entry into the world political and economic arena, transition to a market economy, and programmes for privatization are now familiar in Eastern Europe. In Albania, however, these processes follow some distance behind those of the rest of the region, reflecting the deeper ideological nature and inertia of the old Albanian regime, the country's hitherto relative isolation, and its low level of economic development and technological sophistication. Under these circumstances, Albania could, and should, learn from the mistakes of both East and West: its environment is fragile, but its resources, both physical and human, are potentially great.

Positive environmental benefits will include the closure of heavily polluting and uneconomic plants which have only been developed through a combination of a dogma of self-reliance and poor transport and distribution systems. These closures will involve the Elbasan metallurgical works, lignite-burning power-stations, and at least some chemical plants. An infusion of foreign capital and aid from the EBRD should help establish pollution and waste control technologies and strategies as a high priority with, particularly, 'cleaner' and more efficient processing facilities for the country's mineral wealth (most notably for oil, copper and chromite). Unemployment and emigration may well be considerable in the short- to medium-term, as structural rationalization and technology transfer make their impact.

As subsidies have been reduced and collective agrarian structures broken down, many remoter rural and particularly upland areas are being abandoned, bringing a contraction of cultivable area, collapse of upland terracing schemes, and significant regional, rural to urban and out-migration flows. Environmentally deleterious consequences of the dire food and fuel shortages being experienced during 1991 included the cutting and felling of trees in 'protected' rural areas, urban parks and alongside roads (where lower trunks are whitewashed for night-time traffic guidance in the absence of lighting). Animal hunting in its various forms, including fishing, will have increased considerably, not least again in 'protected' areas.

While the previous unabated pro-natalist approach may well have had a measure of cultural reinforcement, demographic policy needs to be comprehensively reviewed in the light of increasing pressures on domestic food production and other basic needs such as housing and fuel.

Future increased consumerism coupled with an emphasis on tourism development will require substantial infrastructural investments. Inevitably a number of lowland and coastal environments will be further modified, while improved access to upland areas will accelerate processes of change there. Increased waste, a reduction in recycling, improvisation and import substitution, greater use of motor vehicles with attendant atmospheric pollution and congestion problems, will all represent negative environmental consequences of change. However, 'the restructuring economies of east Europe have a golden opportunity to abandon hugely polluting energy intensive industries, and adopt energy efficiency in a big way as they transfer to dependence on solar technologies' (Parkin 1991: 14).

With electricity theoretically available in every village since 1970 (much of it from hydropower sources), Albania could possibly stabilize a reduced rural population and, with Western assistance, undertake a technological leapfrog by introducing decentralized information technology-based employment opportunities which could be provided at home with appropriate training programmes, thereby bypassing environmentally deleterious industrialization processes and minimizing the adverse impacts of rural depopulation. Although there currently exists a low level of technological sophistication within the country, following 45 years of Stalinism, the population's capacity for adaptation and innovation should not be underestimated.

ACKNOWLEDGEMENTS

The author wishes to extend his grateful thanks to Pat Cowell for the cartography, to Ann Howlett for support and advice on earlier drafts, to Petrit Hoda for his field knowledge, to Vasil Trojani and the Geography staff of Tirana University, and to Agim Neza, Nikolla Elmazi and Morna Dewar for their assistance.

3

BULGARIA

F.W. Carter

For many years during the post-Second World War period, Bulgaria's government leaders tended to turn a blind eye to the problems of the environment
believing that they could not occur in a socialist country. This followed from the
rather dogmatic assumption that environmental degradation was only to be
found in the capitalist societies of the West, and that somehow their socialist
economy, as in other Eastern bloc countries, was immune from any impending
ecological catastrophe. Greater awareness by the general Bulgarian public in
recent years has meant that the country's leadership has had to amend some of
these earlier beliefs and revise its views on the growing pollution problems
associated with the country's air, water and soil/vegetation (see Figure 3.1).

HISTORICAL BACKGROUND

During the inter-water period, the Council for the Protection of the Countryside
was founded in 1928; the Council maintained that 'One of the most important, if
not the major question for life in Bulgaria, is an understanding of how to utilize
even the smallest of our resources' (Carter 1978). In 1931 the recreational area
(Silkosija) was established in the Strandzha mountains in 1933 in the Paragalitsa
region (Rila mountains), and in 1934 in the 'Vitosha' National Park, Sofia.

The early impetus of these decisions was lost with the upheavals of the Second
World War, and little preservation of any kind was attempted again until the
1960s. The first state decree for protecting the countryside was passed in 1960
(*Izvestiya* 1960) with further amendments on nature preservation being added in
1967 (*D'rzhaven Vestnik* 1967). In 1969 came the defining of items classified for
protection, with a rather specific list of 'objects of national interest' which
included national parks, recreational areas and observation sites of particular
natural beauty. Towns seen as having historical, scientific or economic value,
together with certain specialist areas of vegetational and animal life were also
listed.

The impact of this law gathered momentum. In 1973, 81 nature reserves, 7
national parks, and 39 places of special interest were designated for protection,
885 sightseeing beauty spots, 712 historical towns, along with several hundred

38

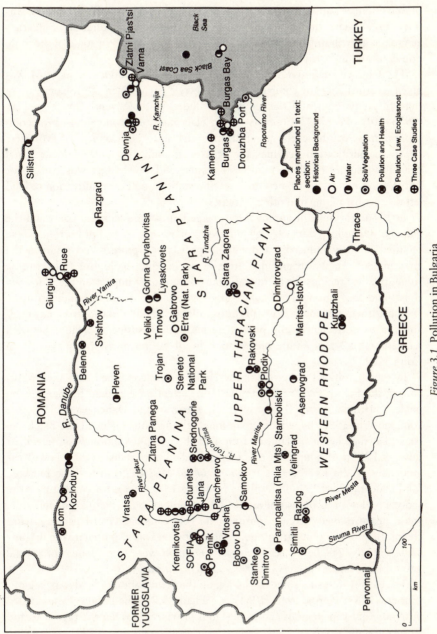

Figure 3.1 Pollution in Bulgaria

species of wild plants and animals. Since 1973, new additions have been added on the preservation list and specific attention has been paid to the protection and preservation of nature reserves and national parks (Popov 1976). Nothing was done to alleviate the growing threat of environmental pollution. Bulgaria's economic growth in the 1950s and 1960s created problems in some of the more populous and industrial regions.

The Council on Environmental Protection and Reproduction in 1974 was given responsibility for combating the country's environmental problems; more than fifty environmental themes were included in the Seventh Five-Year Plan (1976–80); People's Councils had elaborate environmental protection programmes within each administrative district; the Council of Ministers approved purification schemes for areas containing cement works, thermal generating stations and metallurgical plants; and the budget was tripled for purification plants, erosion control, reafforestation and the recultivation of land left derelict by mining and other activities.

In 1976, government approval for protection of the Black Sea coast and the Danube was followed in 1977 by guidelines on environmental protection. These required all new industrial and mining projects to protect and improve the environment. They also included areas of outstanding ecological and scientific value, the best regions for therapeutic purposes, tourist impact on the Black Sea coast, and they initiated a regular survey of erosion and dam building on the Danube. It was now obligatory for the installation of pollution control machinery and the use of scientifically proven norms, thus ensuring a balance between recreational use and nature preservation in all future tourist complexes (State Council 1977).

In theory, official concern for the environment continued into the 1980s. During the Eighth Five-Year Plan (1981–5) attention was paid to the growing nuclear power industry (Khristov 1983, Khristov and Dancheva 1983). At a conference in 1982, papers on environmental questions included the Black Sea coast, Devnja chemical plant, and the Danube region (*Rabotnichesko Delo*, 29 January 1983). Although not officially published, topics discussed followed the guidelines of the 1977 decree (*Ikonomicheska Misl* 1985). They also reflected fund allocation distribution resulting from the 1977 decree for combating more serious pollution in industrial areas (for example, Devnja, Pernik, Srednogorie, etc.); measures were taken to purify water reservoirs and river basins and curb chemical application in agriculture. Reafforestation was also undertaken in some areas. Finally, the Black Sea coast and Danube were singled out for special natural environment preservation (*Ikonomicheska Misl* 1985).

The early 1980s saw some discussion on the delimitation of landscape regions (Daneva 1984; Petrov 1980, 1981) (see Figure 3.2). The country was divided into four landscape regions and twenty-four provinces, sixteen sub-provinces and 127 landscape districts. New nature protection territories (ZPT) were set up (Stoilov 1980 Tishkov 1980). The State Council, in 1983, published further guidelines on basic ecological factors in relation to the food processing industry,

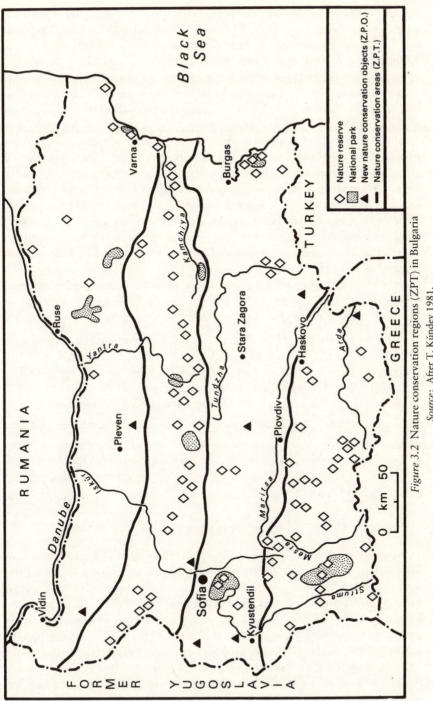

Figure 3.2 Nature conservation regions (ZPT) in Bulgaria
Source: After T. Kündev 1981.

agroecological potential and general development of the economy (D'rzhaven S'vet 1983; *Rabotnichesko Delo*, 29 October 1983). All this gave the impression that the state really cared about the environment.

During the second half of the 1980s, however, all was not well with the Bulgarian environment; an increasing gap between theory and practice was present (Oschlies 1987). The government kept up its facade of environmental care. A national programme for environmental conservation till the year 2000 was adopted in May 1988 stressing the solving of ecological problems at a 'qualitatively new level'. Certain inevitable environmental conservation problems resulting from the country's technical-scientific revolution were admitted, and use of outdated technology for environmental conservation. In future, special attention was to be paid to the rational utilization of natural resources in the long-term interests of society. A new Ministry of Land, Forest and Environmental Conservation was set up, being responsible for a new ecological policy which would be based on better economic and legal management. Finally, a conference on ecological protection in the Balkan Peninsula was planned for later in 1988 (*Tanjug*, 20 May 1988).

This was meant to portray a positive attitude towards the growing threat of environmental degradation. In October 1989 Sofia hosted a meeting on European environmental protection in which it was stated that:

> For many years now we have been trying to solve environmental protection problems, using our economic potential, relying on all public forces and factors, complying with international regulations, we are fully committed to giving the green light to renewal in that sphere as well.
>
> (Press release, Bulgarian Embassy, London, 18 October 1989)

This official attitude towards Bulgaria's environmental pollution problems differed from reality. Here, air, water and soil/vegetation pollution and its effects on health etc., together with case studies, have all contributed to growing dissatisfaction by the people over recent years through the emergence of opposition parties such as Ecoglasnost.

AIR POLLUTION

Throughout much of the post-war period there is a dearth of reliable data on most pollution forms in Bulgaria, especially official government sources. Even with the recent political changes in the country and appearance of environmentally aware pressure groups, there is still little hard information on Bulgarian pollution (Vellev, 1992). Material available supports the view that Bulgaria is probably one of the least serious emitters of sulphur dioxide in the former Eastern block countries. In absolute terms, it has been estimated emissions are between four and five times less than those of Czechoslovakia and the former East Germany (Radio Free Europe Research 1987a); that in itself is no great achievement considering the environmental degradation of the two countries.

Part of the problem is that about half of Bulgaria's primary energy comes from less polluting oil and natural gas resources and their SO_2 emissions per unit of GDP are also the lowest in Eastern Europe. Nevertheless, they remain amongst the 'bad guys' of Europe having consistently increased their sulphur emissions over recent decades (Dampier 1983) and are projected to do so in the future (see Table 3.1). Unfortunately two of the figures in Table 3.1 are predictive, but give some idea of the problem.

Bulgaria is amongst Europe's dumpers in the unpleasant trade of transboundary pollution (that is, one of the states that produce more SO_2 than they themselves receive). Net exports of SO_2 from Bulgaria in 1982 totalled 103,000 tonnes placing them twelfth amongst the European dumpers (Dampier 1983). Even so, in June 1981 Bulgaria was one of the earliest former Eastern bloc countries to sign the Convention on Long-Range Transboundary Air Pollution formulated in 1979; in June 1984 it joined the 30 per cent Club and promised to reduce its 1980 SO_2 level by that percentage in 1993 (Rosencranz 1986). Given that SO_2 emission for Bulgaria in 1980 was 508 kilotonnes of sulphur, the total for 1993 should be 355 kilotonnes (Hordijk 1988). Calculations reveal Bulgaria needs a 16.2 reduction factor if its 1987 SO_2 total of 1,034 kilotonnes attains the maximum permissible level of 518 kg/km^2/year, based on a mean reduction of 90 per cent reduction of its SO_2 (Smit 1988).

In November 1988, Bulgaria also signed the European protocol on nitrogen oxides (NOx) control. This means that after 1994, emissions should not exceed the 1987 level. By the mid-1980s, Bulgaria was about half-way down the European emissions table with 174 kilotonnes of NOx (Slama 1986). Its official total in 1988 was 200 kilotonnes/year. Thus a 3.7 reduction factor is required to reach the necessary permissible level of 6.8 kg/inhab./year if a 75 per cent reduction of the 1980 level is to be obtained (Smit 1988).

Evidence is emerging of atmospheric pollution. In 1987, a Bulgarian journal article examined the general problem of European pollution and then stated, 'Anybody in this country, [Bulgaria] who thinks that this does not concern us because the air we breathe is pure, the water we drink is unpolluted, and there is no danger of acid rainfall as yet, has another think coming' (Radoeva and Ivancheva 1987: 19). It elaborated how between 1978 and 1984 work was temporarily suspended at 133 industrial enterprises and workshops, and at another 62 permanently, because they raised air pollution levels above the norms due to improper use. Of these asphalt manufacture alone totalled 51 and 24 of

Table 3.1 Emissions of SO_2 in Bulgaria (million metric tons)

1972	1982	1992	2002
0.68	0.77	1.18	1.60

Source: N. H. Highton and M. J. Chadwick, 'The effects of changing patterns of energy use on sulphur emissions and depositions in Europe', Ambio XI (6), Table 2, p. 326, Stockholm, 1982.

the erring plants, respectively. It also admitted that one of the first findings in 1977 of the Environmental Protection Committee was that 17 of the country's 983 filtering facilities for trapping substances emitted into the air were out of order, and a further 103 were highly inefficient. By December 1984 filtering devices had risen to 4,100, with only 12 not functioning and another 62 being of little practical use. Further success was acclaimed for preventing harmful substance emissions into the air, which rose from 23–78 per cent between 1978 and 1984. This success (?) was attributed to Bulgaria's signing of the Long-Range Transboundary Air Pollution agreement and the initiation of several projects to trap and recycle the SO_2 emitted by industrial enterprises as well as the national programme to reduce SO_2 emission by 30 per cent.

The same article gave information on cement manufacture. In 1980, the 'Vulkan' cement plant (Dimitrovgrad) deposited between 30–40 tonnes of dust daily, although special electric East German filters were bought in 1968. As production rose these filters worked at abnormally high temperatures (350 degrees C instead of 180) from furnace capacity overloading by 50 per cent and therefore reducing their efficiency. At the Zlatna Panega plant, dust-removing facilities were efficient only from 1966–70. Overcapacity meant white dust clouds covered the area around the plant and local inhabitants suffered from problems of the upper respiratory tract, skin, the subcutaneous tissues, and eyes. Prior to the cement plant, Zlatna Panega had 3,352 inhabitants; 10 years later 44 houses were abandoned and the inhabitants of another 25 were ready to follow. At Beli Iskur cement works, houses were 'stooped like white-haired old men' (Radoeva and Ivancheva 1987: 19).

The misdemeanours of the cement industry illustrate but one of the many air polluters in Bulgaria. Critical in this whole process has been the post-war indus-trialization drive and the country's limited energy resources. Up to the mid-1970s Bulgaria's industrial growth made rapid advances based on a policy in the 1960s which emphasized growth in most major branches. Greatest priority, largely based on Soviet help, was given to the metallurgical industry which now emits 600,000 tonnes of SO_2 and NOx and 40,000 tonnes of dust annually, as well as other dangerous substances. In the first half of the 1970s there was a 9 per cent annual growth in output, higher than the machine-building and chemical industries (Montias 1988). In the 1980s came some refinement with the New Economic Mechanism which was linked to the idea of a 'scientific-technological revolution' (Crampton 1988), involving increased production efficiency and waste elimination to improve quality.

Unfortunately, Bulgaria has very limited energy resources and insignificant deposits of oil and gas. The country had to import 58 per cent of all its energy (crude oil and natural gas) from the former Soviet Union (*Demokratsiya* 2 February 1990: 5). Electricity demand to sustain industry and overall needs comes from a generating capacity estimated at 10,500 megawatts, but perhaps less thanks to antiquated installations working below full capacity (*Rabotnichesko Delo* 23 January 1990; *Vecherni Novini* 1990). In 1988 power-generating capacity came from 8.6 per cent natural gas, 32.5 per cent domestic coal, 17.5 per cent imported coal, 5.8 per cent HEP and 35.6 per cent from the nuclear plant at Kozloduy (Gavrilov 1990a). Thus, nearly nine-tenths came from heavy polluting coal and potentially dangerous nuclear sources. Even more significant, exactly half of the country's electricity is produced from coal, which ejects 2 million tonnes of SO_2 and NOx and 700,000 tonnes of dust annually. A third is poor-quality low calorific domestic lignite. Geological exploration has identified about forty coal basins, with an estimated 4,100 million tonnes of proven recoverable reserves; 99 per cent is lignite with high polluting properties. As a source of electricity, lignite will play an increasing role in thermal power production.

Air pollution from domestic sources come from two major areas, Sofia–Pernik and Maritsa-Istok. The former agglomeration perhaps poses the greatest problem due to its valley location and frequent temperature inversions, preventing pollutants from rising to higher altitudes and escaping (Tishkov 1974). Moreover, prevailing westerly winds carry polluted air from the Sofia-North industrial zone towards residential districts, easterlies blow noxious gases and dust from the Kremikovtsi metallurgical complex towards the city centre and eastern residential areas (Carter 1984). The pollution in Pernik, one of Bulgaria's major industrial areas, comes from a large thermal power-station there which uses low calorific lignite, and from cement and metallurgical industries. By 1970 it was Bulgaria's most polluted town, with prevailing winds carrying polluted air towards residential areas. Since then additional capital for purifying, especially electronic filters, has reduced emissions to more acceptable levels. The other major polluting area is the Maritsa-Istok agglomeration in the eastern Upper Thracian plain (Khristov and Popov 1976). It is based upon large deposits of opencast strip-mined brown coal, which supplies three power-stations. The coal is high in ash, moisture and sulphur (3–5 per cent), and both air and water pollution in the region are excessive in spite of tall chimneys and electric filters.

Other areas of air pollution exist in the country, but these are often in other pollutant forms such as water and soil/vegetation (Vitkov 1991). They include centres like Plovdiv, Gabrovo and the Black Sea coast. The Plovdiv industrial region, which provides a seventh of Bulgaria's industrial output, now suffers from dust pollution 8.67 times above the permissible safety level (PSL = 0.15 mg/m^{-3}) because of its more open site on the Upper Thracian Plain (Zyapkov 1974). Similarly, at Gabrovo, Bulgaria's oldest industrial region, most pollution from post-war plants is quickly dispersed by strong winds in the Yantra valley where water contamination is more serious. The Black Sea coast, with a sixth of the country's production, has some air pollution associated with the Burgas oil-refining complex and various enterprises (for example, chemicals) associated with the Varna–Devnya region. This is critical when onshore winds blow pollutants towards important coastal tourist centres. Finally, there are problems associated with pharmaceutical factory emissions. A special case of transboundary air pollution is at Ruse, across the Danube river from the Romanian town of Giurgiu, site of a chlorine and sodium plant. Chlorine levels are often fourteen times above nationally recognized safety levels. Overall it may be said that it is low calorific coal burning and chemical production that cause continued air pollution problems in Bulgaria, along with other serious pollutants being emitted into the country's atmosphere (Manolov 1985, Carter 1989).

WATER POLLUTION

Bulgaria has a complex drainage pattern characterized, with the notable exception of the Danube by relatively short rivers. There are numerous lakes: coastal (near Varna and Burgas), glacial (the southern mountains), structural, and karstic. Dams have been constructed in mountain areas, whilst there are over 500 mineral springs. Industrial development, however, has been limited by poor hydrological resources, with only 2,490 litres per capita annual volume. The Danube is the only reliable fresh water source attracting chemical and timber processing plants, a nuclear power-station, and hydro-electric power (HEP) projects along its banks. The Danube area provides nearly a tenth of Bulgaria's industrial production and a similar amount of fixed productive assets – for example, Silistra, a complex containing interdependent enterprises for joinery, cellulose, pipes, chipboard and furniture manufacture (Donchev 1983).

The Bulgarian government has always been more open about water pollution than air pollution. By the early 1960s nearly all Bulgarian rivers were polluted. In 1964 an article warned of polluted water problems (Dimova 1964). By 1976 certain positive attempts at combating water pollution were claimed (Dakov 1976). The installation of over 150 purifying plants commissioned during the Sixth Five-Year Plan (1971–5) meant capacity was four times greater than 1971. These included improvements at the Zlata mine, and purification of the River Jablanitsa and Pernik's industrial enterprises. Further, Lobosh dam on the Struma river could now recycle urban effluent; once Samakov's purifying plant was completed it allowed clean water from the upper reaches of the River Iskur into the Iskur dam which supplied drinking, industrial and irrigation water to the Sofia area.

By 1976 measures relating to the pollution protection of the Black Sea provided for collecting about 1,000 tonnes of oil spillage and flotsam from the sea surface around Varna and Burgas. Oil-collecting barges monitored ships' waste (for example, around Burgas), and a helicopter patrolled coastal waters to register ships dumping waste. These measures were Bulgaria's contribution to Black Sea protection from oil pollution by bordering states.

Laudable as all this activity may sound, 60 million tonnes of waste products still enter Bulgaria's rivers annually. About two-thirds comes from livestock effluent. From industry probably the worst offender is chemical production. Intensive industrial concentration in particular areas over the last two decades has inflicted serious damage on the natural environment. Chemical plants have been particularly guilty through leakage of waste products into water courses – for example, large complexes around Stara Zagora and Razgrad. Recent industrial development in Burgas, Pleven and Varna–Devnja poses serious threats to ecological balance. Ten of the country's seventy-five administrative districts periodically suffer from air/water pollution, due to chemical plant emissions (Khristov and Stankov 1982, Khristov and Dancheva 1983). Recent planning has suggested that chemicals, thermal power plants, metallurgy and cement production should be located at least 15–20 km away from the Black Sea coast and the Danube river-bank. There is a need for more careful thought farther inland and locating such industries in the Struma, Mesta, Maritsa and Tundzha valleys.

In 1978, the Environmental Protection Committee's first enterprise to be prosecuted was the 'Verila' chemical plant near Sofia. Industrial waste from the Lesnovska river flowed into Botunets dam, which supplied the Kremikovtsi metallurgical plant. It took seven years to cleanse the river of detergents, lubricants and industrial chemical agents. A purification plant had been planned by 'Verila' in 1974 but was never built; when finally installed in 1980, the initial 10 per cent fine on company annual profits was doubled, indirectly affecting employees' salaries (Radoeva and Ivancheva 1987).

The wide distribution of the country's chemical industry (Dimov 1989) ensures many problems, but other industrial branches are also to blame, including paper manufacture and sugar-beet refining. Industrial agglomerations add their contribution. That covering Gabrovo, Gorna Oryahovitsa, Lyaskovets and Veliki Trnovo in the Yantra valley has added to river contamination from industrial waste (machinery, food, wine and tobacco enterprises), and the construction of biological and chemical purification plants is urgently needed. Plovdiv's industrial region ejects dangerous emissions from a non-ferrous metallurgical plant, a storage battery combine at Pazardzhik, and a cellulose/paper factory at Stamboliski. Attempts at water purification in this important agricultural region have occurred, but recent local protests have demanded the closure of major polluters like the lead and zinc works at Asenovgrad for polluting the River Maritsa, and a chemical works and uranium factory at Rakovski; all are near Plovdiv (*Rabotnichesko Delo*, 4 February 1990). The southern city of Kurdzhali, with its lead and zinc complex, has been blamed for high lead levels in local soils and water supply. Local protesters have succeeded in reducing the output by a third at Asenovgrad and Kurdzhali, with the promise of eventual cessation altogether by 1994 (*Rabotnichesko Delo*, 11 February 1990). At Burgas on the Black Sea coast, the local water supply is unsafe for human consumption (Radio Sofia, 14 March 1990, 3 p.m.). Srednogorie, a town in the Topolnitsa valley, with its copper and sulphuric acid plant, was officially investigated in 1989 to examine treatment of waste water from arsenic contamination (*Sofia News*, 23–9 November 1989: 6). The river's dam (1.1 million m^3) had over 2 mg/litre of arsenic. For years the plant had produced around 35,000 tonnes of sulphuric acid and 10,000 tonnes of sulphur dioxide annually (Manolov 1985).

Even attempts at increasing the country's meagre HEP production have met with difficulties. Two schemes for river diversion threaten the ecological balance of the Rila mountains, whose rivers are already polluted by waste water from hotel complexes. The Rila and Mesta HEP projects hope to transfer water from the West Rhodope mountains to Upper Thrace and from the Rila mountains to Sofia by a complex catchment system. A parliamentary group was established to examine the whole issue (*Sofia News*, 23–9 November 1989: 6), but the whole idea has not been resolved yet (Mihailov 1991).

Finally, and more optimistically, biological and ecological indicators are being used to measure the condition of Bulgaria's rivers (Yaneva, 1986). In 1987, a scheme began in which children collected primary data for the whole country (*Rabotnichesko Delo* 13 February 1987). The results, if and when published, should prove interesting reading (Hristov 1991).

SOIL/VEGETATION POLLUTION

The Bulgarian landscape contains much variety. Open expanses of lowland alternate with broken mountain country, often cut by deep river gorges and upland basins. Over this landscape a natural environment with about twenty different soil types and subtypes has been identified in three main regions. These include fertile black earths (chernozem) and some grey forest soils in northern Bulgaria; the southern more acidic forest varieties provide the most extensive areas, modified with deep, rich humus layers; and mountainous areas with their brown forest, dark mountain forest and mountain meadow soils. Plant and animal life reflects proximity to the Eurasian biogeographic zones, although vegetation displays central European influences. Interspersed between them in the higher areas are arctic/alpine varieties, steppe species in the north-east/south-east of Bulgaria, with subspecies and Mediterranean species in the south.

Post-Second World War economic development has made itself felt on this natural environment. From 1945 to the mid-1960s, Bulgarian agriculture was not only collectivized but also heavily mechanized in order to release labour for industrial development. Chemical fertilizer, not produced in 1939, began to be applied in quantity. Further, by the mid-1970s there was a need for a soil conservation policy to prevent loss of arable land and preservation of its fertility. Soil pollution from the use of noxious chemicals suggested the introduction of less harmful pest control methods.

Former pest control methods have been withdrawn through over use of organic-chloric preparations, as have fertilizers. For example, imported Moroccan nitrates and phosphorous fertilizers were found to contain 100–200 mg/kg of uranium, as well as fluoride. New regulations for the production, trade, storage and use of these highly toxic substances were introduced. Bio pest control using trihogramma was applied instead which reduced costs; by 1974 it was being spread over nearly 50,000 ha for plant protection (Dakov 1976). Bog, salinated and degraded soils were also receiving attention; by the mid-1970s, 264,000 ha of bog and 210,000 ha of salinated soils were mapped for reclamation. Humus collection from construction sites, as well as the recultivation of

11,700 ha devastated by mining and other industrial activity, was undertaken. Arable land (Classes I and II) was restricted in its use for construction purposes.

Unfortunately, these measures were either not strictly adhered to, or they were overtaken by the pace of industrial development. By the mid-1980s there was dissatisfaction with the results. At the 'Man and the Biosphere' UNESCO conference for the socialist states (Sofia 1984), Bulgaria outlined stricter controls on artificial fertilizers and other plant growth preparations, whilst admitting that chemical fertilizer use had continued unabated. Intensive agriculture was the greatest problem coupled with better soil conservation to prevent erosion. In 1983, Bulgaria recultivated 1,960 ha and irrigated a further 6,700 ha of arable land, but wind erosion still proved a problem. In the mining and energy complex of Maritsa-Istok it was admitted that over 16,000 ha of fertile arable soil was lost through opencast mining, with a further 20,000 ha to disappear with further expansion. On a more positive note, Bulgaria had recultivated 3,500 ha over the last two decades, two-thirds in the Maritsa-Istok region; as a result agriculture now covered two-fifths of the region, the rest reafforested (Zabojnikov 1984).

Problems of ecological rehabilitation continued during the mid-1980s, agricultural combines being blamed for decreasing soil fertility (*Rheinische Merkur* 20 June 1986: 4). Further, there was little effort to reverse opencast mining exploitation in the Maritsa-Istok complex; this led to groundwater reduction in an area dependent on irrigation for much of the year (Markov 1978). Recultivation in the Upper Thracian Lowlands is beginning to show some success. A humus layer was placed over the exhausted brown coal beds to reactivate the biological milieu, together with application of azote, phosphorus, potassium etc. Average yields of 350 kg/decar (wheat, barley), and 130–150 kg/decar (sunflowers) have been obtained on this recultivated land. By 1989 9,000 decars of arable land had been recultivated and 13,000 decars reafforested (Kal'chev 1989, Onchev 1991).

Other areas suffered from industrialization, creating a grave ecological crisis. Heavy metals pollute 400–450 decars of arable land with concentrations measured in places at 1,300 mg/kg (lead), 3,000 mg/kg (zinc), 650 mg/kg (arsenic), 22 mg/kg (cadmium), and 3,950 mg/kg (copper). Uranium production (hydrometallurgical process) has affected soils in some regions – for example, Pervomai, Simitli. Around Srednogorie, Razlog, Pernik, Bobov Dol and Stanke Dimitrov, heavy metals have accumulated over 19,600 ha of arable land and a further 13,000 ha have been rendered unusable from dumped industrial wastes (Popilieva 1989). The zinc and lead plant near Plovdiv has been poisoning crops and killing livestock for years (Ekimova-Melnishka 1990); over 300 km² of arable land is unusable after zinc extraction at Plovdiv mines (Dempsey 1990). Heavy metals have polluted some 54,000 ha of farmed land whilst cultivation is banned on another 900 ha (Gavrilov 1990b; Radio Sofia, 27 April 1990, 7.30 p.m.). Ecoglasnost has recently reported that 4.7 million ha of arable land (44 per cent of state territory) was either eroded, polluted or destroyed because 68 million tonnes of dust and toxic gases are emitted annually (Dempsey 1990).

This covers over two-thirds of all agricultural land.

Like soil, vegetation has not escaped this pollution onslaught (Bondev, 1991). Bulgaria has many national parks and nature reserves, and as some are contiguous to industrial areas their vegetation comes under attack. For example, the Planina Mountains, location of post-war industrial complexes (Popov *et al.* 1983) are close to national parks such as Steneto near Trojan (Popov 1989) and Et'ra near Gabrovo. Others include Vitosha (Sofia) and important small Black Sea coastal areas at Zlatni Pjas'tsi (Varna) and Ropotamo (Burgas). By law such parks are not infringed by human activities but their proximity to industrial regions renders them in danger from acid rain. Around lakes (for example, Sreburna Lake reserve) water pollution may affect wildlife such as the Balkan pelican, and a variety of vegetation. The Burgas swamps must be preserved as they are a critical autumn migration point for birds from the Eurasian land mass. The sea lily, a very delicate and fragrant flower, has its natural habitat on the southern Black Sea coast (Dakov 1984); picking it is strictly prohibited as is the picking of water lilies in the Kamchija and Ropotamo rivers, but they are not protected from toxic plumes at nearby industrial complexes (for example, Burgas oil-refinery).

In 1988, 57 per cent of Bulgaria's forests remained undamaged (Painton 1990), but the acid rain threat both from domestic and transboundary pollution remains (Stojanov 1989). Between 26–38 per cent of Bulgaria's rainfall has above average acidity, according to a five-year study completed in 1988. Acid rain has affected more than 520 ha of woodland to varying degrees of which 450 ha are practically destroyed. Government investment in nature conservation (Eighth Five-Year Plan, 1981–5) was increased by a quarter, as well as the construction of green belts around major industrial areas – for example, Sofia, Varna–Devnja, Pernik, Stara Zagora, etc. (Manolov 1985). This is probably too little, too late, but at least showed some official concern for the country's natural environment.

POLLUTION AND HEALTH

Overall, Bulgaria's demographic trends after 1945 suggest general improvement related to social and economic changes. The introduction of free medical care and better working conditions are attributed to a significant decline in death rate; in the 1970s its rise again was largely due to an ageing of population structure. Birth-rate also declined as did infant mortality. In the early 1980s the demographic pattern was characterized by both low birth- and death-rates with a similar trend in natural increase (Mitchev 1983). Profound demographic changes have also accompanied recent industrialization. Large-scale urbanization through rural to urban migration and increased exposure to environmental pollution and noise, accompanied by a growing incidence of stress, shocks and accidents, have all helped damage living conditions and health.

A significant factor in this process is the role played by environmental pollution. Repeated public demonstrations, medical information exhibits, and demands for some investigation, led to official calls for a 'radical restructuring' of Bulgaria's environmental protection policies in May 1988 (Economist Intelligence Unit 1988). A timely study by six Bulgarian polyclinics in 1989 of around a quarter of a million inhabitants found an alarming increase in children's respiratory and circulatory diseases around Sofia and eight other cities – all resulting from high pollution levels (Economist Intelligence Unit 1989). It is generally recognized that over a third of Bulgaria's population is exposed for several days annually to strongly polluted air. By the mid-1980s indications were present that pulmonary diseases were increasing at a significant rate in Bulgaria (Sullivan 1985), but official data was lacking.

Public pressure has recently led to a growing dossier on environmental health hazards. In November 1989, a report on the Razlog valley, which has pronounced temperature inversions, showed that the incidence of respiratory diseases was fourteen times higher in 1984 than in 1970. Further, in 1970 nervous disorders numbered 463/10,000 inhabitants, rising to 1,524 (an increase of 10.6 per cent) by 1984. Blame was squarely placed on the Hydrolysis and Yeast Fodder Factory and the Pulp and Paper Mill for creating two-fifths of the pollution in the Razlog area. For two decades they had been the chief polluters, due to worn-out equipment and no modern waste treatment facilities. Both plants were ordered to close by early 1990 (Popilieva 1989). Even so, much damage had been done; investigations in the valley between 1979 and 1984 showed hydrogen sulphide levels four times above the permitted safety level (PSL), and carcinogenic chemicals forty times higher.

During 1990, pressure from Ecoglasnost and the Union of Democratic Forces revealed that prospects of Bulgaria reaching the 30 per cent reduction of SO_2 emissions by 1993 remain bleak; this is of little comfort to people living in polluted regions. According to doctors in the Plovdiv region the death-rate has doubled over the last two decades. Local residents around the Rakovski uranium plant have the highest death-rate for certain cancer types in Bulgaria, as well as highest mortality. The death-rate of Rakovski's children (39.1/1,000) is treble the national average (Radio Sofia, 29 February 1990, 9 a.m.). In the Plovdiv area, the lead content in children's blood is dangerously high (Gencheva, 1990). In Kurdzhali region, the lead and zinc plant is blamed for the rise in birth defects/stillbirths, together with the number of cancer and cardiac ailments which are well above the national average (*Rabotnichesko Delo*, 9 March 1990).

Other parts of the country have also suffered. In Stara Zagora, dust pollution emissions at 3 million particles/litre are well above permissible levels, causing a high incidence of certain illnesses amongst local children. A similar situation was noted in Vratsa, an industrial town north of Sofia. In the Vratsa region the death-rate is 264/10,000; respiratory diseases total 7,526 (state norm 93), as well as high numbers of heart and brain disease cases. In the Burgas region high dust levels and an unreliable water supply have caused a growing number of illnesses (Radio Sofia, 12 February 1990, 11.30 a.m.).

Another worry for Bulgaria's population is the growing threat from the use of domestic nuclear power generation for electricity production. During the 1970s the country's first nuclear power station was built on the Danube River at Kozloduy downstream from Lom, with a 1,760,000 kilowatt capacity (Carter 1976). By 1981, this station was producing a quarter of the country's electricity needs (Khristov 1983) and by 1987 it was nearly a third. Plans exist for a second nuclear power-station downstream from Kozloduy at Belene, scheduled for the 1990s. Other sites have been planned so that by the year 2000 nearly two-thirds of Bulgaria's electricity supply could come from nuclear energy (Sobell 1988).

Memories of the Chernobyl accident in 1986, and its long-term impact on health, have generated public protest rather than electricity. The Kozloduy plant has a history of accidents and technical problems – for example, in November 1989 an escape of radioactive material (*Otechestven Front*, Sofia, 30 November 1989) and in early 1990 emergency shut-downs (Radio Sofia, 10 April 1990, 11 a.m.). There were three 'near misses' at Kozloduy in 1977, 1982 and 1983, when Chernobyl-scale catastrophes were narrowly averted (Gavrilov 1990b), and plans for shut-down and refurbishment are being planned (*The Times*, 10 July 1991). Public disquiet about nuclear power production in Bulgaria is growing. No reliable nation-wide radiation monitoring service, or necessary health equipment, is available to cope in case of nuclear accident. The planned Belene plant, in an earthquake-prone region (*Trud*, Sofia, 8 February 1990: 3), has received strong objections in Svishtov, a town 10 km downstream, and construction has been stopped (*Narodna Mladezh*, Sofia, 27 February 1990).

Further discussion on nuclear problems continues. Elsewhere in Bulgaria, the problem of nuclear waste disposal has emerged now that the former Soviet Union has stopped its importation. Velingrad, in southern Bulgaria, is a possible site but has met local disapproval on health grounds. Radioactive pollution also worries residents of three villages near Sofia (Jana, Gorni and Dolni Bogrov); a long-established factory using urane and polymetal ores has polluted arable, pasture, and water supplies with Radon 226. More generally, most people's radiation comes from hospital X-rays, where high-dose scopic methods are used due to lack of proper film supplies.

POLLUTION, THE LAW, AND ECOGLASNOST

A number of conservation measures have been introduced since 1945. These include steps to protect air, water, and soil from pollution and to save areas of outstanding natural interest. The Bulgarian Constitution (Article 31), codifies the state's role in nature conservation, rational use of natural resources, and environmental protection (Starzewska 1988). The first ecological law was passed in 1960 against air, water, and soil pollution. This reversed earlier complete disregard for environmental protection through the establishment of heavy polluters – for example, metallurgical combine (Pernik), copper plant (Srednogorie), and ironworks (Kremikovtsi, Sofia). A spate of legislation after 1960 was accompanied by 'Ecology Days' and 'Ecology Weeks'; ecology became a compulsory school subject. A number of official ecological organizations were formed: the largest (National Movement for the Protection of the Environment) with over 7,200 branches (Ashley 1987).

The guidelines of 1977 gave a clear statement on environmental protection legislation. Environmental protection was enhanced by decrees which observed ecological requirements. A plan was approved for the systematic codification, updating and refinement of conservation legislation, based on environmental protection. State legal bodies were instructed to improve legal measures on the environment, provide stricter sanctions and demand full damage indemnity. A new Committee for Environmental Protection was responsible for ensuring that the new laws were observed (State Council 1977). These measures were responsible for curbing environmental degradation up to the year 2000 (Daneva and Vaptsarov 1978).

By the mid-1980s it was evident that this legislation was not working correctly. Formulation of these laws and principles appeared strong on paper. For example, general principles on prevention of occupational chemical hazards were contained in the Labour Code, the Social Health Law, and several additional regulations (Spasovski and Hristev 1983). Complaints were made in 1987 that factories discharging harmful substances above established norms were subject to penalties, but courts often dismissed them thus making sanctions ineffective (*Zemedelsko Zname*, Sofia, 26 January 1987: 2).

A new ecological policy in 1988 improved environmental law but failed to remove earlier legal inconsistencies. Also, the dichotomy between ecological needs and pressure for more efficient, profitable state enterprises failed to be resolved. There was improvement in monitoring/control of foodstuffs and live-stock feed after Chernobyl (Ashley 1988). No mention was made of state funding for its implementation; the end result was like the 1980 programme for minimizing industrial waste and recycling water and effluent: it became less and less effective over time. In theory, the new policy offered stronger and more unified environmental norms, economic incentives for greater environmental protection, and prosecution of offenders breaking these laws. This implied that previous law breakers had been treated far too lightly. Until late 1989, any environmentalist group accusing the state of its failure to apply these laws led to goal sentences. Meanwhile, more ineffectual decrees were passed against environmental degradation (Radio Free Europe Research 1989a, Kovachev 1991).

The demise of the former regime, in November 1989, meant a more favourable milieu for the emergent green parties, the most notable entitled 'Ecoglasnost'. In Bulgaria more than 3.1 million people live in areas where air pollution is above the sanitary norm (Bachvarov 1991). By late 1988 these inhabitants had witnessed the birth of an independent ecological group (Party of the Green Masses) in north-western Bulgaria; it was a response to Danube pollution problems, including Romanian chemical emissions over Ruse. In March 1989 a nation-wide environmental pressure group. 'Ecoglasnost', was founded and received official registration by the year's end (Pehe 1990). 'Ecoglasnost' demanded uncensured environmental information, access to medical statistics, nuclear waste dump locations, evacuation from areas above the permitted safety levels, more protected national parks, free press, arrests for previous environmental crimes, technological innovations, nation-wide ecological reporting network, and ecology to have university subject status. It maintained that 'the degree of an individual's responsibility in decision making is in inverse proportion to his actual suffering caused by ... environmental pollution' (Radio Free Europe Research 1989b). The movement has ninety local branches and was part of the Union of Democratic Forces, which obtained over a third of the parliamentary seats in the June 1990 elections. In December 1989 another ecological movement, 'The Green Party' was founded. It joined the Union of Democratic Forces (UDF) and has fifteen branches in Sofia and eight in the provinces. About twenty other ecological groups with no political affiliations exist in Bulgaria, some with limited aims – for example, cancellation of Belene project (Anon 1990).

CASE STUDIES

1 The case of the capital

Sofia, the Bulgarian capital, spreads out at the foot of Mount Vitosha into a basin plain almost surrounded by mountains (see Figure 3.3). It has a temperate and continental climate. Over 1 million residents live in the Greater Sofia region around a site inhabited since prehistoric times. Since 1945, industrial development, and a large immigrant influx from throughout the country, added to its importance (Carter 1973), but threatened its pre-war 'garden city' image making it essential to preserve areas of greenery (Panev 1972).

Heavy industrial development, and location in a partially closed basin, led to increasing pollution problems. (Hristov 1966). These have been intensified over the last two decades (Carter 1987). A third of the Sofia region has been affected by harmful emissions (1.5 times above national average) and only 15 per cent of the water is treated for drinking purposes (Popilieva 1989). This pollution, including that of the River Iskur, has resulted in the Sofia basin suffering a marked decrease in animal life and vegetation. This was becoming obvious in the 1970s, and in January 1982 a protection programme for the capital was initiated. The result was that 25 air polluting enterprises, 10 garages/vehicle service stations, and 49 warehouses were relocated outside the city. Devices for air and industrial effluent purification and noise reduction began operation. Quarries and infertile/contaminated land covering 312 ha were recultivated. Water conservation through circulation cycles saved, annually, 7.5 million m^3 in the city and a further 5 million m^3 at Kremikovtsi ironworks.

In spite of these measures industrial pollution outpaced conservation. By the mid-1970s there was evidence of serious air pollution in the capital (Bl'skova *et al.* 1979). Dust, SO_2 and NOx data for 1974–8 from various parts of the city

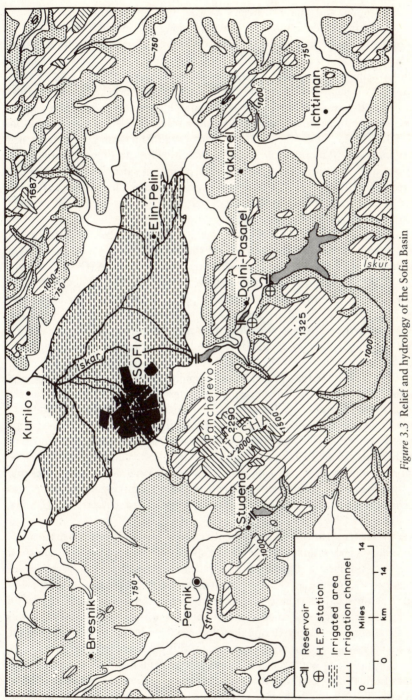

Figure 3.3 Relief and hydrology of the Sofia Basin

Source: F. W. Carter 1973, p. 26.

revealed annual and diurnal changes. During autumn months temperature inversion produced high SO_2 levels, with a longer-term study (1964–78) of atmospheric dust contamination giving results well above the permitted safety level in 1968, 1973, and 1977 (Bl'skova 1983). Another study has shown how industry helps reduce Sofia's winter temperatures (Velev 1986).

The city's main environmental problem is air pollution. A report in 1987 stated that the average dust level was 2.6 times higher than the permitted safety level, hydrochlorine three times higher, and nitrogen oxide double (*Zemedelsko Zname* Sofia, 26 January 1987: 2). By March 1990, Sofia's air pollution reached appalling levels (*Rabotnichesko Delo*, 17 March 1990). At all twelve monitoring stations dust concentration was above the permitted safety level with the city centre highest (quadruple). At Botunets near Kremikovtsi ironworks, SO_2 levels reached 17.3 times the permitted safety level with high nitrogen dioxide, phenol and lead levels around the plant and city centre.

In the early 1960s, the Kremikovtsi metallurgical plant, located in north-east Sofia, became Bulgaria's largest iron and steel works. Locally used ores have a low iron content (28–31 per cent) with a high burn-off rate (Oschlies 1986). Some 200 tonnes of dust are ejected daily, along with 120 tonnes (SO_2), 170 tonnes (NOx), 320 tonnes of carbon oxides, phenols, cyanides, lead aerosols, as well as 2.5 tonnes of mercury annually. As a showplace of Bulgaria's industrial development the Press was not allowed to criticize it until recently (Atanassov 1989). Poor local iron ore should be replaced by better imported varieties (Radio Free Europe Research, 1987b). Logically, the plant is so outdated it should be closed, but instead natural gas will replace lignite for local power generation and certain dangerous alloys will not be produced until filters are installed.

Sofia's water is also a pollution problem. Microbiological and chemical tests on Sofia's reservoir drinking water only recorded an 85 and 53 per cent purity, respectively. Again the north-eastern residential suburbs around Kremikovtsi suffered most with high zinc, copper, chrome, lead, manganese and arsenic concentrations in the water. Also near the works, a large slurry pond stores ore-dressing waste (Gerzilov 1989). Only half Sofia's drinking water undergoes mechanical purification, the rest is chlorine-treated (*Zemedelsko Zname*, Sofia, 29 September 1987: 2).

In the eastern part of Sofia runs the River Iskur, regarded by local residents as the country's largest and most stinking cesspool. The river receives not only most of the city's domestic effluent but also industrial waste; its purity is the lowest category (IV), and is useless for irrigation or industrial purposes. A sewage plant, said to be the Balkan's largest, constructed near the Iskur river, was planned with a 7 m. sec. capacity; due to error it had only 4.5 m. sec. on completion. It was initially planned to supply 40,000 decars of cropland in the Sofia basin, plus the Kremikovtsi works, but is now unsuitable for this purpose. Financial stringencies have prevented any improvements. Water for the capital and these projects now comes from Pancherevo reservoir, which recent dry summers have stretched to capacity (*Trud*, Sofia, 16 March 1988: 2). Soil erosion in the Sofia basin due to insufficient water has limited its growing area.

Health repercussions on the local populace have resulted from all this pollution. The Kremikovtsi plant and nearby uranium mine are the most serious health hazards. Between them lies Jana village, 15 km east of central Sofia. Only 10 per cent of children are healthy, and many other villages in Sofia's northern outskirts are similarly affected (*Narodna Mladezh*, Sofia, 8 March 1990). Noise pollution in Sofia is also a problem. In the mid-1970s, daytime levels in some parts of the city were 10–24 decibels above normal (Dakov 1976). A legacy of small- and medium-sized road junctions also contribute through a confused intra-urban network (Carter 1979); the existing road pattern cannot cope with the present traffic, and noise levels reach 20–40 decibels above tolerable limits at central crossroads, streets and local parks. Further aggravation is experienced from waste processing plant construction as industrial enterprises are reluctant to build their own (*Pogled* 1987). Once the city's new underground metro system has been completed, noise levels may be partially reduced.

2 Ruse, and transboundary pollution

Ruse, located on the Danube, is Bulgaria's fourth largest city with 200,000 inhabitants. It has been subjected to transboundary air pollution for several years from Giurgiu, a city on the opposite Romanian shore. In the early 1980s its chemical plant became operational based on Soviet technology; the production of caustic soda (200,000 tonnes), chlorine, and sodium-based products were exported to the USSR. This plant, together with fourteen others, was planned for the Danube's left bank under a 1986 Romanian–Soviet agreement. Assistance was to come from the Soviet 1986–90 Five-Year Plan. Little attention was paid to safety measures in the plant's drive for increased output. Much of the augmented chlorine emission drifted over Ruse across river. Evidence of this intensified air pollution in recent years is seen in Table 3.2.

No effective anti-pollution measures were made to protect Ruse's residents from the dangerous chemical emissions which affected both air and soil (*Pogled*

Table 3.2 Recorded gas emission levels in Ruse, Bulgaria, 1982–7

Year	No. of Gas Emissions	Times above PSL
1982	26	all 3 ×
1983	33	> 5 ×
1984	56	> 9.2 ×
1985	47	> 6 ×
1986	50	> 6 ×
1987	69	> 9 ×

Source: *Dunavska Pravda*, Ruse, 6 December 1987.

1987). The heavier than air chlorine gas causes nausea, inflammation of mucus membranes, and choking coughs. Public protests in November 1987 followed two days of dense chlorine gas cloud drifting over the city.

Severe chlorine gas emissions from the Romanian plant occurred again in early February 1988. Further public outcry followed with a demonstration by 2,000 Ruse inhabitants. Inquiry evidence revealed that for sixty days annually, average chlorine and mineral acids concentration in Ruse exceeded the permitted safety level. Heavy chlorine gas emission days increased 5 to 8 times, mineral acids 20 times (*Literature Front* 1988). Significant in the longer run, this protest led to the immediate formation of Bulgaria's first independent organization: the Independent Committee for the Protection of the Environment. It was soon followed by others elsewhere, and contributed eventually to the overthrow of the old regime.

Repercussions on health standards caused by these emissions are now clearer. There has been an alarming increase in various lung diseases in Ruse, openly attributed by medical specialists to the Giurgiu plant. Prior to the plant's operation, only 969/100,000 inhabitants suffered any lung disease in 1975; by 1985 this reached 17,386/100,000 (an eighteenfold increase); in 1986, 86,228 children and 62,138 adults attended out-patient treatment with lung diseases, and a further 2,924 children and 1,546 adults were confined to hospital. On average, ten days' employment annually were lost by Ruse area workers from temporary lung disability. (*Starshel* 1987). 'Ruse lung' creates massive changes in respiratory organs and is diagnosed as toxic fibrosis. A third of the elderly die from lung disease, and four-fifths of the region's youth are unfit for military service. Increases in asthma, suffocation, nervous breakdown, and skin and eye inflammation have occurred. Infant mortality, abortions, stillbirths and deformed births are a potent reminder of Giurgiu's chemical plant. Moreover, during humid weather in the vicinity of Ruse, chlorine changes to hydrochloric acid, affecting vegetation, bird life and building façades alike. Efforts have continued to reduce pollution, not only in Ruse but in the whole Danube ecological region (*Rabotnichesko Delo*, 4 April 1990).

3 Black Sea coastal pollution

Bulgaria's coastal zone suffers from serious pollution, especially around Varna and Burgas. An eighth of Bulgaria's total population resides along the Black Sea coast, together with a fifth of basic industrial investment, an eighth of industrial employees, and a sixth of industrial production. Cheap sea transport and increasing economic links with the former Soviet Union, Arab and African countries encouraged development. It proved attractive for industry with access to land and sea transport, and local supplies of coal, copper, manganese, natural gas, marble, quartzite, and rock salt. Two major areas arose at Varna–Devnja in the north and Burgas–Kameno in the south; also, imported raw materials from the USSR and Middle East are processed and re-exported. Harbour deepening at

Varna and Burgas accommodated larger vessels. Varna was linked to the industrial complex at Devnja by canal. Devnja produces industrial chemicals (mainly soda), fertilizers and chlorine. There is a thermal power-station between Varna and Devnja, and shipbuilding on the coast. Farther south, Burgas–Kameno specializes in oil-refining and petrochemicals using (the former) Soviet supplies. Beyond these two industrial areas are important tourist centres (see Figure 3.4). In theory, industrial zones should not be located near tourist complexes; in practice, this would restrict future economic growth of the Varna–Devnja canal area. Stricter protection of internationally significant small nature reserves (for example, the Atanassov Lake bird sanctuary) and other nationally and locally important reserves must be made to prevent trespass and damage by tourists (Carter 1978).

For years this scenario has given rise to ecological concern. A 'basic directive' on environmental protection of the Black Sea and its coastal zone appeared in 1976 (*Rabotnichesko Delo*, 16 March 1976). It stressed the urgent need for installing purification equipment between 1980 and 1985, and for renewal, improvement and enhancement of the environment by 1990. Much of this directive is unrealized, but environmental studies have been made – for example, re-afforestation and forest management in the Varna–Devnja industrial complex area. Its forests cover a quarter of the region and research suggests planting trees more resistant to air and soil pollution (for example, *Quercus cerris*, *Quercus frainetto*, *Carpinus orientalis*, *Rhubinia pseudoacacia*, *Pinus nigra*) will improve the local microclimate (Gorunova and Mourgova 1988).

Both the Varna and Burgas industrial complexes have negative environmental effects. Waste products from them contribute and include noxious gases (hydrogen sulphide, carbon and nitrogen oxides, nitrogen dioxide, etc.), dust particles, polluted water, and middling slime. Factory effluent has meant a decrease in mussels (for example, the *Meretrix rudis*, *Tapes rugatus*, black *Mytillos galloprocimialis* varieties) in coastal areas; inland coastal lakes have a proliferation of poisonous unicellular seaweed (*Prymnesium parvum*), resulting from increased water mineralization and subsequent fish mortality. Some zooplankton (for example, *Copepoda*, *Cladocera*) has disappeared. Dust particles reduce local wool's durability/stretching qualities, and milk quality has declined. Sunflowers, sugar-beet, fruit and vines, all especially susceptible to air pollution, have decreased in output per decar.

The Burgas–Kameno area suffers mainly from local petrochemical pollutants. The refinery's purification equipment is woefully inadequate to deal with the 6,000 m^3/hour capacity, at 0.05 mg^{-2} mg/litre petroleum concentration. Also nitrogenous peroxide, lead aerosols, hydrogen sulphide, formaldehyde and sulphur dioxide, along with high phenol concentrations in the effluent waters of Burgas Bay affect both fauna and flora. Added to all this is effluent ejected by oil tankers unloading in Drouzhba Port, near Burgas (Donchev 1979). Much of the southern Black Sea coast is above the permitted safety level; swimming is forbidden in places and shore birds suffer – for example, swans from eye and beak tumours leading to blindness and death. No funds for improving purification

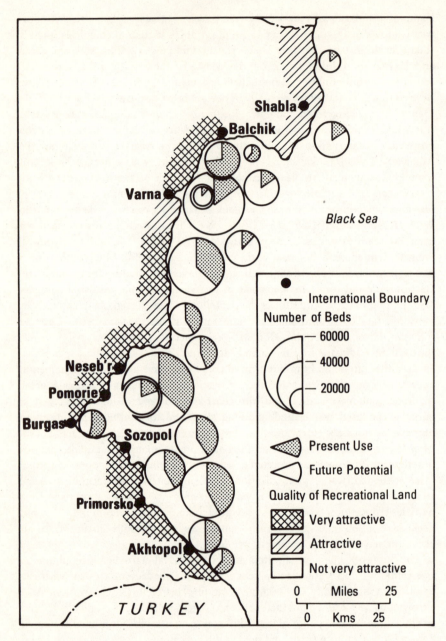

Figure 3.4 Tourism on the Black Sea coast of Bulgaria, 1985
Source: G. Stankov *et al.* 1985.

equipment at the Burgas plant have been provided over the last decade.

There is growing concern about the Black Sea itself. Bulgaria's 378-km coast-line faces on to a near catastrophe. Bulgaria's Institute of Ecology has warned that in a decade the Black Sea could be completely dead. Chemical poisoning from it would then begin atmospheric pollution. It collects ten times more water per m^2 of surface area than other seas, largely from major polluted European rivers. The Danube is worst, carrying water from eight highly industrialized states which also have intensive chemical agriculture. Freshwater influx into the Black Sea has been reduced by 50 km^3 annually from dams and irrigation projects (Hanley 1990). This largely enclosed sea has minimum vertical oxygen flows from the surface downwards, allowing hydrogen sulphide to build up, reducing the oxygenized level to only 60 metres in places. This allowed the proliferation of 'red bloom' and other algae along the southern Bulgarian coast during summer 1990, yet fourteen years earlier the Bulgarian Black Sea project claimed effective plans for coastal shelf use (*Rabotnichesko Delo*, 30 December 1976). Such plans went unheeded, and by the late 1980s Bulgaria's coastal shelf was again demanding attention (Kr'stev 1988). The impact of Chernobyl radioactive fall-out on the Black Sea (Anon 1987), is but one of many problems facing this 423,000 km^2 area of polluted sea.

CONCLUSION: THE FUTURE?

In Bulgaria, environmental destruction may be the former government's most enduring legacy. The country's residents have sound reasons for alarm at the pending ecological disaster. In some parts of Bulgaria environmental degradation is reaching critical proportions and although changing environmental attitudes expressed through the emergence of green parties have created greater aware-ness, the environmental movement has a long, hard struggle ahead. Rethinking of attitudes on economic prosperity, new types of industrial progress, greater attention to material production, and more rational natural resource use all have to be achieved.

A crucial gap exists between various economic sectors and the planners; the latter set targets, often failing to provide means for their fulfilment or appreci-ation of environmental dangers incurred. In the past, equipment and technology lacked environmental considerations and pollution offenders were treated lightly. In the long term, comprehensive solutions to the country's ecological problems have to be achieved, probably through external aid for specific environmental purposes.

Solving environmental problems in the immediate future demands short-term practical answers. Smoke treatment installations at thermal power-stations, reduction in energy-intensive industries, changeover from low- to high-quality steel production, abandonment of uranium extraction using ground-injected sulphuric acid, introduction of catalytic converters, and cancelling propane gas tax on car fuels, may provide some of these answers. Financial external aid for such measures could ease the burden.

Publication of a draft conservation and environmental regeneration programme up to the year 2000 and beyond may give some clues on achievement (Hall 1989). A reduction of dust emissions (by 95 per cent), a 35 per cent drop in gas emissions (over 1987 levels), a 30 per cent cut in sulphur oxides (by 1994), and nitrogen oxides pegged to 1987 levels have an over-optimistic ring. Trans-boundary air pollution must end with international agreement (for example, Ruse). Water pollution improvement through the construction in all urban centres over 100,000 inhabitants of effluent treatment plants, 85 per cent recycling of industrial waste for river protection, and special consideration for the Danube and Black Sea, sound feasible if expensive. More effective use of natural fertilizers (85 per cent), reduction in pesticides, biological pest control, deciduous reafforestation plans, and appraisal of wildlife sources are sensible given the right discipline and will-power. Unfortunately, 3 per cent of national income and up to a tenth of all state investment seems insufficient to achieve all these recommendations given the present constraint on the country's finances. Good intentions and reality have to be linked, but more than forty years of state domination over Bulgaria's society does not augur well for such hopes. It is the only way forward in the 1990s if environmental disaster is to be averted (Bozhanove 1991).

ACKNOWLEDGEMENTS

I would like to thank Delcho Vichev, Secretary of 'Ecoglasnost', for recent environmental data, and Louise Saunders of the Cartographic Unit, Department of Geography, University College London, for the illustrations.

4

CZECHOSLOVAKIA*

F. W. Carter

*Two independent states from 1 January 1993

INTRODUCTION

Czechoslovakia is one of the most polluted countries in Europe. The basis of its environmental problems is associated with the economic and industrial planning of the past, this planning having changed considerably since the Second World War. Whilst Czechoslovakia had an industrially based economy during the inter-war period, the adoption of the Soviet socialist political system in 1948 intensified the situation. Emphasis on large-scale investment in heavy industry from the 1950s onwards largely contributed to the seriousness of the present state of environmental degradation (see Figure 4.1).

Whilst post-war evidence of pollution was known to exist in many parts of the country, until recently attempts to assess the scope and severity of its impact were onerous because data was either inaccessible or at best fragmentary. Such conditions proved confirmation difficult, not only for foreign observers but also for local planners and scientists (Seger 1986). Moreover, the official government line continued to play down the issue of pollution maintaining, as recently as 1985, that under socialism people were cared for and caring, and that this included the environment which was the source of much happiness for those people (*Rudé právo*, 1 March 1985: 4). Reality has proved much different from such idealized statements.

Environmental quality has clearly deteriorated as a result of human activity; the major cause is an excessive and inconsiderate extraction of natural resources, extensive waste emissions and failure to observe ecological and aesthetic laws. These were compounded by an inefficient economy, which consumed inordinate amounts of raw materials and energy, based on outmoded technology which produced manufactured goods with little respect for the ecological consequences. This sad situation was further aggravated by inadequate financial resource allocation for environmental protection, which was of a remedial character rather than one of damage prevention. Much of the blame for this state of affairs must be laid upon the Communist government over the past forty years when legislative, executive and political power was concentrated in the hands of a small controlling group (*nomenklatura*) who did little to correct adverse effects on the

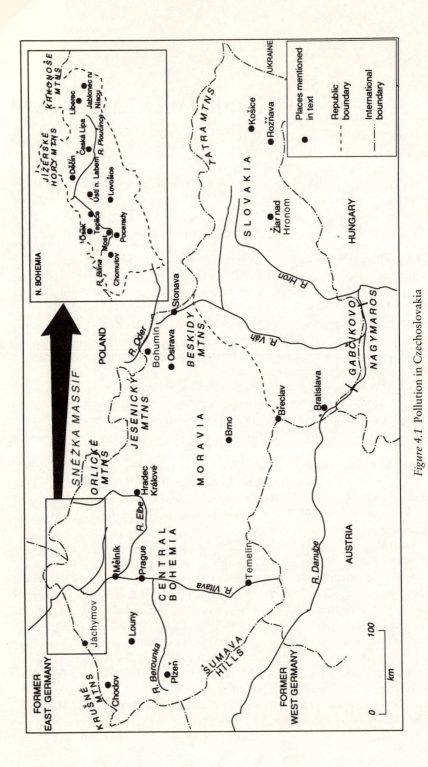

Figure 4.1 Pollution in Czechoslovakia

environment caused by their policies. Added to this detrimental domestic attitude was the significant contribution made by transboundary pollution from neighbouring states, particularly along the northern and western boundaries of the country.

Attempts to assess environmental damage in Czechoslovakia over that forty-year period proved difficult due to the scattered nature of material on the subject in Czechoslovakia and the fairly limited data resources published in the country. By the mid-1980s some appreciation of the problem had appeared in Western literature, including a geographical appraisal of post-war pollution (Carter 1985), and brief surveys on environmental policies and politics (Wolchik 1991: 279–84, Albrecht 1987).

AIR POLLUTION

Air pollution is a serious problem in Czechoslovakia, yet even in the early 1920s it was a topic of serious debate amongst the country's scientists and agricultural-ists (Anon 1988). Two years after the Second World War and before the Communist takeover, the first instances of tree damage by air pollution were recorded in northern Bohemia. The late 1950s saw a sharp increase in SO_2 emission, this reaching 2.50 million tonnes by 1960. The beginning of the 1960s also experienced the first large-scale evidence of forest destruction (Pohl 1985). Thus, Czechoslovakia's air quality has been a national scandal for decades. Currently, the country is blanketed with about 7.6 million tonnes of polluting gases – mostly sulphur dioxide, nitrogen oxides, and carbon monoxide – along with dust, ash and soot from its own industries, power plants, motor vehicles, and residences. Moreover, throughout the post-war period, the Czech Lands (Bohemia and Moravia) seem to have suffered far more than Slovakia. This geographical differentiation is borne out by recently published data (see Table 4.1).

According to Czechoslovakian sources in 1988 the country produced about 8,000 tonnes of gaseous and solid emissions annually (Czechoslovakian TV, 5 December 1988, 3 p.m.) of which sulphur dioxide comprised nearly two-fifths, and carbon monoxide a further sixth, the rest comprising of dust, nitrogen dioxide and other pollutants (*Svobodné slovo*, 10 December 1988: 4). More

Table 4.1 Air pollution in Czechoslovakia, 1986

Region	Area(km^2)	Dust $t/km^{-2}/yr^{-1}$	%	SO_2 $t/km^{-2}/yr^{-1}$	%	NO_2 $t/km^{-2}/yr^{-1}$	%
Czech Lands	78,864	8.7	72	24.0	81	9.0	81.7
Slovakia	49,036	5.5	28	9.1	19	3.3	18.3

Source: Vavroušek, J. and Moldan, B. (eds), *Stav a vývoj prostředí v Československu*, Praha, 1989, p. 5.

recently, it has been stated that Czechoslovakia's SO_2 emission is nearly 3 million tonnes. Of this total, the twenty-five largest sources (most thermal power-stations) produce 1.6 million tonnes, of which approximately half is in the North Bohemian Coal Basin (Klášterská 1991). Furthermore, Czechoslovakia imports nearly 2 million tonnes of transboundary pollution every year (mostly from the former East Germany), whilst exporting two-fifths of its own total to neighbouring countries (mostly to Poland) (*Rudé právo*, 22 November 1988: 4) (see Figure 4.2). Altogether this makes Czechoslovakia one of the worst atmospheric polluters of SO_2 in Europe in absolute terms (along with the former East and West Germany, and the United Kingdom). It is also, after the former East Germany, the worst polluter per unit of Gross Domestic Product (Sláma 1987), the predominantly westerly winds helping to make it one of the worst affected countries for atmospheric pollution in Europe.

Scientists and government planners agree that the worst pollutant is sulphur dioxide generated by thermal power plants and industrial boilers burning poor-quality brown coal (lignite). In the absence of alternative hydrocarbons, national energy policies and tight budgets have tied the country to brown coal, but its low calorific value means that large quantities must be burned. This state of affairs has been further aggravated by regional 'imbalances' due to concentrations of heavy industry with coal mines and polluting power-plants. The worst air quality in Czechoslovakia is found in north Bohemia, where over two-thirds of the country's brown coal is mined and nearly half of its thermally generated electric power-stations are located. These power-plants are found near opencast brown-coal mines which gouge out the material with enormous scoop-shovels. Brown-coal reserves here are predicted to last another half century at the present rate of consumption. Other places seriously affected by poor atmospheric conditions include major cities such as the capital Prague, Bratislava and Brno, along with the industrialized regions of central Bohemia and north Moravia (especially around Ostrava).

Bohemia's urban and industrial areas have been particularly prone to air pollution (Moldan 1990: 54–6, Moldan 1991a). Towns like Most, Chomutov and Ústí nad Labem are infamous for their choking smogs, especially in the winter months when temperature inversions are common. In central Bohemia, Prague's 1.3 million residents suffer from elevated concentrations of sulphur dioxide twenty times above the 'permitted safety level' (PSL = 0.15 mg/SO_2/m³). Plzeň, in western Bohemia, centre of heavy engineering and armament factories (as well as beer) is stricken with similar problems. In Slovakia, Bratislava suffers air pollution from chemical and petro-chemical plants, power-stations, and engineering and cement works, which along with paper mills adversely affects some half million inhabitants.

Motorized vehicles also have added their share of pollution. Czechoslovakia has the largest number of cars per inhabitant in Eastern Europe (one for every six people). Many of these vehicles are highly polluting, such as the Czechoslovakian produced Skoda and Tatra, or the former East German Wartburg and Trabant

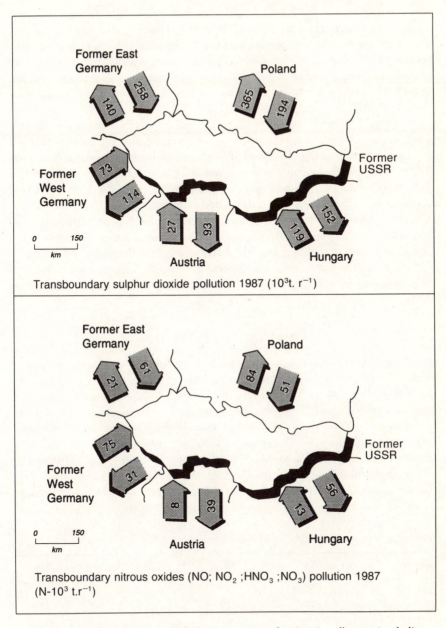

Transboundary sulphur dioxide pollution 1987 (10^3t. r^{-1})

Transboundary nitrous oxides (NO; NO_2 ;HNO_3 ;NO_3) pollution 1987 (N-10^3 t.r^{-1})

Figure 4.2 Czechoslovakia: transboundary movement of major air pollutants (excluding former USSR), 1987

Source: *Životní prostředí České republiky* 1990.

cars, neither known for their low pollution emission, whilst few if any carry catalytic converters and a large number are old and poorly maintained. Passenger cars with two-stroke engines are common, and although they produce less nitrogen oxide than their four-stroke counterparts, they emit large quantities of carbon monoxide, hydrocarbons and various particulates. Some idea of the contributions made by the major pollutants in Czechoslovakia may be gleaned from Table 4.2.

In 1986, Czechoslovakia's government signed the UN Convention on the Long-Term Pollution of the Atmosphere, which committed the country to a 30 per cent reduction of the 1980 level of sulphuric emission and river pollution by 1993. There are some doubts as to whether this can be achieved, particularly for SO_2 levels, which by 1987 had only dropped by 6 per cent (Alm 1989). The key seems to lie in the general restructuring and modernization of Czechoslovakia's industry, which lags behind developments in other industrial countries because it has failed to build enough effective purification plants. Further, there has to be adequate penalties introduced against factories breaking the anti-pollution regulations. In reality, signature to the agreement means the country must reduce its SO_2 emissions by 1993 by about a million tonnes (*Rudé právo*, 22 November 1988: 4).

WATER POLLUTION

Czechoslovakia is often referred to as the 'roof of Europe', as it forms part of the main watershed for the continent. Most precipitation flows out across its borders, while no water flows into the country except the Danube which forms a common frontier with Hungary. Furthermore, Czechoslovakia's hydrological network not only has a low runoff but also suffers from considerable variations in regime through the year. Under such conditions, river pollution over the past decade has become all too visible. This disadvantageous geographical situation has helped threaten the country's surface and underground water resources, further compounded by polluted atmospheric precipitation which is the major source of water supply.

Table 4.2 Major sources of pollution in Czechoslovakia, 1988 (in per cent)

Source	Solid	SO_2	NOx	CO	CxHx	Other gases	Total
Industry (incl. energy)	22	27	10	3	3	3	68
Central and local heating	3	4	1	6	1	–	15
Traffic	3	2	3	6	2	1	17
Total	28	33	14	16	6	4	100

Source: Ceřovský, J. and Podhajská, Z. 'Czechoslovakia', in *Environmental Status Reports 1988–1989*, vol. 1, IUCN East European Programme, Thatcham, 1989, p. 9.

Admittedly, water pollution is not as serious as air contamination, but already in the early 1980s about one-third of the country's stream/river network suffered from some degree of contamination. At the end of that decade over two-thirds of Czechoslovakia's rivers and streams were horribly polluted from untreated industrial effluents (for example, mining wastes, oil spillage), raw sewage and run-off from pesticides and fertilizers (for example, nitrates, liquid manure) used on agricultural fields. A full third of the country's major rivers are so polluted that they cannot support fish life; when caught elsewhere fish are unfit for human consumption.

At present, the most serious problem concerns the pollution impact on drinking water sources. Nearly half the water in the public supply system is below satisfactory consumption standards (Vavroušek and Moldan 1989: 35). Present water demand by the population and economy is around 3.6 m³ annually, of which four-fifths (2.9 m³) comes from surface sources (Kopečny 1988). While there is increasing anxiety about pollution of underground sources, especially in northern Bohemia (*Rudé právo*, 16 January 1989: 3), the critical factor remains the continued contamination of surface drainage. Whilst the average daily water demand per inhabitant in Czechoslovakia (including industry) is 143 m³, in major centres of population and industry it is much higher; for example, in Prague this is about 450 m³ (*Lidová demokracie*, 8 September 1989: 1). If a radical change to water pollution control is not effected during the next decade, a large part of the country's water resources will be unavailable for utilization, except by industry.

The contamination of the drainage network is equivalent to a country with 20 million inhabitants, yet Czechoslovakia has only three-quarters that total. Both surface and underground water is being increasingly polluted by viruses, bacteria, pesticides, and radionuclides. There is also considerable heat pollution of the network from nuclear and thermal power-stations and metallurgical works. However, it is agricultural activities that cause approximately half of the country's water pollution problems through the use of high nitrate concentrations and other chemicals, and the improper disposal of liquid animal manure. In a country with less than half a hectare of agricultural land per person, protection of this resource should be of paramount importance.

Accidental spillage is another contributory factor. In 1988 there were some 500 pollution accidents affecting Czechoslovakia's drainage network, adding to the estimated six cases of water pollution that took place every day somewhere in the country (Radio Prague, DOM/M, 6 March 1989, 9 a.m.). The seriousness of these accidents varies considerably; if they are reported or discovered in time, and the source of the substance known, then prompt steps are taken, but the worst situations occur when offenders try to conceal the accident. For example, in May 1987 a special graphite lubricant was found in Laz water reservoir and the culprit was never found (*Rudé právo* 14 July 1987: 2); in November 1986, 180 tonnes of heating oil leaked from faulty storage equipment at the Ostrava-Kunčice cement plant into the Oder river and thence to Poland (*Neue Zürcher*

Zeitung, 14 November 1986: 4), and the following month 32 tonnes of oil escaped into the same river from a heating plant at Bohumín hospital which led to official government action in drawing up an inventory for all possible pollution sources (*Radio Hvězda*, 16 December 1986, 3.40 p.m.). Even so, in July 1988 30,000 litres of diesel oil leaked into the Nisa river from Jablonec due to insecure railway tankers breaking an oil supply hose (Radio Prague, DOM/M, 23 July 1988, 6.30 p.m.).

The fouling of rivers and streams with untreated sewage from municipalities is a national disgrace. Less than a fifth of the country's sewage is treated properly before being discharged into surface waters. Large cities like Prague, Bratislava, Hradec Králové, and Ústí nad Labem are plagued with inadequate sewage treatment plants, whilst in some popular recreation areas such as the Krkonoše Mountains the government has only recently ordered the building of sanitation systems (Radio Prague, DOM/M 9 February 1989, 5.30 p.m.). All this is made worse by poorly maintained water mains; nearly a third of the total water supply is lost due to defective public drainage systems (*Lidová demokracie*, 8 September 1989: 1).

Damage has resulted also from Soviet troop behaviour over 23 years of occupation in sizeable military compounds. They left behind damage already estimated at about $125 million (*Lidová demokracie*, 23 May 1990: 1), which could be even higher. It included harm done to underground water reserves, these being contaminated by waste oil which in places was several metres deep; in central Bohemia streams glisten in rainbow colours from oil tipped into the ground discarded from military vehicles and aircraft (*The European*, 25–7 May 1990: 5). American technicians inspecting sites in Moravia reported that part of the water supply had probably been contaminated by poisons used in chemical warfare (Pehe 1990).

The Czechoslovakian government's aim to reduce its dependency on brown coal reserves has led to plans for alternative energy sources including hydro-electric power (HEP). Amongst the numerous HEP projects under construction one that has caught international attention relates to the Gabčíkovo–Nagymáros Dam project along 170 km of the Danube, first planned in 1977 (Ostry 1988). It soon became apparent that the new dam would fundamentally alter the eco-systems of the surrounding area; changes to the hydrological system have not yet been fully studied scientifically (Mikloš *et al.* 1989). A government commission was set up in early 1990 to study the environmental impact of the near-completed Gabčíkovo dam (construction on the Hungarian counterpart at Nagymáros was suspended in May 1989 for ecological reasons). It looked especially at its impact on water levels in a known water deficient area, as well as making an evaluation of its effect on the surrounding environment (*Rudé právo*, 23 January 1990: 2). The international implications of the whole project have to be considered by not only Slovakia within Czechoslovakia, but also by Hungary and Austria, the latter being responsible for much of its construction (Vavroušek 1989a). As Slovakia presently only produces four-fifths of its energy needs, importing the rest from the Czech Lands, it is believed that the Slovaks will wish to complete the Gabčíkovo dam in the near future in their drive to be economically independent of the rest of the country (Pomíchal 1990). In July 1991, further attempts were made to end the international deadlock, more with hope than faith in a positive outcome (Davies 1991).

SOILS AND VEGETATION POLLUTION

Perhaps the most evident, and tragic, effect of air and water pollution is the abominable state of the country's forests and soils. In this relatively densely inhabited country there is only 0.44 ha of agricultural land per person. Farming is characterized by concentrated and specialized production in large-scale enterprises in which soil fertilization and irrigation play an important role (Anon 1987). Much of the eastern part of the country is semi-humid with insufficient rainfall unevenly distributed throughout the year. Pressure on the country's 7 million ha of agricultural land (1988) is increasing, especially the 4.9 million ha of arable land. The latter has declined from 0.46 ha per person in 1946 to 0.30 ha in 1988, thanks to growing competition from alternative uses such as urbanization and industrialization.

Soil quality has also suffered from pollution and deforestation. Denuded of vegetation, upland soils can no longer retain moisture; instead they are beginning to erode at alarming rates. Water erosion now threatens 2.9 million ha of land (42 per cent of the country's total area), and wind erosion affects another 870,000 ha. So far only 125,000 ha have been rehabilitated with anti-erosion measures, but it is hoped that by the end of the century some 1.5 million ha will be protected from erosion. At present industrial emissions, largely through acid rain, affect 550,000 ha of agricultural land in the Czech Lands and a further 170,000 ha in Slovakia (Vavroušek and Moldan 1989). Soil intoxication levels have been described as a 'time bomb' (*Podniková organizace*, no. 8, 1986, pp. 377–9) whose fuse is now very short. One major cause of soil contamination is the excessive application of chemical fertilizer and plant protection agents to fulfil the pressure on maximum production; agricultural enterprises also contribute through pollution accidents. In 1984, farms in the Czech Lands were responsible for 74 accidents, 26 of which involved leakage of silage juices into surface and underground water sources. Liquid animal manure caused 17 accidents in 1984, and a further 16 involved oil and oil products. Spillage of artificial fertilizer added 9 more and pesticides 3 (*Zemědělské noviny*, 23 January 1986: 3). More recently, chemicals seeping from waste dumps improperly built by Soviet troops have polluted the soil (Pehe 1990). Livestock and wildlife can also suffer; low-quality fodder leads to serious liver diseases and poor milk standards (*Klub* 1988), whilst near Břeclav in south-east Moravia, 3,000 birds died after picking treated seeds spread in fields to kill mice (Radio Prague, DOM/M, 6 March 1989, 9 a.m.).

Forests cover 4.4 million ha or 35 per cent of Czechoslovakia, but many areas are suffering disastrously from environmental degradation. The 1980s have experienced a rapid growth in areas affected by pollution emissions; in 1982 they damaged 419,000 ha of forest, rising to 691,000 ha two years later and 1.3 million ha by 1986 (*Svobodné slovo*, 29 January 1988: 4). By 1988, surveys indicated that over 30 per cent of Czechoslovak forests were affected by an alphabet soup of air pollutants, stated more recently to have risen to four-fifths (Davies 1991). The surveys revealed the Czech Lands (Bohemia and Moravia) to have been hardest hit, with damage to 58 per cent of its forests compared with 35 per cent in Slovakia (Czechoslovakian TV, 29 October 1988, 3 p.m.; Kopečný 1988). On a regional basis, past and predicted percentage of forest damage is seen in Table 4.3.

Industrial emissions have affected forests in nearly all the northern part of the country. The situation is particularly critical in the mountain regions, where the forests are the most fragile. In the Czech Lands sulphur dioxide fall-out is heaviest in the Ore Mountains (Krušné Hory) where all trees show signs of damage from acid rain. In this region authorities must reforest 1,500 ha annually, but they are fighting a losing battle. Thousands of conifers, weakened by acid rain, topple during heavy snowfalls and gales or succumb to attacks from insect pests and pathogens. Forestry evidence from 1956 reveals that only 40,000 ha had been damaged by then (*Zemědělské noviny*, 25 January 1988: 1) but now the whole area is devastated; today the eastern section of the Ore Mountains resembles a graveyard with few single trees remaining. The heavily wooded region along the former East German–Polish–Czechoslovakian border has also suffered heavily from airborne chemicals. The ecological threat hanging over the Krkonoše (Giant) Mountains, the Sněžka massif (Lokvenc 1988), and the Jižerské Hory Mountains is mainly the result of domestic and transboundary sulphur dioxide emissions. In the Giant Mountains, Norwegian spruce have shed

Table 4.3 Forest damage in Czechoslovakia, 1970–2000 (per cent)

Administrative region	1970	1980	1990	2000
Central Bohemia	–	21.6	47.9	58.4
South Bohemia	–	–	21.7	30.7
West Bohemia	–	19.3	40.4	46.7
East Bohemia	9.5	26.3	100.0	100.0
North Bohemia	33.4	73.7	98.0	100.0
South Moravia	–	–	19.2	21.4
North Moravia	3.6	51.8	96.4	96.4
West Slovakia	3.0	19.1	28.5	42.9
Central Slovakia	4.8	12.2	33.4	50.4
North Slovakia	1.1	12.9	19.5	29.4
East Slovakia	1.6	10.0	15.8	24.4
Average per cent	5.1	22.4	47.3	54.6

Source: Vavroušek, J. and Moldan, B. (eds), *Stav a vývoj prostředi v Ceskoslovensku*, Praha, 1989, pp. 73, 75.

their needles throughout much of the national park's 32,000 ha together with damage to a quarter of all plant life; 90 per cent of mushroom species are now extinct and half of the animal species are endangered (*Svobodné slovo*, 12 December 1988: 3). Many of these are victims of the gaseous pollutants funnelled into the area from steel mills and metal smelters in Poland's Silesian Industrial Zone. In the Orlické Mountains, 85 per cent of the area has been bombarded with airborne pollutants, killing over two-thirds of all the trees (Carter 1989). The southern part of Bohemia is less affected, although here in the Sumava Hills there is increasing evidence of forest damage from pollution (*Jihočeská Pravda*, 23 September 1988: 1). In northern Moravia the Jesenický Mountains, a national park since 1969, have suffered from growing industralization and the resultant pollution of sulphur dioxide (Senjuk 1988).

In Slovakia, forests cover 40 per cent of the territory and also suffer pollution. As in the Czech Lands, there has also been a dramatic increase in the impact of atmospheric pollution. In 1970 it damaged only 2.4 per cent of the growing area, but by 1975 it was 8.5 per cent and a decade later 15.6 per cent, which doubled in three years to 30.4 per cent in 1988 (Fronková 1988a). In northern Slovakia, industrial pollution from the Polish Katowice industrial complex was primarily responsible (Radio Prague, DOM/M, 18 June 1988, 12 noon), but in eastern Slovakia much of the cause is blamed on magnesium dust. The highly toxic dust from factories at Rožňava and Košice has damaged fir trees to such an extent that they have had to be uprooted and new trees planted in their place (Fronková 1988b). Worst polluted forest areas are in the western Beskidy Mountains on the Moravian–Slovakian border, and in central Slovakia areound Ziar nad Hronom (Konopka 1988). Fears also exist of possible forest damage in the Beskidy from a new coking plant at Stonava near the Polish–Czechoslovak border (Radio Prague, DOM/M, 19 May 1990, 3 p.m.). Slovakia is acutely aware of the threat to its forests and a close watch is being kept on contemporary damage (Prochazka 1988).

NUCLEAR POWER

Czechoslovakia is committed to a 30 per cent reduction in the level of sulphuric emission by 1993 according to the Long-Term Pollution of the Atmosphere Convention drawn up in 1979 by the UN Economic Commission for Europe. Whilst 2 per cent of Czechoslovakian investment goes on environmental projects, bureaucracy has prevented progress; sizeable funds are committed to combat pollution but insufficient technology exists to achieve success. This, therefore, has put pressure on the use of alternative energy sources. By the mid-1980s plans for increased nuclear power expansion were approved (Burger 1985). It was hoped that better security regulations would be enforced than had previously been operative. In spite of the Chernobyl disaster, which according to Czechoslovakian authorities mainly affected milk contamination (Medvedev 1990: 204), emphasis on nuclear energy continued (Carter 1986, 1988).

Czechoslovakia's nuclear programme currently has eight reactors (440-mega-watt variety) in operation, which in 1991 produced 27 per cent of the country's total electricity output (Ondráčková 1991). By the end of the century, nuclear-generated electricity should reach half the total demand through the installation of twelve 440-megawatt units and six 1,000-megawatt reactors. The first of the latter is due to be commissioned at Temelín in southern Bohemia in 1992. This was seen by the government as a lasting improvement, allowing the phasing-out of brown-coal-burning thermal power-stations, the installation of purification plants, and a reduction in atmospheric pollution (Brożyniak 1988). In the wake of the Chernobyl accident, nuclear power production continues to be hotly debated in Czechoslovakia. The country's energy needs cannot be met without nuclear power if their polluting thermal counterparts are to be phased out; emphasis must be on greater nuclear safety. Present government feeling suggests that immediate abandonment of the country's nuclear energy programme could be as unreasonable as reverting back to the old coal-burning policy (Sniegon 1990). The problem of improving safety and waste disposal methods still remains (Genillard 1991).

POLLUTION AND HEALTH

By the early 1980s there was increasing public concern in Czechoslovakia about health and the environment. Although the extent of ecological disruption on health was not revealed during the early part of the decade, by 1986 it was admitted what many Czechoslovakians believed – namely, that at least 30 per cent of the population in the Czech Lands and 16 per cent of Slovakia's inhabi-tants suffered constantly as a result of high concentrations of air pollutants. About 90 per cent of these pollutants came from combustion processes and traffic (*Podniková organizace*, no. 8, 1986; pp. 377–9). Even before 1986, the scientific and medical literature in Czechoslovakia had indicated these con-nections (Anon 1986). In 1987 warnings of such links by the dissident Charter 77 group led government officials under the new era of *glasnost* to admit that the state of the Czechoslovakian environment did not correspond to the needs for the preservation of the nation's health.

In August 1988, the Czechoslovakian trade union daily newspaper published a supplement which contained an unprecedented exposé of the nation's health (*Práce* 1988). The main message from the findings revealed that while the state of health in Czechoslovakia had been similar to that of other developed countries up to the mid-1960s it had since declined. Life expectancy between 1960 and 1985 had risen by only thirty-six days to 70.5 years and was even lower than in twenty developing countries. The country's male population had a life expect-ancy of only 67.3 years and their female counterparts 75.9 years, figures that were between 4–7 years below normal average for developed countries. More recently, data published on comparative death-rates in Eastern Europe for 1983–6 reveal Czechoslovakian males with 1,562.3 deaths per 100,000 inhabitants as second only to Hungary, but females fared much better with 933.1, only Poland and the former East Germany coming below this figure (Chobotský 1990). The supplement also referred to the country's infant mortality rate. This has declined from 19.2 deaths/1,000 live births in 1970 to 12.4 in 1990 but is still above the norm (10 deaths per 1,000 births) for developed countries. Moreover, there was a rise in disease among women, and a third of all pregnancies suffered complications.

A United Nations survey of 150 countries, carried out in 1988, revealed that Czechoslovakia was in top place for cancer deaths, coronary diseases (especially ischemia) and strokes (*Hospodářské noviny* 1988a). Even young people are beginning to suffer from serious heart ailments; one in two Czechoslovaks die from cardiovascular disease and one in four from malignant tumours. Czechoslovakia also has one of the highest incidences of rectal and intestinal diseases (*Zemědělské noviny*, 6 August 1988: 3). The sad state of Czechoslovakia's environment is seen as a direct link in this chain. The incidence of disease grew by nearly a third for men and half for women between 1970 and 1985. About one-fifth of the entire country's population suffers from environmentally-related diseases (*Hospodářské noviny* 1988b).

Czechoslovakia is an important producer of various mineral resources besides brown coal, some of which have negative environmental effects on their local areas. Uranium exploitation has led to miners complaining about the effects of high radioactivity levels at the Zadní Chodov enterprise in western Bohemia (*Rudé právo*, 25 July 1987: 2). More recently, inhabitants in Jáchymov near the German border have asked to be rehoused elsewhere due to increased deaths from cancer directly attributable to uranium dust in the air. Although Jáchymov uranium mined closed in 1964 there are still excessive doses of radioactivity in the town because locals were using building materials from the old mine. Houses were recording radon gas levels of 500 becquerels per m^3 (*The Guardian*, 2 November 1989: 10). In Slovakia, the Železorudných bani mine at Rudňany emits 3.4 tonnes of mercury dust annually, a chemical that can affect the central nervous system (Staszova 1988). Magnesium dust problems exist in eastern Slovakia. More serious is the sulphur dioxide emitted from coal-burning power-stations. The World Health Organization has set international norms for clear air and these are accepted by Czechoslovakia; they include a yearly average of 60 micrograms per m^3 for sulphur dioxide. In most parts of northern Bohemia the average annual concentration is 130 mg/m^3, whilst in central Prague it is 136 mg/m^3 (*Svobodné slovo*, 10 December 1988: 4). More generally, over three-quarters of Czechoslovakia's toxic waste is dangerously stored. Toxic metals such as mercury, copper, cadmium, zinc, lead and aluminium occur in industrial waste; in time these enter the food chain causing harm not only to human beings but also to flora and fauna. These are some of the unpleasant realities about the nation's health which have emerged in recent years and demand serious attention in any future planning (Atlas, 1992; Straškraba *et al.*, 1992, 109–15).

POLLUTION AND THE LAW

Czechoslovakia has a plethora of laws concerned with various aspects of the environment. The principle of care for, and improvement of, the environment was incorporated into Article 15 of the Czechoslovakian Constitution and a number of protective laws and regulations have been passed subsequently. Laws were made as environmental problems arose; they have appeared at all administrative levels, from major governmental decrees to local authorities concerned with parochial problems. Constitutional recognition ensured legal protection of the environment; however, laws promulgated were of a general nature. For example, the State Protection of Nature Act, 1956, the People's Health Act of 1966, the Measures for Protecting Air Purity Act in 1967, and Forestry Acts of 1960 and 1977 (Carter 1985: 37).

The federalist structure of the country divides legal powers between the Czech Lands and Slovakia, each being individual republics which since 1988 have had their own ministries of the environment (Ceřovský and Podhajská 1989: 31). These operate within the overall umbrella of the Ministry of Interior which is responsible amongst its varied functions for co-ordinating all state environmental management. Within this framework other ministries find environmental matters impinge on their jurisdiction such as those for Agriculture and Food, for Health and Social Matters, and for Industry, whilst the Czech and Slovakia Geological Offices are responsible for mineral resource control. The Ministry of Culture looks after nature conservation and protection of national monuments, but of the country's six National Parks (only one in the Czech Lands) management in Slovakia is shared with the Ministry of Forest/Water Management/Timber Industry. At a lower level, overall environmental responsibility lay in the hands of National Committees (abolished 1990) who delegated liability to smaller authorities – namely at regional, district, town and village level.

By the late 1980s it was becoming increasingly apparent that many previous laws and their ineffective methods of imposing fines for environmental damage were out of date (Medliak 1990). In 1986, new Acts in both the Czech Lands and Slovakia had been promulgated regarding stiffer fines for offences breaking state nature conservancy acts and decrees, and became operative in 1987. Weaknesses remained: for example, the Criminal Code only protected the environment indirectly through material damage done to state ownership (Nesrovnal 1990). Moreover, ecological aspects were insufficiently covered in the field of economic law (Patakyová 1988), and the legal aspects of nuclear safety within the environment needed revision (Pirč 1988).

Three new Acts entered the Czechoslovakian statute books in 1988: one concerning state enterprises (no. 88), another related to the agricultural co-operative system (no. 90), and the third to the non-agricultural co-operative system (no. 94) (Klapáč, 1989a), with smaller authorities now in operation at the regional, city and ward levels. Each contained specific reference to the environment up to the end of this century; in future activity in every state enterprise, cooperative and corporate enterprise must take heed of any harmful influences that their individual production units may have on people's health. These new laws placed major emphasis on environmental damage prevention rather than liability and reparation – a new approach to an old problem. Specific reference was made to links between exploitation of natural resources and their environmental impact, with special regulations regarding the control of waste products (Klapáč 1989b). The law on state enterprises emphasized use of less wasteful technology, but all three Acts imposed norms on the utilization, disposal and removal of harmful waste substances. Each enterprise or co-operative was responsible for the waste products from its economic activity and must now recycle them for their own use or provide conditions for their safe use by others. A fourth Act was passed in 1990 (no. 105) which placed environmental controls on private enter-prises (Klapáč 1990). All four Acts therefore placed stress on the creation of an effective environmental control system for which management was directly responsible.

CASE STUDIES

1 Northern Bohemia

This region covers only 6 per cent of the country's territory (7,814 km^2), but contains 10 per cent of Czechoslovakia's industrial production and employment (Carter 1985: 23) (see Figure 4.3). The concentration of industry here includes metallurgical, chemical and textile plants, cellulose paper mills, and workshops for glass and ceramic manufacture. Most important, however, is the opencast lignite brown-coal basin which produces 60 per cent of domestic coal supplies (25–30 per cent ash content). Furthermore, a third of the country's thermal power-stations are located in this region and are responsible for producing two-thirds of domestic electricity demand. Coal production from the lignite basin is around 100 million tonnes yearly, and although the mines are slowly becoming exhausted, supplies could last for another 30–50 years.

Signs of pending pollution problems in this region were evident in the early 1960s. Certainly by 1970 the North Bohemian coalfield had the largest pollution per km^2/year in Europe – namely, 86 tonnes of fly ash and 181 tonnes of sulphur dioxide. At this time the Most Basin within the coalfield was producing over half of all solid fuel extracted in the country and three-quarters of all lignite, and also generated 17 per cent of total electricity and manufactured 15 per cent of all industrial chemicals in an area only one-hundredth that of the state. At the end of the 1980s little had changed: of the fifteen worst SO$_2$ polluting industries in the Czech Lands, eight are situated in the North Bohemian coalfield, leading to catastrophic environmental pollution in this region.

Acute air pollution problems in this region originate from both domestic as well as foreign sources. In Czechoslovakia, the emission of solid harmful substances total 1.4 million tonnes annually of which 400,000 tonnes (28.6 per

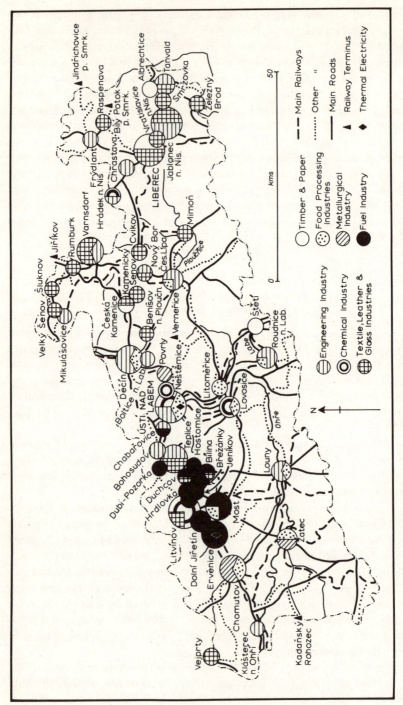

Figure 4.3 The North Czech region: industries and communications

cent) falls on North Bohemia; of the equivalent 3 million tonnes of gaseous emissions, the region receives 700,000 tonnes (23.3 per cent) annually (Ceřovský and Podhajská 1989: 37). Local urban centres within the region have recorded disturbing data; for example, on 19–20 January 1989, the Hydrometeorological Institute in Ústí-nad-Labem recorded an average 24-hour SO_2 concentration of 400 mg/m^3 (*Rudé právo*, 21 January 1989: 2). Part of the problem is geographical as the major industrial towns of northern Bohemia are located in a valley which is prone to temperature inversion, particularly in the winter months. Cold, polluted air is trapped in the valley by warm air aloft which prevents normal daytime winds from dispersing the pollution. For example, at Teplice data from the local health station has revealed that from 1 October to 31 March the average daily SO_2 concentration was 195 mg/m^3 for the period 1975–89; the annual average over the same time gap was 129 mg/m^3. During a period of 102 winter days it reached a diurnal average of 500 mg/m^3 and for a shorter 23-day spell 900 mg/m^3 with a peak of 1,285 mg/^{3} (Anon 1990). The highest concentration ever recorded was at Osek near Duchcov in January 1982 with a daily average of 2,400 mg/m^3 (Dykast 1989). The permitted 24-hour average is 150 mg/m^3 (Klášterská 1991: 3).

The administrative districts of Louny, Most, and Chomutov experience air pollution levels above the permitted safety level for at least half of every week (Cestrová 1989). This is hardly surprising in places like Chomutov, where power-stations within a 35-km radius produce 3,170 megawatts of electricity and utilize 18 million tonnes of brown coal annually (Glenny 1987). The earlier philosophy of planning before people is well illustrated by the town of Most. In 1964, officials discovered it was sitting on a rich deposit of brown coal which could be easily obtained by opencast methods. The town was razed to the ground and a new town for 65,000 inhabitants was built nearby (Schonherr 1988). Problems soon arose when strip-mining released clouds of sulphur water into the air and left a hole several kilometres in length next to new apartments built to rehouse the displaced residents (Echikson 1987).

Attempts to reduce SO_2 emissions have met with limited success. In 1983, Czechoslovakia claimed that much progress had been made in reducing the level of pollution in the region. The chemical combine at Lovosice had perfected a device for collecting nitrogen oxides which reduced emission content by 4,000 tonnes/year, whilst another new development was claimed to eliminate up to 90 per cent of sulphur dioxide from inferior-quality coal through the addition of limestone (Carter 1985). These so-called 'ecological installation projects' received double the previous investment allocation in the 1986–90 Five-Year Plan, but design preparation, delivery shortages and financial support were lagging behind schedule (*Rudé právo*, 19 April 1988: 1). At the Tušimice power-station it was claimed that experimental equipment tested there in 1983 could trap more than 90 per cent of SO_2 and all fly-ash. In 1988 it was reported that desulphurizing equipment would be installed at Tušimice (and other stations like Prunerov, Počerady, Mělnik and Tišova, all north of Prague), based on former

Soviet and former East German technology. Difficulties remain at Tušimice, which uses 17 million tonnes of brown coal to produce 15 milliard kilowatt hours of energy annually, and with construction deadlines taking four to five years it is obvious that effective results will only begin to show after 1995 (*Rudé právo*, 7 April 1989: 2).

Water, like air, is a serious problem in the Czech Lands. The most polluted river in the country (the Bilina) flows through northern Bohemia on its way to join the River Labe (Elbe). Here phenols and other chemical waste are deposited into the slow-moving course, and like other local rivers Bilina has also been affected by mining activity in the area. Radioactive material from uranium mining has also leaked into the Ploučnice river, itself eventually flowing into the Elbe. Forests have likewise suffered: damage accelerated dramatically in the 1960s and 1970s when new heating plants progressively came on line. By 1973, local foresters were felling twice as many dead trees annually as throughout the whole of the late 1960s. A pending ecological catastrophe has been partially averted in the Krušné Hory (Ore Mountains) with the replanting of 1,500 ha annually (estimated at a cost of $15,000 per ha). In the Jablonec district alone 16,153 ha of forest or three-quarters of all woodland vegetation is affected by air pollution. Dying forests also threaten soil quality and water supplies; lack of trees accelerates soil erosion through water run-off, which in the Jablonec district had produced 45,800 metres of erosion tracks by the end of 1988 (Radio Prague, DOM/M, 13 November 1988, 5.20 p.m.).

Attempts at the recultivation of land devastated by mining and industrial emission in Czechoslovakia date back to the late 1950s. Research into the problems of North Bohemia began in 1958, and greatest success has been in its brown coal basin (Škoda 1981). Evidence of recultivation continues – for example, in 1988 1,000 ha were reclaimed around the periphery of Most (Frank 1989a). The territory of the North Bohemian brown coal region is proof that even badly damaged land can be rehabilitated with surprisingly good results, but this assumes that new regulations on industrial waste dumping, estimated to total 2,500 ha by the end of the century, can be applied (Carter 1985).

Finally, some appraisal must be made of all this pollution on the residents of the North Bohemian region. Until recently, the government refused to publish detailed statistics on life expectancy in northern Bohemia but it was generally accepted that it was between three to four years less than the rest of the country. Figures recently made available by the Czech Institute of Statistics now confirm this viewpoint on the half million residents of the brown coal basin (see Table 4.4).

According to the Teplice District Public Health Station's findings the concentration of three of the most carcinogenic polycyclic hydrocarbons is 800 times higher here than the acceptable limit, and winds carry these dangerous substances over the area. It has calculated that for 30 per cent of summer and 75 per cent of winter days, local residents are subjected to carcinogenic and mutagenic air pollutants (Anon 1989: 42).

Table 4.4 Life expectancy in the North Bohemian region, 1989

Administrative area	Male (yrs)	Female (yrs)
Czech Lands	67.1	74.1
Chomutov district	65.2	72.1
Most district	64.2	72.4
Teplice district	65.0	72.1
Ústí nad Labem district	65.8	73.2

Source: Anon, 'Dramatic statistics about the northern Czech brown-coal Basin', *Panos Feedback 90*, September 1990, p. 16.

Children and teenagers are most affected by air pollution, water and soil contamination. Paediatric hospitalization is between 2–12 times higher in the North Bohemian coal mining area than elsewhere in the country. Infant mortality in 1990 was 42 per cent higher than the country's average, and two-fifths more infants suffered from serious illnesses than the Czechoslovakian average. In 1989, the number of bronchial and respiratory illnesses among nursery-age children was four times higher than other regions, skin diseases twice as high and endocrinological problems nearly twice as high. The government's answer has been to organize annual summer camps (Nature Schools) in the High Tatra Mountains for 150,000 children from the North Bohemian region (Frank 1989a), and paying attractive salaries to entice people to reside in North Bohemia for a ten-year period, ironically seen as burial money by some of its inhabitants.

2 Prague

The capital of Czechoslovakia, with over a million inhabitants is, after the North Bohemian region, the most polluted part of the country. The city lies between two of the narrow Vltava river gorges on a wide flat valley (see Figure 4.4) and this physical setting often leads to temperature inversions during winter months which produce fog and smog. These inversions, combined with interference from residential and industrial buildings, lead to poor weather aeration in the lowest parts of the city (Carter 1984).

Air pollution is arguably the most serious aspect of the city's environmental deterioration. In 1946, before the Communist takeover of the country, Prague was suffering from dust fall-out 3–4 times higher in the heavily industrialized city districts than the residential areas. By the end of the 1950s, 50,000 tonnes of dust and ashes were falling on the capital annually, which by the mid-1960s totalled 461 tonnes/km^2 each year (see Figure 4.5). Much of this was attributable to the influence of industrial development in the city, for a tenth of the country's industrial production is located in Prague, especially engineering. Czechoslovakia's Air Pollution Act of 1967 (the so-called 'chimney laws' imposing minimum

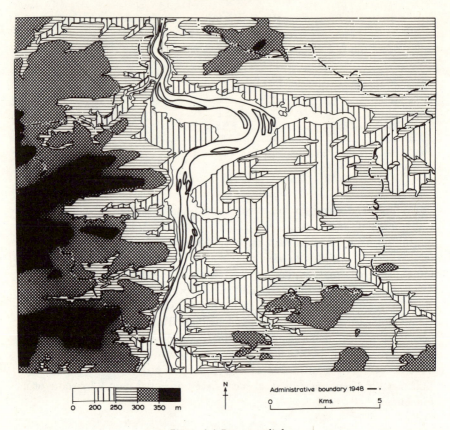

Figure 4.4 Prague: relief
Source: F. W. Carter 1984, p. 260.

smoke stack heights) led to a reduction in dust fall-out reaching a record low of 217 tonnes per km^2 in 1973, but in spite of some early successes in the reduction of SO_2 levels, the increasing use of alternative fuels and number of motorized vehicles have led to a growth of noxious gases over the city. By 1986, the city centre was recording daily sulphurous oxide concentrations of 150 mg/m^3 with a maximum diurnal density of 620 mg/m^3 in Rytířská Street (Brchanová 1988).

Prague's furnaces still use brown coal and release twenty times the permitted amount of SO_2 into the city's atmosphere; this prompted a small public opinion survey (390 respondents) in the capital during 1988, and the results confirmed the growing worries of the inhabitants. Over four-fifths believed the city had ecological problems, with over a third fearing it had reached catastrophic proportions and half that it was getting worse. Most respondents blamed pollution on the poor public transport system which encouraged greater use of private cars; others noted the high intensity of industrial location within the city and inadequate punitive pollution controls against renegade factories (Loudová, 1989).

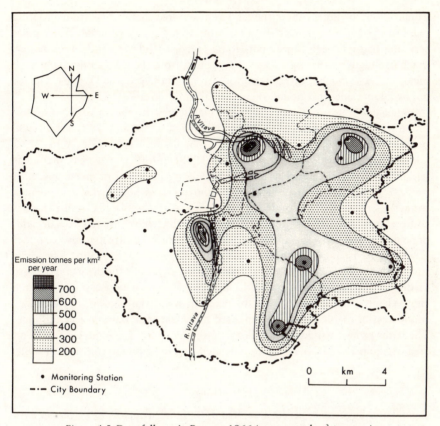

Figure 4.5 Dust fall-out in Prague, 1966 (tonnes per km^2 per year)
Source: *Hygienická Stanice NV Hl m Prahy* 1966.

Meanwhile, it is hoped optimistically to reduce emissions over the city between 1985 and 2010 by 9.2 per cent for dust fall-out, 4 per cent for SO$_2$, and 45 per cent for nitrogenous oxides (*Rudé právo*, 7 April 1989: 2).

The questionnaire also revealed anxiety over water pollution. The main source of water has always been the River Vltava, but increasing demand, particularly for drinking water, has meant tapping other rivers to the north and south-east of the city. Whilst studies of streamwater chemistry have been made in the Prague region (Cerný 1989), the most serious problem remains that of water cleansing. The basic sanitary network was completed in the 1900s, which six decades later was proving entirely inadequate. New sewage plants have been constructed but pollution continues to aggravate the system. In 1987 there were over forty cases subjected to fines (Vondráčková 1988); large amounts of untreated sewage are released into the Vltava river and from 1986–8 oil-spills increased by 50 per cent (*Večerní Praha*, 20 April 1988: 1). For example, mazout was detected in the central sewage treatment plant, brought by waste-water canalization. By going counterflow it was discovered that it came from a heating plant in the Michle suburb, which on being found guilty admitted that regular servicing had not been carried out (*Rudé právo*, 14 July 1987: 2). Fortunately, much of Prague's drinking-water supply comes from the Zelivká river, but such incidents cause much local concern; today none of the water in Prague is safe for use by infants since it contains too many harmful substances.

Other environmental problems were highlighted by the survey. Noise pollution was a topic of discussion already in 1967 (Carter 1984: 268), yet in 1988 between 40 and 50 per cent of Prague's inhabitants were said to suffer from above average noise levels (Vondráčková 1988). Insufficient recreational areas within the city were noted in spite of increasing provision of green spaces. A pressure group has been formed to try and preserve Stromovka Park against further encroachment from other land-use forms (Stursa 1988), whilst there is concern about the 2,000 historical monuments which have suffered from decades of pollution and decay in the city. The 270-ha historical centre may also be under attack from visual pollution by property developers (Frank 1989b). The city's surroundings also demand attention (see Figure 4.6). Calls for the increasing use of desulphurizing equipment for the numerous thermal power-stations in the city's vicinity, and abandonment of plans to flood the Berounká river valley south-west of the city with the proposed Křívoklat HEP dam project continue. The latter project would destroy flora and fauna extant since the Ice Age in an area which in 1977 UNESCO recommended should be preserved (Merta 1988).

3 Bratislava

Capital and largest city in Slovakia, had 440,000 residents in 1989. Situated on the Danube it has considerable riverborne commerce and to the east its industrial quarters spread out over the Danubian plain (see Figure 4.7). Large thermal power-plants utilizing brown coal, together with engineering, petro-chemicals, cellulose-paper plants and cement manufacture, create considerable air pollution. Atmospheric deterioration during the 1970s meant that the city's lower air layers were seriously affected by dispersed solid particles, particularly during the winter months; in 1975, 110,000 tonnes of pollution fell over the city, nearly two-thirds SO_2. Moreover, vehicle registrations doubled during that decade, further contributing a quarter of all emissions making the city centre permanently above the permissible maximum concentration for average daily CO levels (Hesek *et al.* 1979). Water pollution resulted from the discharge of raw sewage into the Danube and its local tributaries.

In the 1980s, the overall pollution situation in Bratislava further deteriorated. Much of the blame was placed on the 'Slovnaft' petro-chemical combine, which covers much of the city with fly-ash and chemical waste. Furthermore, although the combine was built in the late 1960s, it was not fitted with a water purification plant until the end of 1985; thus, for a decade and a half, 15–30 tonnes of oil waste products were leaked annually into the Danube daily. It is not

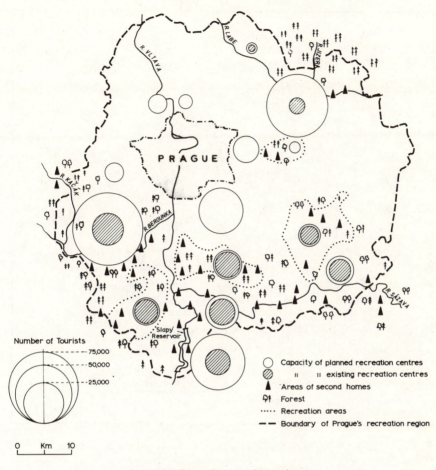

Figure 4.6 Prague's recreation regions
Source: F. W. Carter 1984, p. 271.

surprising, therefore, that concern for the local environment led to the publication of a *samizdat* report on the city in an attempt to alert its inhabitants to the dangers they faced in their daily environment.

The appearance of this report made disturbing reading. The report was written mainly by members of the Slovak Academy of Sciences (Budaj 1987). It suggested that Bratislava was the most polluted city in Europe, a fact played down by Czechoslovakia's official media. The critical report, divided into three parts, gave detailed information on the pollution of the natural environment (air, water, soil/vegetation, noise, radioactivity), urban situation (city centre, suburbs, industry, transport, building construction), and the social environment (cultural infrastructure, services and health/hospital facilities). It claimed that the major cause of increased cancer deaths (up by a third from 1980–5), heart diseases (up

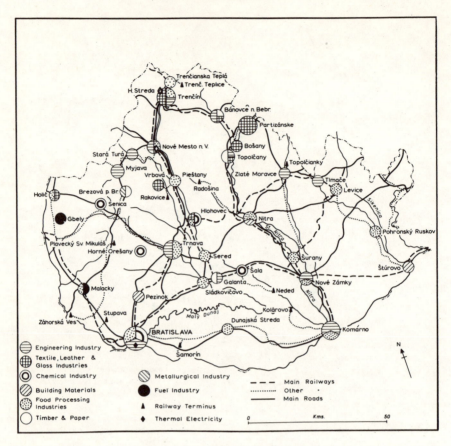

Figure 4.7 The West Slovakian region: industries and communications

by four-fifths since 1970) and infant mortality (up by two-thirds since 1960) was directly related to official disregard for the environment. Main criticism was levelled at the dumping of untreated industrial waste into local water courses (especially by chemical combines), and concentration of toxic materials in the air exceeding even those in Prague; it called for improved hospital facilities and radical anti-pollution measures to be taken.

Yet damning evidence continues to emerge: air, water and soil pollution in the city have worsened with children there three or more times likely to suffer from respiratory diseases than their counterparts in the rest of Slovakia (Radio Prague, DOM/M, 2 December 1988, 9 a.m.). Air quality over the city suffers from increased quantities of sulphur oxides, aerosols and vehicle exhausts (Rak 1989), whilst forecasts indicate that the number of Bratislava's inhabitants will rise to 545,000 by 2010 (Barták 1988).

GREEN POLITICS

Perhaps more than in most east European countries, Czechoslovakian environmentalists played a critical role in the eclipse of Communist rule. Their *samizdat* publications enabled the general public to become more aware of the atrocious conditions in their country, a factor which state-sponsored environment organizations had failed to do. The latter was best represented by the 'Brontosaurus' ecological movement of Czech youth which was established in 1974 as part of Czechoslovakia's national ecological programme. Much emphasis was placed on recreation and ecological education, accompanied by musical and literary evenings with lectures and debates by state experts – but the movement was financed by the government and therefore reluctant to provide open criticism (Kulich 1989).

It was not until the emergence of Charter 77 that more negative attitudes to state environmental policy were revealed. Although the Czechoslovakian political system suppressed possibilities for open popular environmental protests, the Charter organization complained about nuclear power-plant safety only a year after its inception. The early 1980s saw other letters/reports emerge which were critical of the overall environmental situation in the country, especially with reference to northern Bohemia (Waller 1989). The Charter group, therefore, was fundamental in providing alternative environmental information contrary to official state sources. Their cause was greatly enhanced by the Chernobyl disaster, as they provided a platform for anti-government protest on its dangers to the environment (Charter 77, Document no. 15, 1986). Other groups took courage from their lead such as 'Antiatom' which protested against nuclear safety at the Temelín power-station under construction in Bohemia.

During 1987 and early 1988 further damning reports on the country's serious ecological situation, notably the previously discussed *samizdat* on Bratislava, together with evidence on meat and dairy contamination and dissenting views on the Gabčíkovo–Nagymáros Dam scheme, increased public anger (Ashley 1989). The appearance of a *samizdat* ecological bulletin provided a forum for renewed unofficial debate, but nothing on this activity was mentioned in the state media. The emergence of a new government after the so-called 'Velvet Revolution' was to change the whole approach to information dissemination in the country.

Ecological pressure groups were no longer illegal and a plethora of new nongovernmental organizations (NGOs) appeared including the 'Prague Mothers', 'Children of the Earth', 'The Ark', the 'Rainbow Movement', the 'Independent Ecological Society' and 'Ecoforum' (Doležal 1990). A group of environmental activists formed a Greenpeace organization 'for a greener Czechoslovakia' (Anon 1989); such activity soon led to political involvement with the emergence of a Green Party (Wolchik 1991: 282). It was becoming increasingly apparent that if things were to change, the new government had to include environmentalists amongst its members, and given the popular support for anti-pollution measures such candidates would stand a chance of election. In the elections of June 1990, however, the 'green vote' went to Civic Forum which, like Public against

Violence in Slovakia, supported the country's main environmentalists for the posts of Czech and Federal environment ministers.

CONCLUSION

The depressing state of Czechoslovakia's environment is seen as a direct link with the centralized nature of authority during the period of Communist control. Economic growth took place at the expense of the environment; present thinking stresses the opposite view – 'what is not ecological is not economic' (Dočekal *et al.* 1989). It is now realized that growth must be accompanied by more efficient use of natural resources if further pollution is to be prevented. Greater investment alone will not suffice to halt the destruction; there must be greater awareness of the problems both by officials and the general population. Ultimately, the key lies in the general restructuring and modernization of the country's industry if environmental pollution is to be reduced. Progress will be slow, but already the former Communist adage that heavy industry is the measure of success has been discredited.

The increase of democracy in Czechoslovakia means more public knowledge on the environment, and the new government supports groups willing to criticize present inadequacies in the milieu in which they all live. There has also been a shift of emphasis by environmentalists in the country – they no longer concentrate on the weaknesses of specific projects but instead appraise overall government strategy. The increasing appearance of hard data is now transforming the former situation, where information on environmental quality was difficult to assess (Vavroušek 1989b). Ecological damage is now being taken more seriously within the country's legal framework (Zeman 1989), but it remains to be seen how effective this attitude will be. While the government is suggesting the phasing out of large thermal electricity power-stations, especially in northern Bohemia, the alternative is to replace them with nuclear counterparts. Little emphasis has yet been placed on energy efficiency or consumer reduction (Vavroušek 1990, Moldan 1991b).

Ultimately, the answer seems to lie in greater international environmental cooperation in central Europe. It is easy to blame neighbouring states for some of the emissions and effluent deposited within your borders, but perhaps 'cleanliness should begin at home' (one such domestic meeting took place in Dobříš in May–June 1991). Only then should international meetings be held to discuss a more effective multilateral approach to central Europe's ecology. In this way we may eventually see some improvement in the Czechoslovakian environment.

ACKNOWLEDGEMENTS

The author would like to gratefully acknowledge help given by Peter Brod, Prague Correspondent, Radio Free Europe, and Louise Saunders of the Cartographic Unit, Department of Geography, University College London.

5

HUNGARY

D. Hinrichsen and I. Láng

INTRODUCTION

'Everyone knows Hungary stands on two feet,' goes a current joke making the rounds in Budapest, 'but the problem is one foot is in the twentieth century and the other in the nineteenth!' This has been the reality for Hungarians over the past forty years of socialist rule and examples abound. On the edge of the charming university town of Pécs a huge thermal power-station burns polluting brown coal to produce electricity for the southern part of the country. The plant has two tall stacks: one belches out clouds of thick black smoke while the other emits next to nothing. The clean stack has been fitted with an efficient Swiss filter which removes most particulates, the other has no pollution controls at all.

There have been dramatic changes since the late 1980s. Following Gorbachev's restructing of the Soviet economy and his policy of *glasnost*, Hungary, along with the rest of Eastern Europe, was told, in effect, that she could go her own way. The socialist regime crumbled in 1989 in favour of a new coalition government (led by conservatives), and all remaining Soviet armed forces in Hungary were withdrawn by the end of July 1991. But times are still tough. Hungary, like all the rest of what used to be called 'the Iron Curtain countries', has been left with an economy in shambles, a horribly outmoded industrial infrastructure, critical housing shortages in most cities, and a dreadful legacy of over-used resources and widespread environmental degradation.

An impression of the extent of the problem may be gained from a selection of extracts from recent reviews (mainly Hinrichsen and Enyedi 1990). Thirty per cent of the country's agricultural land is endangered by wind and water erosion and over almost half of it the soils have inherent nutrient limitations. Around the country some 700 small settlements, amounting to 300,000 people, must obtain their drinking and household water from tanker trucks or have it piped in from neighbouring areas. Their own wells are too contaminated with nitrates to be usable. Over 40 per cent of the population, or 4.7 million people, live in highly polluted industrial zones or cities where they are exposed to elevated levels of sulphur dioxide, nitrogen dioxide, dust, soot and heavy metals. The State Statistical Office estimates that 35–40 per cent of Hungary's 2,300,000 cars, trucks and buses should be banned from the roads for pollution and safety reasons. The National Institute of Public Health estimates that every twenty-fourth disability and every seventeenth death in Hungary is caused by air pollution. By 1990, 22 per cent of the country's forests (including both conifers and broad-leafed trees) were suffering from die-back and death brought on by an assortment of air pollutants.

The country's population now totals around 10,375,000 (based on 1990 census), roughly half the number of people living in the much larger state partitioned under the Treaty of Trianon after the First World War. But even within the present frontiers the natural increase is now negative as birth-rates continue to fall. There are actually 7,000 more deaths than births in the country every year, a fact attributed to the very low fertility levels of Hungarian women, not poor medical facilities (United Nations Population Fund 1990). Before the arrival of Soviet troops at the end of the Second World War, Hungary had enjoyed several decades of independence and cultural revival. The country's economy blossomed in the 1920s and 1930s, buoyed up by the expansion of both heavy and light industries such as electronics, metal working, engineering, chemicals and agro-technologies (Nemeth 1981). But pollution from industries and urban areas was a continuing problem from the 1870s onwards. József Fodor, a physician, pioneered a series of studies on the quality of Budapest's air in the 1880s, discovering that it was full of sulphur and particulates. The situation in the capital continued to deteriorate over the years. By 1929, when the Capital Institute of Hygiene and Bacteriology conducted more air pollution studies, the pungent odour of sulphur 'permeated the city center and accumulated around the capital's railway stations, due in large measure to the use of high sulphur coal for heating purposes and in steam engines' (Várkonyi and Kiss 1990: 49). The measured values of various pollutants continued to increase up to the Second World War.

The quality of environmental data has always been of high standard in Hungary. But under socialism much of the accumulated data on the state of the country's environment were not published in easily accessible form (in some cases data were considered too politically embarrassing to be released at all). Nevertheless, beginning in the 1970s, the Hungarian Academy of Sciences initiated a number of important interdisciplinary studies (such as the Lake Balaton project) and helped to set up a nation-wide environmental monitoring system in co-operation with the State Statistical Office and the National Office for Water Management (now merged into the Ministry for Environment and Regional Development). The Academy established a number of permanent committees to consider environmental questions of national importance. These committees also issued regular expert reports and studies on various environmental problems confronting the country and what needed to be done about them (Láng 1978). Hungary now has a solid international reputation for its integrative approach to environmental science and there is no lack of authoritative data. But overcoming the environmental neglect of recent decades remains a major problem, as it is in other East European countries, all the more so because the 'greens' in both the government and scientific establishment are still constrained by political inertia and slim budgets.

AIR POLLUTION

With the advent of socialism (or what the Soviet's called 'goulash Communism'), Hungary's economy (largely dependent on agriculture) was forced to give more emphasis to heavy industry: steel mills, metal smelters, metallurgical plants, chemicals, cement and mining. An ugly industrial crescent was created across the northern part of the country from Ajka, Györ and Tatabánya in the west to Miskolc and Ozd in the east. Within this industrial corridor, most of the country's heavy, polluting industries were concentrated. But some heavy industry appeared in other provincial cities and virtually all industrial complexes were built without regard for pollution control measures or environmental repercussions. At the same time pollution in the main urban centres was increasing due to motor traffic and central heating systems. This was evident immediately after the Second World War and became more pronounced in later years.

In 1948/9, data from Debrecen (Hungary's eastern-most large city) painted a portrait that became typical for many Hungarian cities. Dust levels were found to be lower in those areas of the city protected by forests; there was a close relationship between dust levels and bacteria content in the air; and the city centre was found to be two to three times more polluted with soot and dust during the winter heating season than in the summer. Studies carried out in 1959 in the university town of Pécs (surrounded by a ring of heavy industries by this time) showed that on average 847 tonnes of dust per square kilometre fell on the city. At Tatabánya in the north-west, researchers found the dust load per year

Table 5.1 Areas of polluted air in Hungary (determined by measurements of sulphur dioxide, nitrogen dioxide and dust) and the population affected

	Km^2	*Inhabitants (in thousands)*
Greater Budapest	1,946	2,527
Borsod-A, -Z. County region	2,225	512
North Trans-Danubian region	1,369	277
Middle Trans-Danubian region	1,283	310
Baranya County region	1,304	271
Nógrád and Heves County region	1,165	221
Other towns	1,136	613
Total	**10,428**	**4,731**
From total:		
Budapest	525	2,064
Other polluted towns	3,218	1,821
Other polluted settlements	6,703	846

Source: Várkony and Kiss 1990, p. 51.

averaged 368 tonnes per square kilometre, while at the city's lime works, dust levels averaged 1,250 tonnes per square kilometre throughout the year. At the same time Budapest's air was also becoming increasingly fouled with pollutants, generated mostly from industries and home heating appliances burning brown coal. In March 1959 an unusually thick smog fell over the city, reducing visibility in the centre to 80 metres. Investigators discovered that sulphur dioxide levels exceeded 4 mg/m^3 of air for long periods of time, while the soot concentration averaged 5.4 in the worst affected areas (Várkonyi and Kiss 1990: 50). Pollution was so pervasive in the 1950s and 1960s that pollution plumes over Budapest, Miskolc, Pécs, Tatabánya and the Mátra Mountains often extended upwards to between 2,000 and 3,000 metres. The main polluters were identified as power-plants, industrial enterprises, home heating units, iron and aluminium production, and the cement and chemical industries. By the late 1960s private vehicles, trucks and buses also contributed to the deteriorating quality of air in cities and towns.

Between 1960 and 1980, the air quality in a number of urban areas improved significantly when most home heating units switched from coal to oil. In 1974, the National Ground Level Concentration Monitoring Network was established to provide pollution data on a continuous basis. However, a profile of the country's airshed made with more recent data (collected up to 1988) revealed a lingering legacy of highly polluted air over industrial areas and large cities and towns. Data for three pollutants (sulphur dioxide, nitrogen dioxide and dust) analysed by the Department for Air Hygiene at the National Institute of Public Health in Budapest, found that the most highly polluted areas were Greater Budapest and the Borsod County Industrial Region. Overall, the total polluted area of the country amounted to 10,400 km^2 or 11.2 per cent of its total territory, affecting 4,700,000 people, or 44 per cent of the total population (see Tables 5.1 and 5.2). Figures 5.1, 5.2 and 5.3 show areas affected by high concentrations of sulphur dioxide, nitrogen dioxide and dust, respectively. In all cases the heavily polluted areas, with the exception of Budapest (which contains

Table 5.2 Areas of polluted air in Hungary as percentages of the total area and population of the country

Size of the country	93,030 km^2	100%
Total population	10,679,000	100%
City dwellers	5,982,000	56%

	Territory %	*Inhabitants %*
Total polluted territory	11.2	44.3
Territory polluted by sulphur dioxide	9.7	41.4
Territory polluted by nitrogen dioxide	7.0	40.4
Territory polluted by dust	5.4	31.5

Source: Várkony and Kiss 1990, p. 51.

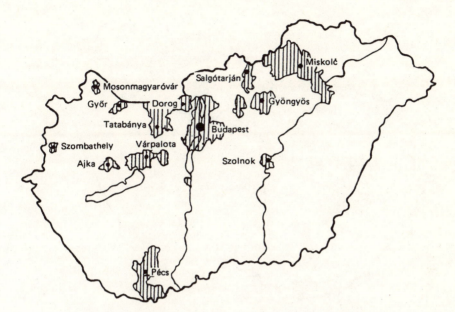

Figure 5.1 Areas of Hungary affected by sulphur dioxide pollution
Source: T. Várkonyi and G. Kiss 1990.

Figure 5.2 Areas of Hungary affected by nitrogen dioxide pollution
Source: T. Várkonyi and G. Kiss 1990.

Figure 5.3 Areas of Hungary affected by dust
Source: T. Várkonyi and G. Kiss 1990.

2.2 million people, 21 per cent of the country's entire population), were heavily industrialized zones dominated by antiquated smokestack industries (Várkonyi and Kiss 1990: 50–1).

Virtually two-thirds of all urban dwellers are exposed to high levels of air pollution, whereas only a small proportion of the rural population is affected. High levels of sulphur and nitrogen dioxides affect 4,300,000 people (40 per cent of the total population) while excessive dust pollution affects 3,400,000 (32 per cent of the total). About 3,200,000 people (30 per cent of the total) are affected by all measured pollutants, but only 3 per cent of these people live in small villages or on farms. Of course, the urban population is exposed to other harmful pollutants as well. In terms of its prominence and its danger to health, carbon monoxide is in the same category as sulphur and nitrogen dioxide. Based on measurements carried out in Budapest (using the telemetric system) virtually the entire population is exposed to high concentrations of carbon monoxide: 2,064,000 people covering an area of 525 km^2 (Várkonyi and Kiss 1990: 52).

During the winter months Budapest is often swaddled in a blanket of smog from vehicles, industries and home heating appliances. On some days pollution is so bad that it is virtually impossible for anyone in Buda (on the west bank of the Danube) to see anyone or anything in Pest (on the east bank). Both parts of the 'Pearl of the Danube' disappear under a pall of pollution. Even during the summer months, pollution from Budapest's enormous fleet of cars, trucks and buses can turn the city's air into a photochemical soup of oxidants. Ozone

94

episodes are on the increase. Fearing the worst, air pollution experts are beginning to monitor ozone events on a regular basis and have set up an early warning system for the city, which will alert motorists to stay clear of areas with high concentrations of oxidants (Hinrichsen 1989).

The terrible state of Hungary's vehicle fleet is also a cause for concern. Every year 1,000,000 tonnes of carbon monoxide is emitted along with 130,000 tonnes of hydrocarbons, 120,000 tonnes of nitrogen oxides, 36,000 tonnes of solid particulates, and 510 tonnes of lead and lead compounds (Vukovich 1990: 29–30). Until 1989, nearly all of Hungary's private cars came from other East European countries (mainly Czechoslovakia, the GDR, Poland, Romania, and the Soviet Union). None contain any pollution control devices such as catalytic converters or lean-burn engines. To make matters worse, the two cheapest and most popular private cars in Hungary (the Trabant and Wartburg from the GDR) run on highly polluting two-stroke engines. Although two-stroke engines produce less nitrogen oxide than four-stroke, they emit huge amounts of carbon monoxide, hydrocarbons and heavy metals. Hydrocarbon emissions from these vehicles exceed permissible health standards by 8–10 times. No one knows exactly how many are on the roads, but some estimates are as high as 40 per cent of Hungary's cars (Hinrichsen 1989). In addition, many vehicles remain in service far beyond their useful lifespans. The State Statistical Office estimates that 35–40 per cent of Hungary's 2,300,000 cars, trucks and buses should be banned from the roads for pollution and safety reasons. It may take two decades to phase in more non-polluting vehicles (Láng 1991).

The health costs of air pollution

Air pollution creates crop damage over large areas of prime agricultural land, but the urban areas face the greatest threats in terms of chronic health hazards which are now taking an increasing toll on Hungarians. Statistical surveys of death and illness related to air pollution have revealed an estimated annual cost of 3.7 billion forints (616,000,000 US dollars) covering hospitalization and medicines, lost work hours, and other related factors. Furthermore, the percentage of medical cases attributable to air pollution is rising. In 1988, it was estimated that every twenty-fourth disability (requiring absence from work) and every seventeenth death in Hungary was caused, directly or indirectly, by air pollution (Várkonyi and Kiss 1990: 56).

Transboundary pollution and acid rain

In 1980 Hungary generated about 1,600,000 tonnes of sulphur dioxide, ranking the country fifth in sulphur emissions for Europe. In 1984/5, emissions dropped to 1,400,000 tonnes. Since then, the amount has come down even more, to around 1,200,000 tonnes. But according to computer models, Hungary is a net exporter of sulphur, sending 45 per cent more abroad than it imports (Soloman and Kauppi 1990). Although there is some dispute over this figure, there is little disagreement over the fact that the northern part of Hungary, especially the Mátra Mountains, is affected by acid deposition (both wet and dry). The average pH level for Hungary as a whole is 4.67, but those areas of the north not adequately buffered by calcium suffer from the depredations of 'acid rain' (discussed below under 'Soils and Land Use').

Regulation of air quality

The comprehensive regulation of air pollution began in 1971. In 1973 the Council of Ministers passed regulations on thirty-one air pollutants in an effort to improve air quality. And in 1976 the Parliament framed the Act on the Protection of the Human Environment, in which air quality was an important component. However, up to 1986, Hungary's air pollution regulations contained a number of serious flaws. They were too 'flue oriented' (ignoring other sources); no provisions were made for controlling non-point sources of pollution; and the end values for primary pollutants were not sufficiently distinguished. Moreover, regulations were unable to control polluting emissions from vehicular traffic; air pollution fines were insufficient to prevent or discourage polluting activities; and legislation was too compartmentalized, lacking integration into a comprehensive package.

So the 1986 Ministerial Decree stipulated that air quality control must be taken into account when planning new industries and it set pollution limits on a plant-by-plant basis (both new and old plants), following the 'bubble concept' popularized by the US Environment Protection Agency. The Decree also distinguished between harmful and dangerous pollutants. Under this definition, pollution is considered harmful if it exceeds permissible air quality standards, and it is dangerous when it reaches levels that jeopardize human health. Budapest has now introduced a three-stage smog alarm system.

Altogether, some 327 air pollutants are regulated, with different values set depending on the area in question and the time-scale. The ten most common pollutants are then regulated according to values set over specified periods (a year, a full day or 24 hours and 30 minutes). This has brought Hungary's air pollution legislation more in line with Western European standards, but enforcement is still slack and fines are not nearly stiff enough to halt serious offenders. In addition, funds to finance improvements (including badly needed retrofitting of highly polluting industrial plants) are in short supply. Over the next five years Hungary will invest only 6 billion forints (120,000,000 US dollars) in air pollution control measures. At the same time, air pollution is expected to result in a loss of 18.7 billion forints (374,000,000 US dollars) to the national economy because of illness and premature death.

THE FUTURE OF AIR POLLUTION

During the 1990s, Hungary's emissions of sulphur dioxide are expected to drop as the country continues to implement commitments made under the Convention on Long-Range Transboundary Air Pollution, administered by the United Nations Economic Commission for Europe (ECE). This agreement calls on all signatories to reduce their emissions of sulphur dioxide by 30 per cent by 1993 (using 1980 levels as the baseline). Between 1980 and 1985, Hungary was able to reduce its sulphur dioxide emissions by 13.5 per cent, but reaching the 30 per cent level by 1993 may prove difficult. Hungary's new energy policy calls for continued reliance on lignite or brown coal to fire its power-plants and industrial boilers. Without proper controls on these plants, air pollution will continue to be a severe problem in Greater Budapest and the industrial belts. At the same time the urban areas will be inundated by more private vehicles than ever before, making the air in cities like Budapest and Miskolc even worse than it is now, despite controls. Without proper controls on vehicle emissions, including lorries and buses, little real improvement can be expected in the quality of air in urban areas.

WATER RESOURCES AND POLLUTION

Water managers in Hungary joke that there are really only two kinds of water in the country: too little, or too much (and usually at the wrong time of the year or in the wrong place). As in much Hungarian humour, there is some reality in this statement. Hungary's total water resources average about 120 km^3 a year, but only half of this arises from direct precipitation. Moreover, the bulk of the country's water (94 per cent) falls within the international water basin; so that the availability and quality of water in Hungary is determined, to a large extent, by the country's neighbours. Water resources within the country are also very unevenly distributed. As a result, Hungary has been forced to develop intensive nationwide water management programmes. Indeed, Hungary's water managers have been coping with this situation since the nineteenth century, building countless kilometres of flood protection dams and dykes, as well as canals and irrigation systems.

Water resources and demands

The total amount of water available in Hungary amounts to roughly 20 billion m^3 a year; with 67 per cent of this supplied by surface water and 33 per cent by groundwater. Annual water use amounts to about 20 per cent of the total resources available (Hock and Somlyódy 1990: 67). It is the surface water available in August that is used in water management planning because this is the driest month of the year. This policy makes good economic sense considering that only small sections of the country's two largest rivers (the Danube, with just 2 per cent, and the Tisza with 37) lie within Hungary. Most of the groundwater resources consist of bank-filtered waters, followed by artesian waters, unconfined groundwaters and karstic waters. The most important bank-filtered groundwater is stored in alluvial formations along the banks of the Danube and Tisza rivers. Taking exploitable resources into account, theoretical supplies of groundwater from this one source are estimated to be around 7,000,000 m³ per day, with 85 per cent coming from the Danube's catchment and 13 per cent from the Tisza's (Hock and Somlyódy 1990: 68).

Water use

This has increased sharply over the last few decades. The growth of water demand by industry, agriculture and municipalities is illustrated in Figure 5.4. In terms of water use, the most important sector of the economy is industry (accounting for 73 per cent), followed by agriculture (about 13 per cent) and finally municipalities (8 per cent). The energy sector, most notably coal-fired electricity generating plants, account for most of the water used by industry. Water use by agriculture varies considerably from year to year, depending on the amount needed for irrigation purposes. However, in a normal year, irrigation and aquaculture account for 76 per cent of the water used in the agricultural sector. Domestic, or municipal, water demands have increased steadily as well. The supplies needed to meet this rising demand come from bank-filtered water (46 per cent), artesian and karstic waters (44 per cent) and from surface water (10 per cent) (Hock and Somlyódy 1990: 69).

A simple comparison of aggregate water demand and availability gives the impression of a huge untapped surplus but the reality is more complicated. When

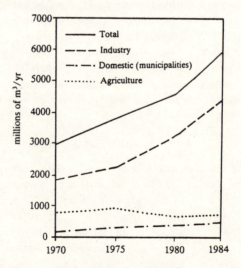

Figure 5.4 Water use in the Hungarian national economy, 1970–84
Source: Central Office for Statistics 1986.

taking into account future demand for water resources by the industrial and agricultural sectors (not to mention the rapid growth of urban areas) several factors conspire to undermine Hungary's reserves. First, the monthly balance of use and availability reveals some limitations, a point which is very significant in drought years when irrigation needs may rise precipitously. Second, the distribution of water resources and demands is far from uniform. The percentage of available, but so far uncommitted, water resources (after subtracting the base flow in the river bed) is 85 per cent for the Danube but only 15 for the Tisza which flows through Hungary's mercurial Great Plain. Hence, during drought, serious water shortages can be expected during the growing season in the latter area. Third, since Hungary is essentially a downstream country with respect to its surface waters, it is safe to assume that as upstream countries increase their water use and consumption less will be available for Hungary. The Tisza River's flow, for example, is expected to decline due to the construction of large reservoirs in the former Soviet Union. In the future it will be necessary to set more realistic prices for water because in the absence of such a policy all water available cheaply will be used. On this basis there will be enough water available to meet Hungary's future demands up to the turn of the century. However, difficulties may arise in allocating water use along the Tisza River system and perhaps in other deficit areas. Because of this potential shortfall in water, authorities will have to implement an effective and carefully worked out water management policy for the entire country.

Water pollution

Pollution is another reason Hungary may have trouble meeting future water needs. Not surprisingly, since industry is the largest water user, it is also the largest source of water pollution. Despite the fact that untreated industrial effluents discharged into Hungary's surface waters have been reduced significantly over the past two decades (due mainly to the construction of treatment plants and the reuse of waste waters) some 84,000,000 m³ of untreated industrial waste water still ends up in rivers and lakes every year. Consequently, the discharge of organic pollutants, heavy metals, oil and its derivatives and dissolved salts has increased accordingly (Hock and Somlyódy 1990: 70). The majority of COD loads (chemical oxygen demand) into the Danube come from Budapest; all other sources combined account for only two-thirds of the communal and industrial load from the capital.

Waste water and effluents from the country's thousands of farming operations are insignificant when compared to industry's contribution. Most of the water harnessed for agricultural purposes is used to irrigate crops and supply aquaculture ponds. Large-scale 'industrialized' farming operations produce liquid manure at the rate of 48,000,000 m³ a year. Of the total volume produced, 42 per cent was spread on fields for fertilizer, 40 per cent was discharged into designated areas, and 12 per cent was pumped into rivers and streams (creating localized problems of eutrophication). But another source of pollution to surface waters comes from the discharge of untreated or partially treated municipal wastes (mostly sewage). Some 1.3 billion m³ of untreated sewage is dumped into the country's surface waters every year. In total, however, two-thirds of Hungary's waste waters are discharged into water courses after mechanical treatment only. Over the last five years, some of this untreated or partially treated sewage has been leaching into groundwater aquifers, fouling drinking water (Hock and Somlyódy 1990: 71).

The results of an extensive evaluation of surface water quality carried out over the 1970s and 1980s revealed some startling trends. In a number of important rivers, including the Danube, Kapos, Tisza and Zala, nitrate and phosphate concentrations have risen significantly (Figure 5.5). Consequently, eutrophication is a growing problem throughout the country. Another finding is that lead and mercury concentrations in sediments increased dramatically during the 1980s, particularly in the Danube. There are fears that heavy metals may eventually leach into the bank-filtered groundwater reserves found along this river.

The quality of bank-filtered groundwater is another matter of great concern since such water is the most important source of supply for communities along the Danube River. Budapest alone withdraws some 312,000,000 m³ of bank-filtered water every year for municipal use. But quality is dependent on the condition of water in the Danube, and as the quality of the Danube's water has deteriorated so too has the bank-filtered groundwater along its banks. In particular, nitrate concentrations in some areas are now reaching dangerously high levels. On Csepel Island, in the southern part of Budapest and the site of the city's largest harbour, nitrate levels in the groundwater have reached 40 mg/litre over half the island, while 5 per cent of the island's wells contain nitrate values of 200 mg/litre. Since the standard for nitrate in drinking water has been set at 20 mg/litre, municipal authorities are increasingly alarmed by these trends and a

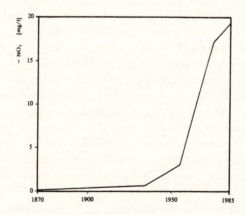

Figure 5.5 Nitrate concentrations in the Danube downstream of Budapest, 1870–1985
Sources: M. Balló 1975, B. Hock 1987, J. Lesenyei 1975.

number of wells on the island have already been closed. Elsewhere along the Danube, groundwater is increasingly polluted with ammonium, iron, manganese and lead. In many areas surveyed, both north and south of Budapest, the quality of bank-filtered water is deteriorating. The picture for the rest of the country is not much better. Because of the growing acidification of soils in Hungary as well as increased use of fertilizers and pesticides, nitrate pollution is on the increase across wide areas of the country. Of Hungary's 3,000 large settlements, around 700 (containing 300,000 inhabitants) now rely on bottled water brought in by lorries, or water piped in from neighbouring communities because their own wells are too contaminated to be used (Stefanovits 1984).

Future trends

The deteriorating quality of Hungary's water resources is largely due to the rapid urbanization of the country over the past 30 years, along with a tremendous increase in industrial and agricultural activities. Although nearly 95 per cent of Hungary's population is now connected to a communal water supply and more waterworks are expected to come on stream by the turn of the century, sewage treatment lags far behind the provision of piped water – a trend that will likely prevail well beyond the year 2000. Meanwhile, Hungary's consumption of water is expected to rise dramatically. If current trends continue, per capita consumption of water will increase fourfold (over 1960 levels) by the year 2000 from 50 to 200 litres per day (Hock and Somlyódy 1990: 82).

Yet at the same time both surface waters and groundwaters are becoming more polluted. This 'scissors effect' on water resources will have to be confronted by water managers. There is little doubt that if current trends continue unchanged, the country's water resources will be squeezed between rapidly escalating demands and deteriorating quality, prompting increased costs to consumers and utilities. In particular, nitrate pollution will likely remain a serious problem well into the next century. This situation is further complicated by the fact that the biological method of treating sewage is ineffective in removing nitrates. Only expensive, three-stage sewage treatment plants are capable of removing nitrates. And aside from a few projects around Lake Balaton, Hungary has no plans at present to build more. Moreover, if water quality in rivers and streams is not improved, the amount of water available for irrigation and even for some industrial uses may shrink. The sorry state of Hungary's water resources could, if not improved, limit future economic growth. It is a prospect that planners and economists are beginning to take seriously.

SOILS AND LAND USE

Across Hungary's vast agricultural lands, the backbone of its economy, a 'quiet crisis' has been brewing. Studies carried out by the Research Institute for Soil Science and Agricultural Chemistry in Budapest, found that 30 per cent of the country's land was endangered by wind and water erosion. Nearly 50 per cent of the soil had inherent nutrient limitations, being too acidic, too alkaline, or too sandy for large-scale crop production. This translates into the need for ever-increasing doses of fertilizers and a constant outlay of funds to rehabilitate degraded agricultural land. Foreign-exchange earnings depend to a large extent on maintaining high agricultural output, and farmers are encouraged to use more and more agro-chemicals (pesticides and fertilizers) in desperate attempts to maintain yields. By 1980, farmers were dosing their land with 60,000 tonnes of pesticides and 1,300,000 tonnes of fertilizers – over 300 kilos of active agents per hectare of land (Vukovich 1990: 25–6). Driven by misplaced government incentives, Hungarian farmers often seem to be waging a chemical war on their land. Increasingly, over-fertilization and a failure to match crops with soils types, is giving farmers a harvest of degraded soil and falling productivity. Fertilizers replace only nitrogen, phosphorus, and potash, thus reducing the lime, humus and microelement content of soils. Intensive fertilization on unstable soils with few reserves of nutrients also reduces soil pH and increases salt content. Because of soil acidification and the leaching of nutrients, the salt content of groundwater is also increasing in some regions and surface waters are becoming eutrophied (Vukovich 1990: 24).

Overall, some 10 per cent of the country is affected by salinity and/or alkalinity. And some 120,000 ha of soils suffer from secondary salinization and alkalization. The quality of irrigation water is fairly good but the water quality in unlined irrigation canals deteriorates from time to time; and seepage from unlined canals and water reservoirs, as well as filtration losses from irrigated fields, causes the water table to rise, bringing saline groundwaters to the surface. As can be seen from Table 5.3, land use patterns have changed considerably in Hungary during the post-war period 1950 to 1986. Nearly 10 per cent of the total arable land (some 815,000 ha) has been lost, mostly through urbanization. Grasslands have also diminished along with vineyards. But there has been a notable increase in the amount of land used for orchards, gardens, forests, reed beds and fish ponds (Várallyay 1990). Land rehabilitation is needed to keep agricultural land productive. Every year some 40,000 to 50,000 ha need some remedial measures to make worn-out and degraded soils more productive. This is a problem that may get worse over the next ten years.

Forest death from air pollution

Most important at this moment is the fate of Hungary's conifer and oak forests along the northern border with Czechoslovakia. *Waldsterben* (forest death) affects some 22 per cent of Hungary's forests (Lean *et al.* 1990). The causes of *Waldsterben* are not entirely clear but it is thought that a virulent assortment of air pollutants (mostly sulphur dioxide, nitrogen dioxide, ozone and other oxidants, acid deposition, heavy metals and organic chemicals) working together with natural stresses such as insect pests, wind and snow damage and

Table 5.3 Land use patterns in Hungary, 1950–86 (per cent)

Year	Arable land	Gardens	Orchards	Vineyards	Grassland	Forests	Reeds	Fishponds	Uncultivated area
1950	59.4	1.0	0.6	2.5	15.9	12.5	0.3	0.0	7.8
1955	58.1	1.1	0.7	2.2	15.8	13.5	0.3	0.1	8.2
1960	57.1	1.1	0.9	2.2	15.4	14.1	0.3	0.2	8.7
1965	54.6	1.6	1.8	2.7	14.0	15.3	0.3	0.3	9.4
1970	54.2	1.6	1.8	2.5	13.8	15.8	0.4	0.2	9.7
1975	53.5	1.6	1.7	2.2	13.7	16.6	0.4	0.2	10.1
1980	50.9	3.1	1.5	1.8	13.9	17.3	0.4	0.3	10.8
1985	50.5	3.6	1.1	1.7	13.4	17.7	0.4	0.3	11.3
1986	50.6	3.6	1.1	1.6	13.3	17.8	0.4	0.3	11.3

Source: Várallyay 1990, p. 114.

pathological factors, are at the root of the problem. Hungary's oak forests are particularly hard hit by this puzzling dieback. Many stands along the border with Czechoslovakia have been damaged and killed off over the past six years. Hungarian scientists studying the phenomenon think that transboundary pollution, especially acid rain from their northern neighbour, may be an important agent in the decline of the country's oak forests. This northern band of dead and dying timber corresponds fairly well to the area of Hungary susceptible to the ravages of acid deposition. In general, pH values tend to be lowest in this region and since the soils are poor in buffering agents, damage to vegetation has been a problem for some time. The acids in the rain cause nutrients (such as calcium, potassium and magnesium) to be leached out of thin soils, depriving trees and other vegetation of these needed growth agents. It may also make the trees more susceptible to damage from insect pests and disease pathogens (Hinrichsen 1986: 203–24). Combating *Waldsterben* and keeping agricultural land productive will be two of the biggest challenges facing land resource managers over the next ten years.

NOISE POLLUTION

Noise pollution is a problem only in Greater Budapest and around some industrial centres. About 80 per cent of all complaints about noise in cities, however, involve vehicle traffic (especially heavy lorries and buses). Vehicles made since 1980 generally comply with noise regulations. However, there are many old, out-of-date vehicles on Hungary's roads. The majority of passenger cars operating in the country have outside noise levels of around 80 decibels, lorries average 88, garbage trucks around 90, and buses between 87 and 90. Noise pollution from air traffic is only a problem for those Budapest neighbourhoods located close to the capital's two air terminals at Ferihegy Airport. Passenger airliners like the TU-134 and 154 have noise levels at 115–118 and 110–115 dB, respectively, from a distance of 100 metres. Some 17,000 aircraft use Budapest's airport annually, creating a regular pattern of noise pollution in its environs. With Malev (Hungary's state-owned airline) no longer required to buy Soviet-made airplanes, one hope for cutting down on noise pollution from the airport is the prospect of buying more efficient, safer and quieter airplanes from Western suppliers.

VISUAL POLLUTION

In spite of the shambles the Germans and Soviets made of Budapest after fierce house-to-house fighting at the end of the Second World War, the gutted capital was quickly restored to its former glory and is now one of the most attractive capitals anywhere in Europe. Most smaller Hungarian towns and cities (those fortunate enough to lie outside the grimy industrial belt created by state planners in the 1950s) also escaped much of the Stalinist block-house architecture which dominated so many other East European cities after the war. Unfortunately, when Budapest began to expand in the 1960s and 1970s the new suburbs built by the state planners were horrors of mediocrity. Parts of Budapest are still surrounded with these crumbling, badly built, high-rise 'slums': monuments to the corruption which allowed them to be built in the first place. Now, with the economy in a state of disarray and reorganization, the building boom of the last decade has stalled. After the dust settles, perhaps the government will make it easier for more people to build their own homes or restore older houses left to rot. With most cities experiencing severe housing shortages, some creative scheme must be developed which will help solve Hungary's urban housing crisis.

TECHNOLOGY TRANSFERS

One of the benefits of the painful economic reordering now going on in Hungary will be greater access to Western technologies, particularly telecommunications systems and computers. Parts of Hungary's telephone network dates from World War One, so annoying 'party lines' and constant breakdowns in service are a regular feature of everyday life. Stories are told of couples searching for flats for years because the husband or wife absolutely had to have a home phone in working order! In some of the older sections of Buda, trunk lines are so overloaded that no new phone numbers have been issued since the 1970s. Needless to say, one of the country's top priorities is a modern telephone and communications network. Out-dated phones and long delays in placing calls has been one of the main drawbacks for doing business in Hungary. Many foreign firms operating in Budapest have only a storefront office; most of the real work is done out of Vienna where phones and faxes are cheap to install and operate efficiently.

The other big drawback, preventing foreign firms from investing in the country, is the lack of computerized offices and trained personnel. This is being tackled at the present through the importation of thousands of office computers and retraining programmes for those office workers who want to become proficient on computers. But it will take time before the proper infrastructure is in place. Nevertheless, foreign investment in Hungary has soared since the 'year of change' (1989). Tourism has always been a heavy investor in Hungary. With 20,000,000 foreign visitors a year (nearly double the entire population of the country) tourist facilities are being expanded at a breakneck speed.

On the manufacturing front, one of the most promising signs is the deal, struck between General Electric of the US and Hungary's old 'Tungsrum' company, to manufacture light bulbs for the US and European markets. With wages low (by Western standards they are considered rock bottom), Hungary's skilled work-force is still a bargain for Western companies seeking to establish manufacturing and distribution outlets to service consumers in both Eastern and Western Europe. In order for the country to continue with its development programme, exports will be emphasized for the foreseeable future. This will mean that Hungary's limited stock of natural resources will come under heavy pressure. In order to avoid the Third World syndrome (exporting more and more agricultural produce and commodities just to pay off bank loans) Hungary will have to evolve a strategic development plan that will allow it to build on the country's strengths, while continuing to attract foreign investment.

CONCLUSION

The next ten years are likely to be critical ones for Hungary's social, political and economic structure. The country is under new management. But with the dismantling of the state apparatus, and the transformation of the economy back to a market-oriented one, there are many perils in the short run. Unemployment rates, especially in heavy industries, will probably be very high. At the same time, the country wishes to retain its generous social security system (including national health insurance, state-run old-age pensions, unemployment and housing benefits). But some difficult compromises will inevitably have to be made. There is little likelihood that the country can afford to do everything it wants to on the social front. Hospital benefits might be severely slashed, along with unemployment benefits and other social services.

While attempting to get the country on its feet again, the state of Hungary's environment may well get worse before it gets better. There are simply too many other pressing matters to contend with and little time to sort things out. In its rush to join Western consumerism, Hungary may put environment and resource issues on the back burner.

In periods of great uncertainty and readjustment, there is a strong temptation to place development concerns over issues of environment protection. But hopefully cooler heads will prevail. Hungary badly needs a national plan for proper resource management and environment protection, similar to the one that has evolved over the years to regulate its water resources. If the new regime is to avoid the mistakes of the past, it will be compelled to manage the country's resources, not just exploit them.

The country has many strengths to build on, not least the fact that the agricultural sector has been run, more or less, by consumer demand and market forces for the past twenty years. Hungary can still feed itself at a reasonably efficient level of production. It is the industrial disaster left by forty years of 'socialist reconstruction' that has left Hungary (indeed the whole of Eastern Europe) hopelessly out-of-step with the rest of the developed world. That gap cannot be closed easily. The next ten years will be full of challenges, but they will also be ones full of hope. After all it was citizens' environmental movements (such as the Danube Circle, led by Janos Vargha) which played major roles in preparing the ground of political change. Given the emergence of a 'new environmentalism' in Hungary, and the upsurge of citizens' action groups, there is reason to believe that the excesses of the past will not be tolerated in the future (French 1990).

6

POLAND

F. W. Carter

INTRODUCTION

It is generally accepted that Poland, along with Czechoslovakia and the former East Germany, is amongst the most polluted countries in Europe not only for air and water but also its soil and vegetation. Yet nearly two decades ago the eminent Polish environmentalist and expert on nature conservation, Władysław Szafer, seemed optimistic when declaring:

> It may be stated objectively that ... during the last twenty five years, Poland has attained a level equal to that of only a few countries, in which nature conservation has passed from the conservation stage to a new one, becoming a factor which exerts its influence on man's management of all natural resources, as well as a popular social and cultural movement modelling man's attitude towards nature.
>
> (Szafer 1973: 24)

With the demise of the former political regime in Poland a more accurate extent of environmental pollution has been revealed.

More than forty years of central planning policy and extensive industrialization meant Poland's national economy was directed towards heavy industry and energy production, in most cases based on coal as the main fuel source. The use of old ineffective technology, together with lack of a modernization policy, resulted in Poland consuming twice the amount of energy comparable with other industrialized nations. Further, poor control of pollutant emissions from badly located state enterprises and thermal power-plants has led to severe environmental degradation. During the 1950s and 1960s little attention was given to environmental protection, but the 1970s revealed a greater need in this direction. Unfortunately, this coincided with an investment boom so that the call for much needed environmental protection policies was left unheeded. Only in the 1980s were there changes in official attitude thanks in part to the efforts of independent organizations like the Polish Ecological Club; yet still in the mid-1980s the government continued to deny that pollution was a problem (Rich 1985).

By the late 1980s such official indifference could no longer continue; in

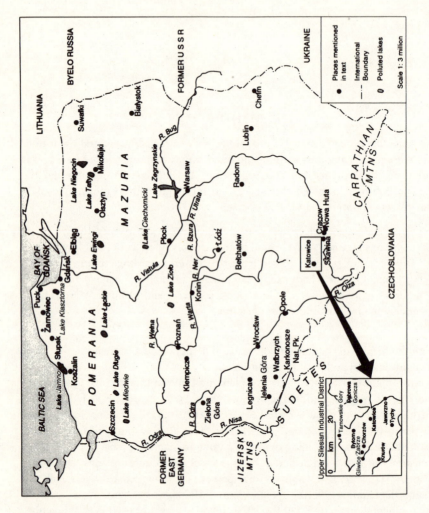

Figure 6.1 Pollution in Poland

February 1988, state environmental inspectors admitted that there were severe problems with Poland's ecology (Radio Prague, DOM/M, 1 February 1988, 3 p.m.), and by the end of the year the Polish Academy of Sciences issued a report which maintained that twenty-seven areas covering over a tenth of the country's land area had suffered 'a complete breakdown of the state of natural balance' (Simmons 1988: 3). Thus it became apparent that Poland, a country of 312 km^2 and a population of 37 million, found itself with a third of its inhabitants (that is, 12.3 million) living in areas threatened by ecological catastrophe.

Poland emerged from the Second World War devastated by foreign occupation and a sizeable population loss, with its agriculture and industry in ruins, a westward movement of its state boundary and a loss of a third of its former territory in the east. The advent of a new socialist regime based on Marxist principles faced the problems of reconstructing this war-torn country and improving its standard of living to that of more fortunate European states. This was to be achieved through an ideology that favoured industrialization and urban growth, and which promised to rapidly create the material abundance enjoyed by Europe's capitalist countries. Modernization was therefore the key to Poland's success, accompanied in the early years of socialist government by strong denials that any environmental problems could be associated with this drive. Unfortunately, for any industrialization programme to succeed, there must be some creation of waste products and occurrence of environmental hazards which became increasingly evident during the 1960s and has continued unabated (see Figure 6.1). An attempt is made to examine Poland's misuse of three major resources – namely, air, water and soil/vegetation, and to assess the role of the legal system, environmental health problems, the rise of green parties, and an in-depth appraisal of two case study areas.

Data have been utilized from varied sources; official Polish statistics on the state of the environment have been included in the country's Statistical Yearbook (*Rocznik Statystyczny*) since the 1970s but it was often difficult to verify true validity since no alternative data sources existed until the early 1980s. Besides the Polish Ecological Club, the Roman Catholic Church in Poland began to provide certain data as well as conclusions that differed from those based on official Polish statistics (Perłowski 1988). In 1986, the 'Atlas of natural resources and degree of contamination of the natural environment' was sent to press ten years after its initial conception and is due to appear in the early 1990s (Kondracki and Leszczycki 1989). Meanwhile, other interesting data is emerging from Poland's main Statistical Office (Główny Urząd Statystyczny), including a report on the threat to and protection of Poland's environment in 1990 (Grzesiak 1990). It provides very detailed information on various pollution aspects, but is weak on very important topics like underground water pollution; the reason is probably high financial research costs, rather than secrecy, which previously bedevilled many of the early publications on environmental degradation.

AIR POLLUTION

The first disturbing realization of air pollution came in 1959, when it was revealed that the average daily deposition of particle matter in the Upper Silesian District equalled the combined total of London and Berlin – namely, 3,000 tonnes over 24 hours (Paszyński 1959). Further analysis of Poland's air pollution problem up to the mid-1980s (Carter 1989), confirmed the deteriorating state of the country's atmosphere. By the early 1980s it was becoming increasingly clear who were the main culprits. It was obvious that thermal electric power-stations were playing an important role; in 1980 they ejected well over a third of the gas and nearly half the dust emitted into Poland's atmosphere. The Bełchatów industrial region and its dependence on thermal power generated from local brown coal supplies supported this contention (Liszewski 1985). Metallurgical plants were another serious contender, emitting well over half the gas and nearly a seventh of the dust, whilst cement works provided a further fifth (Carter 1989: 167). Even more worrying, when the country's twenty-nine major industrial dust and gas polluters (a quarter of Poland's air pollution, 1980), were located spatially, a third were found to be in the Upper Silesian industrial region; they supplied 34 per cent of the country's total dust and 38 per cent of its gas emissions. Furthermore, a line drawn from Szczecin in the north-west to Chełm in the south-east proved that all major emitters were located to the south and contained some of the country's largest urban agglomerations. The situation since the mid-1980s has shown a little improvement (see Table 6.1).

During the 1980s there was a noticeable decline in dust particles from nearly a third of total emissions to just over a fifth (see Figure 6.2); since 1985 the trend has shown a steady decrease. More seriously, gas emission rose accordingly from just over two-thirds to well over three-quarters of the total, over half of which consisted of sulphur dioxide. The emission of gaseous pollutants into the atmosphere provides a serious threat to the natural environment as a result of their rapid diffusion and direct effect on living organisms and buildings. In the peak

Table 6.1 Dust and gas emissions in Poland, 1980–90 (in thousand tonnes)

Category	1980	1985	1986	1987	1988	1989	1990
Dust	2,338	1,788	1,821	1,803	1,615	1,513	1,163
(per cent total)	31.3	26.6	25.5	25.0	23.7	22.8	22.0
Gas	5,135	4,932	5,323	5,399	5,193	5,112	4,115
(per cent total) of which:	68.7	73.4	74.5	75.0	76.3	77.2	78.0
SO_2	2,755	2,652	2,824	2,903	2,827	2,790	2,210
(per cent gas)	53.6	53.7	53.0	53.7	54.4	54.6	53.7
Total	7,473	6,720	7,144	7,202	6,808	6,625	5,278

Source: *Rocznik Statystyczny 1989* (GUS), Warszawa 1989, p. 13; Grzesiak, M. (1990) *Raport o stanie, zagrożeniu i ochronie środowiska 1990* (GUS), Warszawa, 1990, p. 25, ibid. (1991); *Ochrona środowiska 1991*, (GUS), Warszawa, pp. 138–9.

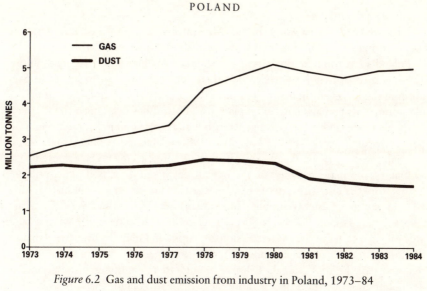

Figure 6.2 Gas and dust emission from industry in Poland, 1973–84
Source: *Rocznik Statystyczny*, relevant years.

year, 1987, about 1,200 industrial enterprises in Poland emitted nearly 5.4 million tonnes of these pollutants into the atmosphere, of which nearly 54 per cent constituted sulphur dioxide, 28 per cent carbon dioxide and 14 per cent nitrogenous oxides. On top of this domestic total, a further 900,000 tonnes of SO_2 emitted from other states, especially Czechoslovakia and the former East Germany, pour into Poland on the dominant westerly winds and form about two-fifths of the sulphur dioxide deposited on its territory (Nowicki 1990a). Meanwhile, the provision of dust-collecting equipment remains woefully inadequate to deal with this problem, largely resulting from the lack of appropriate purification techniques through inhibitive costs.

The significance of SO_2, a gas emitted from burning sulphurous coal and oil, may be appreciated when it is realized that in 1989 average emission was 8.9 tonnes per km^2, compared to the European average of 2.3 tonnes. These figures, however, only take into account the main industrial polluters, but if broader energy use data are applied, SO_2 annual emissions result in country-wide depositions of 38 tonnes per km^2 (Grzesiak 1990: 129). This places Poland in a less favourable light than either Czechoslovakia or the former East Germany, with totals of 35 and 22.6 tonnes respectively. In Poland, the worst affected area is the Upper Silesian Industrial District which accounts for nearly a third of all the SO_2 generated (Pudlis 1989), whilst in Cracow concentrations of SO_2 in the atmosphere regularly exceed legal norms by 700–800 per cent. Much of this deposition results from coal, which in 1989 accounted for 78 per cent of primary energy consumption (OECD 1991: 7); half of the SO_2 is generated by large power-stations (Cofala and Bojarski 1988) running over 7,000 hours/year in which most have only dust control systems and burn not only hard but also brown coal. The latter, with a 5 per cent sulphur content, produces on average

111

35 kg of SO_2 for every tonne burnt (Marcinkiewicz 1987). Power generation emissions may be reduced by half in the year 2000 (Nowicki 1991b).

Poland's environment ministry has threatened to close down the eighty filthiest enterprises if they do not clean up their act in an attempt to halve the SO_2 norm in the air, even though a quarter of the country is still unable to attain the old norm. This perhaps helps to explain why Poland has not signed the Protocol to the Convention on Long Range Transboundary Air Pollution, requiring a 30 per cent reduction of SO_2 by 1993 over the 1980 level. It is most unlikely that present levels in the country would anyway allow the 1993 objective to be achieved.

One of the main reasons for the overall detrimental air pollution situation results from the intensity with which the economy uses energy, its industrial mix, and the nature of energy supply. Comparison with the former West Germany revealed that Poland generates 80 per cent more SO_2, although having a land area only a quarter larger; SO_2 emission per capita in Poland is over twice that of the former federal republic, whilst SO_2 emissions per km^2 are six times greater than those of the USA (Kabala 1989). Although there are fifteen installations for desulphurizing flue gases in operation at present in Poland, they are only pilot plants. The 1990s will hopefully see an increase in numbers with special preference being given to thermal power-stations and metallurgical plants (Osterberg and Teknik 1989).

A survey on industrial pollution and atmospheric protection in 1981 carried out by the food and agriculture department of Poland's statistical service designated three major pollution groups: first, sulphur dioxide (SO_2), nitrogenous oxides ($NxOy$) and carbon monoxide (CO), which the survey stated had wide dispersion and high concentration in Poland; second, lead (Pb), cadmium (Cd), arsenic (As) and mercury (Hg), also with wide dispersion rates, high toxicity and persistent throughout the environment. Third, cancerous highly dispersive hydrocarbons ($CnHm$) were admitted as 'dangerous to people's health' (Carter 1989: 167). After sulphur dioxide, perhaps the most serious atmospheric problem is linked with the emission of nitrogenous oxides. Between 1980 and 1989 this pollutant more than trebled from 187,000 to 781,000 tonnes or 11.8 per cent of all emissions; carbon monoxide provided a further 1,335,000 tonnes or a fifth of total pollution in 1989 (Grzesiak 1990: 138). Increasing car ownership during the 1990s will double NOx and create a 70 per cent increase in carbon dioxide emissions (Agren 1991). All these cancerous hydrocarbons and heavy metals will effect Poland's water, soil/vegetation, and health.

WATER POLLUTION

Degradation of Poland's surface and underground water supplies is the most serious factor threatening the natural environment after atmospheric pollution. As with air pollution, greater awareness emerged in the late 1960s. Increased industrial demands from a country with low water resources soon aggravated the situation. Industrial location along Poland's major river courses intensified the pollution problem in the early post-war period. For example, between 1945 and 1970, 238 large industrial plants were built along the Vistula, Poland's major river. Their siting meant easy water access, sewage disposal and presence of nearby urban centres for labour supply. Some enterprises were cellulose and chemical plants, together with thermal power-stations – all major polluters.

During the 1970s a dramatic fall in Poland's water quality occurred; by 1977 an official government report admitted that Poland's water courses were 'very highly polluted' with only a few rivers in the far north-eastern part of the state and in western Pomerania that could still be regarded as relatively uncontaminated (*Nowe drogi* 1978). Part of the reason was related to an intensified planning policy between 1971 and 1975 to locate industrial enterprises, especially the power industry, along or near such rivers as the Vistula and Odra. Progressive decline in river water quality resulted (see Table 6.2).

Biological criteria has only been applied to Poland's rivers since 1984, but even the physiochemical evidence is quite dramatic. The cleanest rivers with water fit for human consumption (Class I) declined from nearly a third in the mid-1960s to below 4 per cent by the late 1980s; Class IV, which is water unfit for either agricultural or industrial use, rose from just over a quarter to two-fifths, whilst biologically three-quarters of Poland's rivers were dead by 1988 (Kacprzyk and Żurek 1990: 21). There are also many lakes, marshes and waterlogged areas in northern Poland, where research has shown that only about 1 per cent of the water is Class I whilst over half is in Class IV (Klimek 1990).

Much of this deterioration is associated with increased municipal and industrial sewage disposal. By 1979, over two-fifths of all sewage from these sectors was completely unpurified, much of the rest receiving only mechanical treatment which fails to dissolve soluble impurities in the water. By the late 1980s purification of industrial and municipal waste was totally inadequate; for example: of the 4,639 large industrial plants responsible for producing over four-fifths of all

Table 6.2 Poland: river pollution classification, 1964–89

Year	Class I (%)	Class II (%)	Class III (%)	Class IV (%)
1964–67	31.6	25.6	14.0	28.8
1968–70	24.8	29.2	17.6	28.4
1971–73	23.4	32.2	18.0	26.4
1974–77	9.6	30.7	26.7	33.0
1978–83	6.8	27.8	29.0	36.4
1984–88:				
physiochemical	4.8	30.3	27.8	37.1
biological	0.0	3.9	20.3	75.8
1989:				
physiochemical	3.4	28.3	27.6	40.7
biological	0.5	3.9	17.0	77.7

Source: Grzesiak, M. (1990) *Raport o stanie, zagrożeniu i ochronie środowiska 1990* (GUS), Warszawa, p. 98, ibid. (1991); *Ochrona środowiska 1991* (GUS), Warszawa, p. 122.

waste, 1,750 (that is, 37.7 per cent) discharged straight into the urban sewage network. Furthermore, of the 2,889 enterprises discharging waste into surface water areas, 508 (17.6 per cent) did so without any purification. Municipal sewage disposal also has its inadequacies. By 1989, 366 Polish cities, including four with 200,000 or more inhabitants – Warsaw, Łódź, Białystok and Radom – still had no municipal sewage treatment plants (Grzesiak 1990: 90). In 1989, out of 825 towns, 775 now possess a water supply network whilst 459 contain sewage processing facilities; of those only 425 have a waste purification unit, up to two-thirds of which work inefficiently.

It is not surprising, therefore, that by the end of the 1970s Poland's surface water courses suffered from intense degradation. Amongst the worst were the rivers Ner and Wełna (which flow into the River Warta), and the Bug, Bzura and Utrata (flowing into the River Vistula), all figuring below the state BOD5 norm throughout their lengths. Others less affected included the Odra (64 per cent), Vistula (56 per cent) and Warta (51 per cent) (Mańczak 1979, Anon 1981a). Lakes were also becoming increasingly polluted. Of the 500 large lakes in Poland, about 300 had suffered degradation and pollution from municipal/ industrial sewage disposal by 1979. Lakes in the provinces of Suwałki, Szczecin, Olsztyn, Koszalin and Słupsk were the worst affected and amongst those with critical pollution levels were lakes Mikołajki, Niegocin, Tałty in Mazuria, along with Jamno (Koszalin), Długie and Ewingi (Olsztyn), (Giercuszkiewicz-Bajtlik 1987), Klasztorna (Gdańsk), Zajczierze (Szczecin) and Zioł (Poznań). Moreover, in the provinces of Olsztyn and Suwałki, twelve lake beaches were closed because of bacteriological pollution (Kassenberg and Rolewicz 1985).

During the early 1980s water pollution continued unabated. Every year it increased, particularly as rivers throughout the country became more degraded. The consequences of river pollution on drinking water were becoming increasingly apparent by 1982 when only 18 per cent of mains supply was officially considered to be safe hygienically. A year later about two-thirds of all rivers in Poland were classified as Class III or worse. Heavy metals continued to pose a threat. The mercury content of the Vistula river below Cracow during the second half of 1980 was more than 200 times above its permitted norm; an excess of chrome, lead and iron was also noted. It was estimated that over half the running water used by Cracow municipal water supply system fell short of state norms and was threatened with pollution (Survey of World Broadcasts 1983).

Underground water supplies have also been affected. For example, Szczecin has three main drinking water sources, one underground and the others the River Odra and Lake Miedwie. As in Upper Silesia, Łódź and Warsaw, underground sources in Szczecin are almost exhausted, whilst much of the Odra's water is undrinkable. Finally, Lake Miedwie has been undergoing eutrophication over the last two decades due to high nitrogen and phosphorous levels, which create mass organism development. As a result the lake water is being poisoned by hydrogen sulphide; aquatic life has decreased, and the spontaneous self-cleansing process in the lake reservoir has continually been reduced. Through pollution, this combination of events has meant that Szczecin now has a critical drinking water supply problem (Survey of World Broadcasts 1983). Other lakes are suffering also, particularly in Zielona Góra province (thirteen lakes now have inferior water purity), largely from urban and holiday centre effluent as well as industrial waste, so that water pollution in them is continually rising (Survey of World Broadcasts 1983).

Over 90 per cent of Poland's water supply for industry came from surface sources in 1989 (Grzesiak 1990: 75), yet these sources were constantly being polluted by industrial accidents and waste dumping – not all through Poland's own endeavours. A lot of river pollution comes from Czechoslovakia; for example, in November 1986, a spillage of 190 tonnes of heating oil from a factory in Ostrava in northern Moravia eventually reached the Odra river in Poland, causing one of the largest disasters in the region (Gesing 1986); in 1988 Czechoslovak authorities admitted leaking 30,000 litres of diesel oil into the Nisa river (Radio Prague, DOM/M, 23 July 1988, 6.30 p.m.) for which Poland demanded compensation. Poland's main river, the Vistula, is now so poisoned and corrosive-laden that stretches are considered unusable even for factory coolant systems, let alone for drinking purposes. This has prompted objections by the Greenpeace organization (Papuzińska 1990) whilst the Pope has lamented that the rivers and lakes of his native country 'are now largely dead' (McCarthy 1990: 3).

The Vistula river, which runs through Cracow and Warsaw, was known as the 'Queen of Rivers' but is now a potent symbol of environmental degradation. The stretch which passes through Cracow is devoid of biological life and the concentration of sodium chloride (salt) is higher than in the Baltic. Every river in the Katowice province is polluted beyond all acceptable limits. One of the major sources consists of 9,000 tonnes of salt which are discharged daily from coal mines in Upper Silesia into the rivers Vistula and Odra, which in turn create drastic changes in the biological life of water plants and algae, besides its corrosive nature for industrial use and drinking water supplies (Lemiński 1990). While Silesia only processes 3 per cent of its sewage, there are promises of 3,000 waste-water treatment plants to be built throughout the country (Kamiński 1990). Meanwhile, as it flows north the Vistula collects effluents, chemicals (Pawłowski and Kozak 1988), and untreated sewage from industrial plants and urban centres along its route to the Baltic Sea.

The Baltic is rapidly developing into one of the world's most polluted enclosed seas, the northward flow of Poland's rivers adding severe pressure to this situation (see Table 6.3). Clearly, Poland is responsible for over a fifth of the BOD5 and more than a third of the nitrogen and phosphorus deposited in the Baltic, a worrying factor for the Helsinki Commission responsible for the Convention on the Protection of the Marine Environment of the Baltic Sea Area in its attempt to halve toxic discharges by 1995 (Hagerhall 1990).

The Vistula river contributes 34 km^2 of liquid waste annually (Pudlis 1989). Agricultural run-off from Poland's heavily fertilized farmlands amounts to nearly 100,000 tonnes of nitrogen entering the Baltic yearly, whilst the entire ecosystem

Table 6.3 Poland's contribution to Baltic pollution, 1989 (thousand tonnes/year)

State	BOD5	Nitrogen	Phosphorus
Soviet Union	504.7	130.3	5.9
Poland	358.4	231.8	18.0
Sweden	361.6	129.5	6.9
Finland	249.8	69.5	4.4
Denmark	159.7	69.1	9.5
West Germany	22.7	16.4	2.4
East Germany	13.3	3.6	0.4
Total	1,670.2	650.2	47.5
Poland's per cent	21.46	35.65	37.89

Source: Calculated from Grzesiak, M., Raport o stanie, zagrożeniu i ochronie środowiska 1990 (GUS), Warszawa, 1990, p. 101.

of the Bay of Gdańsk is under threat because pollution is killing off the phytoplankton and rockweed, which form the food-chain base, whilst contamination prevents swimming along much of Poland's Baltic coast (Ponard 1987, Żyskowski 1988). Along this coast, 82 places were classified as sewage outlets, 48 of them spill effluent directly into rivers, and the rest straight into the sea (Praca zbiorowa 1981: 25). Around the Bay of Gdańsk and Puck, chemical industries have developed, including oil refineries, factories for producing artificial fertilizers, sulphur purifying plants, thermal power-stations, docks, etc., the latter containing oil tankers which spill huge amounts of oil into the bay thus harming bird, fish and plant life along the shore (Marcinkiewicz 1987).

SOIL/VEGETATION POLLUTION

Environmental disruption in Poland created by air and water pollution has affected parts of the country's soil and vegetation cover. Yet this is no new phenomenon; in 1890, scientists demonstrated that a forest near Katowice, in the centre of the Upper Silesian industrial region, had been damaged by SO_2 emissions (Schreiber 1985: 3). Up to the Second World War the only areas seriously effected by any pollution form were the coal basins of Upper Silesia and Wałbrzych, but the rapid industrialization programme after 1945 meant many more areas were subjected to environmental degradation.

During the 1960s enormous amounts of toxic industrial dust and gases began seriously to affect the soil. By 1970 the total surface area with a disturbed biological equilibrium was estimated at over around 300,000 ha, of which a sixth came from dumped industrial waste (Greszta 1974: 396). In Upper Silesia, spoil heaps covered 7 per cent of the surface area, whilst vegetable garden soil samples from there had a lead content which in 1979 ranged from 42 mg/kg (ppm) to 8,890 mg/kg; cadmium content varied between four and sixteen times the prescribed national limit, and nearly a fifth of the farmland should not have been cultivated as a result of substantial heavy metal presence (Timberlake 1981).

Other areas suffered too, including the soils around the Nowa Huta Metallurgical Works, Cracow, which contained high dust deposition of zinc, lead, copper, molybdenum, aluminium, iron, manganese and barium (Strzyszcz 1982). In agriculture, increased use of herbicides, pesticides and artificial fertilizers filtrated through the soil and eventually entered watercourses, the high nitrate content augmenting the already contaminated water supplies. Livestock breeding and industrial farms also added their contribution; for example, in 1976 two 600 m³ animal manure storing tanks near Łąck (Płock province) burst, leaking effluent into lakes Łąckie and Ciechomicki, the dangerous pollution ensuring that swimming was forbidden there for nearly a decade (Kowalczyk 1984).

During the late 1970s most attention was focused on air pollution's impact on forest vegetation. Over a quarter of Poland's surface area is covered by forests; nitric and sulphuric acids, through acid rain, weaken plants and trees which are then killed by low winter temperatures, high summer ozone formed by sunlight acting on hydrocarbons, together with other factors like pests. Heavy metals also took their toll. Coniferous trees that grew near zinc, lead, iron or coking plant complexes – for example, in Chorzów Park of Culture, Silesia – were found to lack resistance to SO_2 (Białobek and Tachwal 1981). Scots pine needles near a zinc smelter in Katowice province reflected the 'unfavourable influence of these metals' on growth (Niemtur 1982: 26).

This pollution of the soil and vegetation continued during the 1980s. By 1982, some 26,000 ha of Poland's forested area suffered severe pollution damage (Category III), which by 1989 had increased by two-fifths to 37,000 ha (Grzesiak 1990: 161). Poland's forested area covers about 28 per cent of the country's total surface area (Hasiński 1987); according to a study in the mid-1980s, between 5 and 10 per cent of its forests were damaged (United Nations 1985), but unofficial estimates suggest that the figure has been nearer three-quarters, partly as a result of imported acid rain from Czechoslovakia. Clearly the south-west of the country is worst affected. Pollution coming from Czechoslovakia and the former East Germany has had a distinct impact on the Sudetes and Carpathian Mountains (Mazurski 1986). For example, in 1975 the Jizerský Mountains forest in the western Sudetes contained extensive stands of spruce; by the end of the 1980s over 100 km² of forest had died (Mazurski 1989a) and reafforestation with young saplings has failed (Marcinkiewicz 1987).

Four-fifths of Poland's forest are located in national parks where similar devastation has been noted. In the Sudetes, the Karkonosze National Park previously suffered most damage above the 800-metre level, but this is now extending to below this level (Andersson 1987). In March 1988, a protocol on co-operation between the former East Germany, Poland and Czechoslovakia agreed to work together in reducing air and water pollution, and to protect the border forest areas; government officials were warned that unless current levels were reduced, between 30 and 50 per cent of forests in the border regions could be dead by 1995 (Radio Warsaw, DOM/M, 3 March 1988, 7 p.m.). Ecological damage to forests is expanding over increasingly larger areas, and in particular is devastating

mountain woodlands in south-western Poland. Evidence suggests the phenomenon is moving south-eastward largely due to pollution from Silesian industries
(Mazurski 1989b, 1990: 73).

Industry also continues to pollute the soil; in 1984, already over a quarter of
Poland's forty-nine administrative provinces recorded totals of 2,500 ha or more
of devastated land. Predictably these were provinces with high pollution from
mining, thermal power-stations, etc.; the highest in 1988 was Katowice province
with 10,865 ha, followed by Konin (6,222 ha), Jelenia Góra (5,049 ha) and
Opole (4,841 ha) (Mazurski 1988a, 1991a). Refuse storage adds a further
dimension to land devastation. Public outrage erupted in 1988 when it was
learned that an agreement had been signed to store some 900,000 tonnes of
highly polluted Rhine river sludge from Germany in the Bogatynia region of
Upper Silesia (*Der Spiegel*, 23 January 1989: 3), whilst depleted shafts in the
Bełchatów and Konin lignite mines in central Poland were also being considered
as possible dumping places for imported waste (*Życie Warszawy*, 17–18
December 1988: 2). Local pressure may have prevented both cases taking place,
but it leaves the question as to how many others escaped public attention.

Attempts at the rejuvenation of land devastated by mining and industrial
emissions in Poland dates back to the early 1960s. By 1970, over two-thirds of
Poland's post-industrial waste land had evolved from mining activity, the rest
from rubbish dumping by steel works, the chemical industry, thermal power-
stations and other industrial sources. A decade later, degradation of the soil cover
totalled about 100,000 ha, whilst only a quarter of despoiled land had been
restored by the late 1970s. Changes during the 1980s are seen in Table 6.4.
Whilst the devastation of land reached its apogee in the mid-1980s from which
there has been a steady improvement, the overall percentage change from 1975
to 1990 is less encouraging. Devastated land increased by 2 per cent, whilst that
reclaimed dropped by two-fifths and the hectarage back in cultivation declined by
over a third (Mazurski 1991b). These findings also support Malisz's view that by
1990 there would be an increase in unreclaimed land (Malisz 1978: 84).

Intensive agricultural production has also added to soil degradation. In the
densely populated Polish countryside the growing demand for cultivable land has
led to an even greater fragmentation of fields and to a transformation of forest
grass and pasture land into arable, often to over 900 m above sea-level. The

Table 6.4 Land reclamation in Poland 1975–90 (in hectares)

Category	1975	1980	1985	1990
Land devastated	91,660	109,260	103,724	93,679
Land reclaimed	4,480	4,424	4,690	2,665
Land returned to cultivation	3,535	4,332	3,604	2,264

Source: Grzesiak, M. (1991), *Ochrona środowiska 1991* (GUS), Warszawa, p. 41.

growing problem of soil erosion (Mazurski 1989c) combined with greater use of artificial fertilizers, etc, has led to ecological catastrophes in such places as around the Vistula estuary where some of the country's best agricultural land is located (Ponard 1987). In 1989, Poland received a European Community 'gift' of pesticides valued at 6 million dollars; of the thirty-two varieties, twelve had already been banned as unsafe in Western Europe (Pudlis 1991)! Moreover, many Polish industrial workers grow their own fruit and vegetables on allotments near their homes in soils highly contaminated with lead and other pollutants (for example, in the Katowice and Cracow regions); such produce, though eaten, is really unfit for human consumption. On the positive side, however, long-term damage from the more intractable herbicides and pesticides in Poland is not so great as in some West European states, judging by the varieties and numbers of wild flowers still in evidence as one travels through the Polish countryside.

NUCLEAR POWER POLLUTION

Outside the former Soviet Union, Poland was more seriously affected by the Chernobyl accident than any other country (Medvedev 1990: 202). Between 27 and 29 April 1986 almost the whole of Poland was covered by the ensuing fall-out from the accident and the long-term effects of this event can only be conjectured. The impact of the event can be judged by the comparative annual average fallout of caesium 137 and strontium 90 for the 1980s (see Table 6.5).

Besides the obvious implications to health, and the well-being of Poland's inhabitants, the disaster clearly affected the Polish economy which lost some £4 million in tourist revenue, and an estimated £17–25 million as a result of the food ban imposed by the European Community (Boyes 1986).

Nevertheless, Poland's nuclear programme was said to be proceeding according to plan in 1986 – namely, to complete a reactor on the Baltic coast and

Table 6.5   Recorded caesium 137 and strontium 90 in Poland, 1980–90 (Bq/m^2)

Year	Caesium 137	Strontium 90
1980	17	5
1981	10	3
1982	6	2
1983	6	2
1984	5	2
1985	6	2
1986	1,511	22
1987	22	4
1988	12	4
1989	8	2

Source: Grzesiak, M. (1991), *Ochrona środowiska 1991* (GUS), Warszawa, p. 218.

build three other plants to produce 15 per cent of the country's electricity by the end of this century. In fact, Poland had been constructing a 440-megawatt reactor at Żarnowiec near Gdańsk since 1983, which elicited considerable protest after the Chernobyl accident (Leszczyńska and Vetter 1987.) Maybe as a result of these protests, the commissioning of the first unit at Żarnowiec was postponed until 1991 (and is still not operational), with the other reactors coming annually on line thereafter. A second plant (Warta) with four 1,000-megawatt units was planned to be built at Klempicz, near Poznań, and would be operational in 1997 to produce 20 per cent of the country's electricity; yet another station similar in size was to be constructed on the Vistula river (Anon 1987).

During 1989 there was a growing groundswell of anti-nuclear generation protests in Poland (Orval 1989). Demonstrations were held in such cities as Lublin (von Urban 1989), which had been closer to the Chernobyl fall-out than many Polish places, as well as protests in and around Gdańsk (Anon 1989a). Activity against the Żarnowiec plant continued in May 1990 with an unofficial referendum held in Gdańsk only 50 km away which declared an overwhelming opposition to its construction (Paczka 1990). Its future remains shrouded in doubt, and despite some opposition to the project at ministerial level a definite decision still awaits clarification (Pape 1990a, Albinowski 1990).

POLLUTION AND THE NATION'S HEALTH

In theory, Poland should be a healthy society; the nation's well-being during the post-war period of socialist rule has been secured by a free and universally access-ible health service both for treatment and prevention. In reality, evidence from the early 1980s suggested this was not the case, when a report drafted by a group of Polish physicians revealed that 'neither the health of the population is good, nor does the health service function properly' (Anon 1983: 488). Part of this situ-ation was related to 'the impact of the environment on human health' (Żmuda 1980: 126). Unfortunately, less is known about the nation's health before 1980 due to a state ban on all published data concerned with many health issues, including the noxiousness of environmental pollution. Media information on such problems remained fragmentary, whilst there was a prohibition on data concerned with employment health hazards and building materials and other substances readily available on the open market.

Since 1980 political changes have allowed some relaxation on this control, and by 1983 it was officially admitted that 2 million people in Poland were employed in conditions that were harmful and arduous, and sick rates from occu-pational illnesses were proved to be closely tied to working conditions (*Trybuna Ludu*, 4 January 1983: 2). Certainly there was a sharp increase in all-cause mortality for Polish males between 1972 and 1982, with rises of a quarter or more in all five-year age groups between 40 and 59 years, although females recorded little change. The main causes for male deaths appear to have been cardiovascular diseases and lung cancer, which have been attributed to more generalized changes associated with the country's rapid industrialization; but results remained difficult to evaluate in environmental terms (Millard 1982; Cooper *et al*. 1984).

Nevertheless, hints at pending disaster were emerging in the early 1980s; in 1982, Poland's State Planning Commission admitted that the Upper Silesian Industrial District, the copper mining region around Legnica, and the Cracow region, were areas of ecological catastrophe, whilst twenty-three other areas were recognized as under such a threat. These areas were stated to cover about 27,000 km^2 of state territory, where 12 million people resided – that is, a third of Poland's population (Marcinkiewicz 1987). Details for 1984 began to emerge on the high levels of toxic pollutants recorded on vegetables grown in allotments near to industrial complexes (Maziarka 1987, Mazurski 1987), whilst there was increasing criticism on the availability of health services in rural areas and low ratios of doctors per rural inhabitant (Kowalczyk 1987).

Obviously, the Chernobyl disaster heightened people's awareness of links between pollution and health. Immediately after the accident, restrictions were imposed on grazing cattle, together with milk consumption, but these were later lifted after the panic had receded, although the long-term effects remain in question. Food pollution levels from radiation and other sources were still high a year after the accident but were stated to be no longer a threat to the population's health (Radio Warsaw, DOM/M, 4 July 1987, 10 a.m.). However, suspicions about links between pollution and health remained amongst the population. Ecology experts called for the closure of dangerous industrial plants and enterprises and for a radical restructuring of the country's economy if ecological targets were to be met (Radio Warsaw, DOM/M, 24 March 1988, 5 a.m.), a factor becoming particularly serious in built-up urban areas (Mazurski 1988b). Meanwhile, fears about the unhealthy state of many Polish food staples continued (Radio Free Europe Research 1988).

The mid-1980s appears to have reached a nadir in life expectancy in the country (see Table 6.6). There has been a slight improvement over the 1985–90 period for females, a trend which one hopes will continue, although the male population still has a life expectancy below that of 1965. Of course,

Table 6.6 Poland: life expectancy, 1965, 1985, 1990

| Age | Male | | | Female | | |
	1965	1985	1990	1965	1985	1988
15	55.7	53.3	53.1	61.8	61.3	61.8
30	41.7	39.2	39.1	47.2	46.7	47.2
45	28.2	26.0	26.0	32.9	32.5	33.0

Source: *Rocznik Statystyczny 1989* (GUS), Warszawa, 1989, Table 29(97), p. 58; Grzesiak, M. (1991), *Ochrona środowiska 1991* (GUS), Warszawa, p. 253.

these figures disguise the fact that incidents of cancer and bronchial illnesses are higher in polluted areas, where life expectancy and infant mortality are higher than the national average. In 1989, the Polish Health Ministry predicted a devastating increase in cancer cases by the year 2000, with one in every four Poles suffering from the disease. In that year it is estimated that there will be 125,000 cases and 90,000 deaths from cancer; over a third caused by bad diet, slightly less than a third from smoking, and 15 per cent as a direct result of pollution (Clough 1990). By 1990, a third of Poland's population lived in areas of ecological degradation (Anon 1991). Noise pollution is greatest in urban areas; a fifth of Poland's area has levels periodically exceeding the PSL (60 dB and affecting a third of the population (Kacprzyk and Żurek 1990: 32). However, the infant mortality (IMR) has revealed a more positive trend; in 1970 the IMR per 1,000 live births was 33.4, which had dropped to 19.1 in 1985 and 16.1 three years later. The 1988 IMR figure was therefore less than half that of 1970 (but still nearly double the rate of a country like Spain), with highest incidence around Poland's industrial cities.

POLAND'S POLLUTION LEGISLATION

There is considerable evidence of Poland's environmental degradation since 1945. Government reaction to the problem and methods of improving it left much to be desired, the response to combat pollution mainly being concerned with preventive legal measures. Theoretically, the principle of care for, and improvement of, Poland's environment had existed since the late 1940s through various legal measures (Brzeziński 1974: 47). In 1949, the Nature Protection Act stipulated afforestation and tree planting on shifting sands, waste land, steep mountain slopes and the creation of green belts to prevent continued soil erosion; unfortunately this law was more a declaration of intent than action. The State Inspectorate for Water Conservation, established in 1954, was placed under the jurisdiction of the Communal Economy Ministry; by 1957, over 100 minor regulations referring to water pollution had become law but few had practical application.

An encouraging move in the late 1960s was the prospect of combining various consultation/research groups with state administrative offices to analyse Poland's pollution problems. A Committee on Environmental Protection was created in conjunction with the Polish Academy of Sciences. Most research emphasis was on water pollution, but there were few air pollution studies. This was more remarkable considering a major law, passed in 1966, which was officially described as 'the best formulated and most progressive law of this sort in the world' (*Trybuna Ludu*, 25 March 1969: 1). In 1970, a forestry protection law was promulgated which, in theory, made it impossible to establish industrial enterprises in any part of Poland's forested region. Furthermore, state funding of environmental projects seemed more than adequate; for example, sufficient finance to build over a thousand purification plants was allocated in the 1971–5 Five-Year Plan (*Sztandar Młodych*, 25 March 1970: 3; Jastrzębski 1990: 22).

By the end of the 1970s Poland had amassed a plethora of legal measures adopted by the government to combat pollution. Throughout the 1970s, in line with other CMEA countries, laws were passed to assist environmental protection. The central planning system was seen as providing a coherent framework for environmental impact assessment procedures, with a checklist based on components (for example, atmosphere, biosphere, soils, phytosphere, etc.) and used as a basis for positive and negative feedback on human/economic activity and pollution. Environmental protection amendments were added to the Polish Constitution in 1976 with statements like 'the Polish People's Republic secures protection and rational development of the natural environment' and 'its citizens have a duty to protect it' (Kormondy 1980: 35). Yet, despite the profusion of regulations, administrative offices/ministerial responsibilities, and being armed with sufficient funding, results were disappointing – particularly for water control. Unfortunately, the gradually worsening economic situation in the country in the late 1970s meant that in spite of new environmental regulations Poland's financial support for reducing pollution considerably diminished.

Nevertheless, laws continued to appear, accompanied in the 1980s by a significant change of attitude towards the environment (Kassenberg 1988). This was manifested in several ways, beginning with the Statute on the Protection and Development of the Environment in 1980 (Polish Government Decree 1980). The Act was divided into eight parts, each concerned with some specific environmental aspect, and contained 118 regulations; in theory, the Act was a model answer for solving contemporary pollution problems but, in reality, there was little chance of necessary funds for such reforms at that time. Furthermore, the 1980 law failed administratively to place environment problems under the aegis of a national environmental council, free of government interference. Instead they were linked to innumerable ministries (Mining, Metallurgy, Administration, Regional Economy and Environment Protection) each one carefully guarding its own political/economic interests. Similarly, the problem of effective fines remained unsolved; for example, the Soil Protection Act of 1983 meant that penalties so far imposed have totalled only 2.5 per cent of most company profits (Mazurski 1990).

Other evidence of change came with the setting up of the State Council for Environmental Protection in 1981, the establishment of the Office of Environmental Protection and Water Management in 1983, and a year later the creation of a State Environmental Protection Inspectorate. Finally, increased public awareness was demonstrated by the founding of the Polish Ecological Club (1980) and in 1986 the Ecological Social Movement. Even so, the 1980s have been bedevilled by many shortcomings related to the legal aspects of environmental protection. Amongst these there still exists a lack of cohesion and uniformity within the legislation, this still placing emphasis on the encouragement of economic growth (Klimek 1990: 111). There are still inadequacies. Different attitudes exist towards the environment as revealed in the civil and criminal codes; marine environment laws need tightening, and more consistency must be practised regarding the management of economic environment controls.

The late 1980s have witnessed increasing legal concern for the environment. A decree on the Protection of the Land Surface Area in 1987 (Polish Government Decree 1987) made specific mention of waste storage pollution control, sewage disposal, protection of flora/fauna, and harmful application of chemical and biological substances which may alter the balance of nature. This act was not only aimed at the 27 regions where the natural balance has been destroyed, but also preservation of the country's 15 national parks, 988 nature reserves, 43 landscape parks and 165 landscape-protection areas (Anon 1990a: 73). Environmental protection institutes exist in Warsaw, Katowice, Wrocław and Gdańsk to support these aims, whilst discussions were under way in 1990 on ecological priorities over the next three years; these include limiting dust and gas emissions, providing more clean water through a revitalized pumping system and canalization, revision of water charges, closer supervision of the 500 worst-polluting industrial plants, coal enrichment, recycling toxic waste, limited food production from toxic soils, devising an anti-pollution transport programme, and increasing the ecological education of society (Council of Ministers 1990).

Poland also has international legal commitments as a result of signing various conventions, declarations and agreements on environmental control. It signed the Helsinki Convention on the protection of the Baltic in 1974 (Łukaszuk 1989) and in the early 1980s was signatory to the World Charter for Nature Protection adopted by the United Nations in 1982 (Klimek 1990: 113). More recently, the Czechoslovakian, former East German, and Polish premiers signed an ecological agreement on co-operation in July 1989, pledging to reduce gaseous substances (especially SO_2 and NOx) in their border areas through a system of control. In the case of pollution accidents the three states would offer mutual assistance and compensation (Anon 1989b). The Polish government has discussed an international bilateral agreement to limit bilaterally the danger of polluting each of its neighbouring states (Council of Ministers 1990).

POLAND'S ENVIRONMENTAL ORGANIZATIONS

The change in Poland's attitude towards the environment, which occurred in the early 1980s, resulted from rather exceptional political circumstances. The emergence and success of the independent Solidarity trade union encouraged the founding of several non-government environmental organizations worried about the state of the country's pollution. One of these organizations was within the Solidarity movement, another linked to Rural Solidarity, and the third and probably best known – the Polish Ecological Club (PKE) – was associated with it. The latter was formed spontaneously in the heavily polluted city of Cracow in September 1980 and managed to secure the closure of an aluminium plant at nearby Skawina; the following May it was officially registered as an independent body with seventeen branches throughout the country. This unique position was soon lost with the return of martial law in December 1981 and it went underground; but gradually, as conditions became more tolerant, it re-emerged with renewed vigour (Fura 1985).

In 1985, another independent group – the Freedom and Peace movement (WiP) – was established and was particularly active in arousing considerable public sympathy for espousing 'green' issues (Rich 1987). At the same time, the state sponsored League for the Protection of Nature developed with its local branches opting for increasingly independent views; this attitude challenged the original state emphasis on environmental education rather than fostering independent views. Also in 1985 the Committee for the Protection of the Environment (KOS) was established by members of Solidarity, and the ubiquitous Roman Catholic Church also joined in the debate stressing the moral and spiritual side of environmental awareness as exemplified in 1986 by the establishment of the Franciscan Ecology Movement (Perłowski 1988).

The Chernobyl disaster was to add further impetus to these organizations. For example, KOS initiative gathered momentum after April 1986 with the formation of a working group for environmental protection, published its own bulletin *Zagrozenie* ('The Threat') in June of that year (Radio Free Europe Research 1986). In response, the government tried to gather all the various groups under one umbrella – the League for the Protection of Nature, but this broke up into independent parts; the government then established an ecological social movement working under the patronage of the Patriotic Committee for National Rebirth (PRON), dating from the martial law period, and aimed at co-ordinating the work of clubs, committees and individuals involved in protecting the environment and cultural monuments.

Meanwhile, independent activity continued; by 1988 the Polish Ecological Club had 2,500 members with seventeen branches and eighty-five sections in Poland, swelled no doubt by the success of its second conference held in 1987 which severely criticized the government's environmental policies and attempts to hide the true facts (Radio Free Europe Research 1987). The Freedom and Peace group became involved in securing official approval to launch the country's first 'Green Library' (Salter 1988) and raise money to provide centres in a number of cities with information on a range of 'green' ideas and issues (Schreiber 1989). In the autumn of 1989, the Polish Ecological Club established an office in Katowice to advise on environmental education, organizing seminars and guiding foreign visitors interested in air pollution problems (Pape 1990b). It also played a leading role in the round table meeting on ecology attended by Solidarity members and the government in that year.

All this activity led to discussions between Solidarity, The Polish Ecological Club, the Freedom and Peace group and other movements, about forming a Green Party (PPZ). This finally crystallized in December 1988 to become the first of its kind in Eastern Europe (Parkin 1989). It now has about 3,000 members in thirty-five provinces, but due to its fragmentary nature has not become a major political force (Delorme 1989, Pehe 1990). Meanwhile, new groups continue to be formed such as the EcoBaltic Foundation in July 1989 with specific interest in the future of

Poland's northern waters (Partek 1990), whilst foreign organizations such as 'Greenpeace' take an active interest in the Polish environment (Rosenbladt 1988, Anon 1990b), as do local representatives (Kalicka 1991; Poskrobki 1992).

CASE STUDIES

1 The Upper Silesian Industrial Region (GOP)

This is the most polluted part of Poland, 6,650 km^2 (2.1 per cent of Poland's total area) and containing ten towns with a combined population of 4 million. Four-fifths of Poland's coal is mined here, and it produces nearly a third of the country's industrial output. A long history of industrialization made this area the obvious choice for intensive development after 1945. By the mid-1960s, the amount of dust deposition varied between 500 and 1,200 tonnes per km^2/year in the ten largest cities of this region, whilst the SO$_2$ air content frequently breached the accepted safety limit of 0.25 mh/m^3 (Żmuda 1967: 125–6, Dienes 1974). This resulted from intensive use of local brown coal and lignite supplies that contained a high sulphur content, often further aggravated in winter by tempera-ture inversion. A high concentration of thermal power-plants here during the 1960s, especially in the Katowice province, produced large quantities of fly-ash, whilst the location of electric furnaces for steel production added both to dust and gas emissions. The 1970s experienced further intensification of industrial production so that by the end of the decade two-thirds of all dust and gaseous matter came from six provinces in south-east Poland, of which Katowice provided a third (Carter 1989).

The 1980s have seen little improvement in the situation, particularly in Katowice province (Żmuda 1980). Here, by 1980, nearly a million people suffered from intensive fall-out of poisonous and strong toxic dust/gases, with permissible safety limit (PSL) barriers exceeded several times over – a situation which continued to worsen. By the mid-decade the province had dust fall-out 40 times above the highest annual concentration PSL mean at some locations, with nitrogenous oxides 7 times higher, carbon oxides 50 times, lead and other car-cinogens by 60 times (Zięba 1986). In fact, a Katowice city councillor commented that the air was so contaminated that apart from the health hazard it also enhanced the corrosion of machinery and equipment (Reeves 1986). Indus-trial dust falling on the towns of Bytom, Gliwice and Zabrze was 1,000 tonnes/km^2, or 4 times higher than permitted health and safety standards (Marcinkiewicz 1987). Part of the problem lies with the high sulphur content of coal mined in the eastern part of the region (that is, 1.55–2.19 per cent sulphur), and utilized by five large thermal power-stations there which emit 40–45 per cent fly-ash content but produce only 20–28 MJ/kg^2 of energy (Luchter 1988).

Sulphur emissions in Upper Silesia are, for example, 5 times higher per km^2 than in Germany's Ruhr valley, and such levels are found nowhere in Europe except the north-west region of Czechoslovakia. In fact, about a fifth of the SO$_2$

falling on Upper Silesia originates in Czechoslovakia or the former territory of East Germany. It is not surprising that the Freedom and Peace Movement chose Gliwice to house a seminar in 1988 calling for greater co-operation between the governments of East Germany, Poland and Czechoslovakia to work together in combating industrial pollution. Meanwhile, Upper Silesia continues to receive a third of all Poland's gas emissions, a similar proportion of SO_2, and two-fifths of its dust, but produces two-fifths of Poland's power generation. In 1989, Katowice province received 2.9 million tonnes of SO_2 on its territory (more than three times that of its nearest rival province Cracow, with 895,000 tonnes), with a staggering 443 tonnes/km^2/year (Grzesiak 1990: 129).

Water, soil and vegetation naturally suffer from this pollution onslaught. In the early 1960s, a hydrological study of the Upper Silesian Industrial Region – an area of small outflowing rivers – revealed serious problems associated with large quantities of coal mine drainage. This, together with urban and industrial sewage, was being discharged through natural run-off into the Vistula and Oder rivers (Gilewska 1964). The situation was further compounded by chronic supply shortages in this watershed area, combined with heavy pollution. Continued environmental abuse of the hydrological network during the 1970s and 1980s has meant that now nearly two-thirds of the rivers and streams in the Katowice province are no longer usable, even for industry (Górny 1991). A quarter of them have been degraded irreversibly due to their heavy metal content. The ejection of 6,000 tonnes of coal mine salt daily into rivers like the Olza give chlorine readings 8 times higher than the permissible standard (Zięba 1986).

Soil and vegetation equally suffer; industrial development quickly led to landscape devastation in the 1960s through spoil heaps and disruption from mining and manufacturing. In the region, waste dumps covered 2,000 ha and were growing by 67 ha per annum: three-quarters from coal mining and a fifth from burning due to spontaneous combustion (Gilewska 1964). During the 1960s soil over an area of 100,000 ha was subject to the effects of toxic industrial dust and gases, whilst tests recorded the growing presence of heavy metals like lead and zinc near industrial plants. The 1970s saw an intensification of this process, with spoil heaps now covering 7 per cent of surface area and in some urban centres 12 per cent (Greszta 1974). The process has continued into the 1980s with 17 per cent of the land surface beyond reclamation, and the lead content of vegetable garden soils exceeds permissible standards by 150 to 1,500 per cent with values ranging from 42 mg/kg (ppm) to 8,890 mg/kg (PSL = 20 mg/kg) (Marcinkiewicz 1987). A tenth of all gardens are in areas very unsuitable for growing crops. Likewise, the impact of air pollution on forests in the region has been noticeable. About four-fifths (150,000 ha) of forest in the Katowice province is endangered by pollution – about a sixth heavily and a further third moderately.

Finally, all this has taken its toll on human health (Budnikowski 1991). The Katowice province has one of the poorest health records in the country with 15 per cent more circulatory diseases, a third more tumours, and nearly 50 per cent more respiratory complaints than the Polish average. Moreover, recent research has proved that air pollution in the province has led to negative gene mutations linked to chromosome changes amongst its inhabitants (Chorąży 1990). Life expectancy in the province is at least two years shorter than elsewhere, whilst children appear clearly disadvantaged. Oxygen deficiency to the tissues (anoxia) is the major cause of infant mortality for babies aged over one month. Older children suffer higher than usual effects of coal-dust, creating anthracosis and fibrous tumours (fibroma) in the lungs and lymphatic glands, whilst half of Poland's schools for disabled children are located in Katowice; some experts believe this is due to high heavy-metal levels in local air and water supplies together with above average concentrations of lead and cadmium concentrations in the soil (Kabala 1991a).

Overall, this devastating picture of the Upper Silesian Industrial Region is perhaps best summed up by a local saying which states that here is a place where even the Devil says goodbye.

2 The Cracow region

This region and the Upper Silesian Industrial Region occupy less than 3 per cent of Poland's territory but manage to produce half the nation's gas pollution and a third of its dust. Cracow is the third largest city in Poland (after Warsaw and Lódź) with 750,000 inhabitants and is the centre of a province that suffers severe pollution problems. The three main sources of pollution are industry, domestic heating, and vehicles – the latter being of least importance (Kabala 1991b).

In the early post-war years, the government wished to shift the economy away from a strong agricultural dominance in the south and south-east, leading to a growth in industrial investment to provide alternative employment. One result was the planned location in 1949 of a huge metallurgical works at Nowa Huta, a new town 2 km north and 8 km east of the old medieval city centre. By the early 1970s its impact was beginning to be felt; ostensibly its location benefited from being half-way between Ukrainian ore supplies and Silesian coal, and would use the ample water supplies from the Vistula river. In reality cynics believe it was deliberately placed in the Cracow conurbation to modify the conservative, traditionally intellectual and bourgeois outlook of Cracow society; the new plant would infuse an industrial, proletarian community into Poland's former capital city (Dawson 1984).

Cracow and its surrounding province occupy a low basin of the Upper Vistula river valley, which sucks in polluted industrial air from the Nowa Huta metallurgical plant to the east (annual steel production 11 million tonnes), from the Upper Silesian belt to the west and, during the 1970s from the aluminium works at Skawina to the south. As with the Katowice province, one must add sources emitted from residential apartments in the central city area heated by low-grade coal. Pollutants from these sources often hang over the city, particularly during winter due to temperature inversions. Sulphur dioxide is the predominant threat, both to the local inhabitants' health and the city's historic buildings (Carter

1982). During the late 1970s pollution evidence from the region was dramatically illustrated by the high concentration and deposition of SO$_2$ in the central city and industrial areas (Kasina 1978, Carter 1987: 4) (see Figure 6.3). Such views were confirmed by the Cracow Provincial Atlas in 1979 which showed large areas of the region above the PSL (0.15 mg/SO$_2$/m^{3}t), (*Atlas* 1979: 19) (see Figure 6.4).

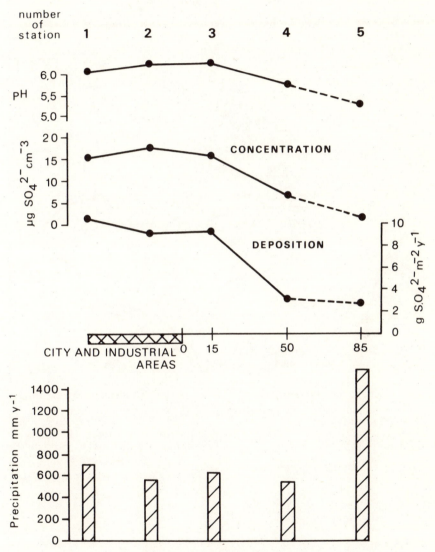

Figure 6.3 Cracow: spatial distribution of mean pH and SO$_2$ values, January 1975 to December 1977

Source: S. Kasina 1978.

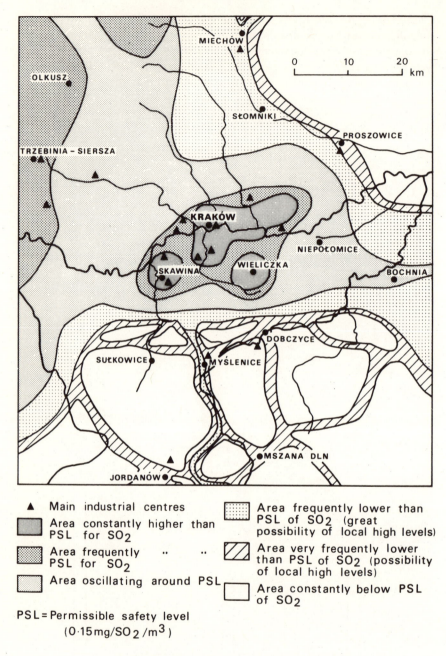

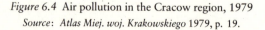

Figure 6.4 Air pollution in the Cracow region, 1979
Source: *Atlas Miej. woj. Krakowskiego* 1979, p. 19.

130

Research efforts by the Polish Ecological Club in 1980 revealed that the city and its immediate environs were seriously threatened by industrial pollution. Their findings indicated that dust fall-out was 9 times above the national limit over the city, and that annually 170,000 tonnes of dust and 1.2 million tonnes of gas (mostly SO_2 and CO) fell on Cracow. The main source was the Nowa Huta complex (four-fifths of all dust and gas), while other contributors included the Skawina aluminium plant, the 'Solvey' soda factory, and the 'Bonarka' cement works; the steel works alone gave off 7 tonnes of cadmium, 170 tonnes of lead, 470 tonnes of zinc, and 18,000 tonnes of iron in its annual dust emission (Polski Klub Ekologiczny 1981). The club's efforts led to the closure of the Skawina plant in 1981 (Anon 1981b).

By the late 1980s the Cracow region contained 1 per cent of Poland's surface area, 3 per cent of its population, produced 4 per cent of the state's industrial output, but emitted 17 per cent of Poland's toxic gases and 7 per cent of its dust (Hess *et al.* 1989: 21). After Katowice, Cracow province had the highest SO_2 emission in Poland in 1989 of 895,000 tonnes (Grzesiak 1990: 129). The Nowa Huta plant annually emits 400,000 tonnes of CO, 50,000 tonnes of SO_2 and 60,000 tonnes of other particulates (Rosenbladt 1988). It is therefore not surprising that within the city three-quarters of the residents live in an area of very heavy air pollution where the average annual deposition is 123 kg of dust and 659 kg of gas per inhabitant (Horwath 1988). Here people reside in an area where SO_2 emissions are continually above 64 mg/m^3/year and dust over 250 tonnes/km^2/year (*Atlas* 1988, map 16). It is not surprising that Cracow residents consistently showed the most negative attitude to industry and the highest environmental sensitivity of the twenty-five cities investigated in Poland, four-fifths of Cracovians believing that industry was the main danger to health (Domański 1989), this being further fortified by a devastating environmental account of the whole appalling situation in the region (Polski Klub Ekologiczny 1990).

Water pollution fares little better. As with air pollution, greater awareness emerged in the late 1960s when increased industrial demands on low water resources soon aggravated the problem. One of the worst affected rivers was the Przemsza, which spilled 40–50,000 kg of soluble compounds per hour into the Vistula river in the mid-1960s; this was soon felt downstream at Cracow (120 km away) where sewage analysis revealed evidence of ammonia, nitrates, sulphates (SO_4), chlorides, iron compounds, phenols, heavy/light oils and tar. During the 1970s and 1980s the situation was intensified; for example, the mercury content of the Vistula river below Cracow during the second half of 1980 was 200 times above its permitted norm; an excess of chrome, lead and iron was also noted. Further, it was also estimated that over half the running water used by Cracow's municipal water supply system fell short of state norms and was threatened with pollution (Polski Klub Ekologiczny 1981). By the late 1980s drinking water was regarded by many inhabitants as risky due to the infiltration of industrial waste into the system.

As elsewhere in parts of Poland, the 1970s saw an intensification of soil and vegetation pollution in the Cracow area. Research on the heavy metal content of soils around Nowa Huta Metallurgical Works in 1970 recorded high dust deposition of zinc, lead, copper, molybdenum, aluminium, iron, manganese and barium (Strzyszcz 1982: 50). These tests were taken nineteen years after the plant came into operation and show a high intensity of certain metals (for example, manganese). Also, great variance was registered between maximum and minimum intensity for all the heavy metals listed. Even minimum values for some metals (for example, lead) were high down to a soil depth of 25 cm and well above the state agreed maximum of 20 mg/kg/ppm. In the 1980s little improved; at the village of Pleszów in 1983 unwashed lettuce recorded values of 8,700 mg iron/kg, 16.8 mg/lead/kg and 1.64 mg cadmium/kg (Curzydło 1990: 180). Trees and forests also suffer; in Cracow, a city previously noted for its greenery, trees have suffered from a two-pronged attack: dust and gases in the air and communication cables under the pavements (Wachacki 1989). Within Cracow province over a third of the forested land has suffered damage (20,000 ha: 54,000 ha) from air pollution through acid rain.

Health hazards of living in and around polluted Cracow are noticeable. There was an alarming increase of malignant cancers in the city during the late 1970s, which by 1981 was six times higher than the world average (Jędrychowski et al. 1990). Employees at the local Nowa Huta Metallurgical Works suffered considerably during the decade 1971–80; over that period only a fifth of the employees left the combine due to normal retirement but over two-thirds received some form of disability pension. In 1980 only an eighth left under normal retirement conditions, whilst four-fifths claimed state disability pensions. Many of these workers had their own allotments on land around the plant, which have since wisely been abandoned. Meanwhile, Cracow has the dubious distinction of having amongst the highest infant mortality rates in Poland – 285 per 100,000 inhabitants as compared to the national average of 184 per 100,000. Unfortunately, it has yet to be proved how many infant deaths are due to pollution, or which contaminants are responsible (Hinrichsen 1988).

Finally, the very fabric of the city is also under attack, including the 72 churches, 43 monasteries and 6,000 classic houses. Acid deposition is causing brickwork to crumble and roofs to disintegrate in a city which the United Nations have classified as 'O' category – that is, top priority for conservation. Ironically, the city escaped damage during the Second World War, but the post-war effects of metal and chemical manufacturing have ruined many of the historic treasures including buildings from the medieval, Gothic, Renaissance and nineteenth-century period. Concerted efforts are now being made to try and save the city's heritage, not only for its value to European civilization but also as a wealth of interest for the 3 million tourists who annually visit the city.

CONCLUSION

The main theme has been to outline some of the effects that severe and prolonged pollution have had on the Polish environment – particularly air, water, soil and vegetation. The evidence for environmental misuse amongst Poland's major resources is irrefutable; in turn its impact on health problems, governmental reaction through legislation, and the rise of 'green' opposition groups reflects its significance. Two case studies have highlighted these problems within the confined polluted parts of the south/south-west.

Four general conclusions can be made. First, the whole question of attitude towards the environment over the post-war period was critical. This was largely a governmental responsibility, but to a lesser extent involved ordinary people. Old attitudes towards the environment are gradually being forsaken; these were closely linked with the belief that socialist societies were incapable of committing any ecological atrocities. Since the early 1980s such views were slowly replaced by a more coherent, logical approach, which admitted that any true society cannot afford to neglect its environment. Fortunately, believers in the latter idea are now in a majority and able to influence important decision-making processes (economic/political/social), which augurs well for the future.

Second, the role of government is paramount if environmental degradation is to be curbed. Legal measures and national pollution norms must be rigidly and effectively applied; collusion between the state's environmental protection agencies and the country's polluters must be strictly prevented, because pollution laws are meaningless if they are not strongly adhered to and applied. In this context, Poland's inhabitants now have greater access to results obtained from pollution data and made widely available by the government. The problem now is less one of secrecy, but rather one of lack of finance to complete intensive research on such topics as underground water contamination.

Third, there must be more cohesion between planners and the planned. The urge for economic development through industrial growth, whether under centralized control or a market economy, has to be understood as being extremely dangerous for the environment because of Poland's natural conditions and energy base. Decades of environmental neglect and heavy reliance on solid, massively polluting fuels means greater financial support from the state to lessen the impact of this legacy. In the early 1980s this was demonstrated by the publication of environmental impact assessment reports on particularly affected areas (Polish People's Republic 1983), which then remained under wraps in some government office. Admittedly, fiscal limitations provide a heavy brake on any government labouring under enormous international debts like Poland (Pape 1991). However, international realization that an improvement in Poland's environment would be beneficial for other European countries, especially in Scandinavia, may create possibilities for closer co-operation.

Fourth, evidence presented here has shown that in post-war Poland, despite public ownership of the means of production (including natural resources), the government was unable to prevent pollution and wastage of state resources in spite of elaborate planning methods. This resulted from governmental decisions based on economic growth at the expense of all else. Now there is a drive towards more efficient use of natural resources and greater awareness of what has to be done. Attempts to catch up and compete with the world's more advanced nations should not inevitably be at the expense of safer living conditions in Poland. The present situation demands an air of optimism, and let's hope the Polish nation responds in a positive way.

ACKNOWLEDGEMENTS

The author would like to thank Professor Zbigniew Wójcik of the Museum of the Earth (Muzeum Ziemi), Warsaw, for his help, and Louise Saunders of the Cartographic Unit, Department of Geography, University College London, for the illustrations.

7

ROMANIA

D. Turnock

INTRODUCTION

Perusal of most of the literature published in Romania before the revolution would convey an impression of a socialist paradise engineered by the genius of the Romanian Communist Party (Partidul Comunist Român – PCR) and its erstwhile leader, Nicolae Ceauşescu (Romanian Communist Party 1988). It was claimed that one important characteristic of this new society would be its development in ecological harmony with the country's Carpathian–Danubian environment; especially in view of the legislative programme of the 1970s. However, since the revolution of 1989 there has been universal condemnation of the 'Ceauşescu Epoch', not least for its indifference towards ecological problems. This chapter must try to strike a balance between the two perspectives and assess the real extent of the ecological disasters which have occurred.

Some problems have certainly been inherited from the past because Romania's modernization started along capitalist lines after independence in 1878, and the origins lie even deeper in the past in the northern and western territories which enlarged the Romanian state after the First World War (Turnock 1977). The metallurgical complex of Reşiţa, which might be regarded as the first industrial growth centre on the present Romanian territory, originates in investment decisions made in 1768. Therefore, air and water pollution problems are long established. The greatest difficulties seem to have arisen from the growth of the chemical industry, but here again there are origins in the pre-socialist period, with the expansion of oil drilling and refining in the late nineteenth century and the exploitation of the methane gas deposits of Transylvania during the inter-war years. On the land the harmful consequences of sustained cropping and deforestation can be linked with the generation of large grain surpluses and the cutting of the Carpathian forests in the late nineteenth century. However, the scale of industrialization before 1945 was distinctly modest by comparison with the present. The growth of production from 1938 to 1985 may be quoted for electricity (1,130,000,000 to 71,820,000,000 kWh), steel (280,000 to 13,790,000 tonnes) and sulphuric acid (440,000 to 1,830,000 tonnes) to substantiate this point. And while most of the inherited industrial centres have expanded many

135

new locations have been developed (see Figure 7.1). The legacy of the capitalist past cannot therefore be regarded as a major factor in the present situation.

Data sources

There is very little published data dealing with pollution nation-wide. The statistical yearbook (*Anuarul Statistic*) offers no details whatsoever apart from basic information on the economy, although the professional literature may occasionally contain useful material (such as the data on green spaces in towns used in Figure 7.4). However, while the publication of data on the precise scale of pollution problems has been discouraged, it is evident that the monitoring of pollution has increased very greatly over the last twenty years. Monitoring of water quality dates back to 1960. In particular the Institut de Meteorologie si Hidrologie, in collaboration with local offices for water management (Gospodărirea Apelor) has had more than two decades of experience, and a water quality control system (Sistemul Naţional de Supraveghere a Calităţii Apelor) was inaugurated in 1978 (Mărcuţă and Serban 1990). All industrial estates are checked while water quality is assessed at 271 principal control sections and at 91 other places of local importance: in all 5,000 water sources are controlled (Ionescu 1988). Air pollution checks were introduced in some of the industrial centres (Hunedoara in 1969, and Baia Mare and Copşa Mică in 1971) prior to the establishment of a comprehensive monitoring system by the health ministry in 1973 (overhauled under the UN Environment Programme in 1978). Particular attention was given to monitoring in Bucharest in that year and at Giurgiu from 1984.

It has been established that all but 3 per cent of the pollution affecting Romania is generated within the country (Vădineanu 1991: 113). Estimates of the total amount of pollution are put at 138,000,000 tonnes of emissions into the atmosphere. Much of this is dust but there are also 1,760,000 tonnes of sulphur

---

KEY TO FIGURE 7.1

Key to other towns mentioned in the test:
1 Anina; 2 Băile Herculane; 3 Bicaz; 4 Bîrlad; 5 Borşa; 6 Cavnic; 7 Cîmpulung; 8 Copşa Mică; 9 Făgăraş; 10 Gh. Gheorghiu-Dej (now Oneşti); 11 Giurgiu; 12 Haţeg; 13 Hunedoara; 14 Mediaş; 15 Năvodari; 16 Ocnele Mări; 17 Reghin; 18 Slănic; 19 Sulina; 20 Techirghiol; 21 Turnu Măgurele; 22 Valea Călugărească; 23 Zlatna.

Key to other locations mentioned in the text:
a Băbeni; b Belceşti; c Berbeşti; d Căpuş; e Caraorman; f Coştiui; g Crucea; h Hoghiz; i Iacobeni; j Maliuc; k Ocna Şugatag; l Ostrov; m Racoviţa; n Răstoliţa; o Roşca-Buhaiova–Letea; p Roznov-Săvineşti; q Săsar; r Slătioara; s Ştefăneşti; t Voineasa.

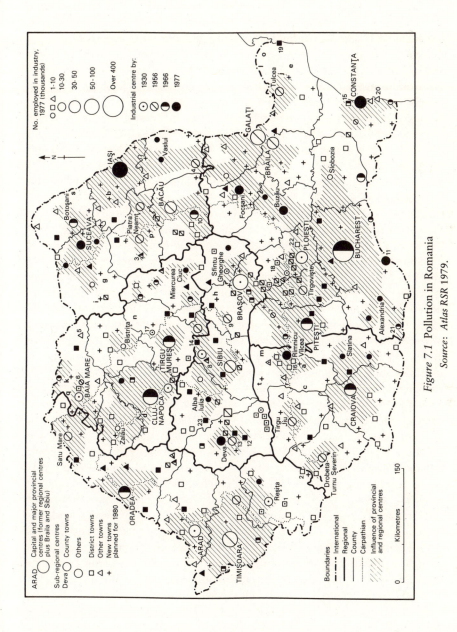

Figure 7.1 Pollution in Romania
Source: Atlas RSR 1979.

dioxide, smaller quantities of ammonia (8,500 tonnes), carbon disulphide (5,450 tonnes) and phenols (2,300 tonnes). Sulphurated hydrogen emissions are locally serious (Brăila, Galaţi and Suceava) and fluorine is a problem at the aluminium smelter in Slatina. It is also estimated that some 6,870,000 tonnes of pollutants are discharged into rivers (including ammonia, chlorine, nitrates and phosphates) (Hâncu 1990b). The quantities are large in relation to the size of the economy and reflect not only unsatisfactory control but also an excessive energy consumption (three times Western levels) because of out-dated technology. They are doubly wasteful because in addition to causing damage to the environment the economy is denied the materials which could be made available by the better processing of waste. Pollution affects some 10 per cent of Romanian territory and 20 per cent of the population (that is, 4,500,000 people). Around 1,500,000 people are suffering from the effects of pollution on a permanent basis, and this means not only unpleasant living conditions but also lower life expectancy and increased risk of serious illness.

Monitoring facilities have been more powerfully activated since the revolution though the equipment is not altogether adequate. Moreover, the shortcomings of the past are being fully exposed. At the Giurgiu chlorine plant, a major source of pollution which was opened in 1977, a laboratory to monitor pollution was only made ready in 1986. Moreover, scientific research must go a good deal further than mere study of pollution levels. Useful work has been done on atmospheric pollution to examine the washing effect of rainfall at Piatra Neamţ (Apostol 1985, 1987). There have been several other studies of air pollution in the context of atmospheric dynamics (Pîrvulescu 1985, 1986). Research on the Baia Mare depression demonstrates how the highest pollution levels correlate with temperature inversions and long calm periods: so chimneys should be taller than 120 m, enabling pollutants to be dispersed by higher altitude air currents, and the city should expand in a westerly direction where the physical conditions are most conducive to self-purification of the air (Erdeli *et al.* 1987). Biogeographers and geomorphologists have contributed to the understanding of soil erosion (Bălteanu *et al.* 1985, Calin 1988). Correlations exist between vegetation cover and the thickness of the moving material: a well-developed plant cover prevents serious landsliding (Muică 1987). Road-building in the mountains can produce problems of erosion (Nicolescu 1987). But afforestation can restore a measure of stability, as in Sibiu (Voicu-Vedea *et al.* 1986). And such protective work has been proceeding in Vrancea since 1950, with research to check the effectiveness of different species (Untaru *et al.* 1986). The plains suffer from problems of both drought and excess water but remedial work to provide irrigation or drainage must be very carefully related to local conditions to give good results (Bogdan 1987). Far more research of this kind if needed.

AIR AND WATER POLLUTION

Sources of air pollution are numerous and they range from large stock-raising units, responsible for offensive heaps of manure, to major factories like cement works and lignite-burning power-stations which discharge huge quantities of ash and dust into the atmosphere. The capital city is badly affected as are the various county administrative centres (see Figure 7.2). In Bucharest the total emissions are estimated at 4,400,000 tonnes (1991) compared with 6,400,000 tonnes in 1984, but there are great variations within the city. The main intersections register high pollution levels due to the density of traffic (involving a high proportion of old and poorly maintained vehicles), but the highest recorded levels have been found on the industrial estates. The legal maximum for dust is 0.15 mg/m^3 but in the Panduri industrial area the level reaches 1.5. This same area registers 9.0 mg for carbon dioxide compared with the legal maximum of 2.0. Another pollution black spot is the Obor industrial area which records the highest levels in the city for ammonia, hydrochloric acid and phenol: 0.90 (0.10), 0.60 (0.10) and 0.10 (0.03) mg/m^3, respectively (with the legal maxima in brackets). To the north-east of Obor lies the small Acumulatorul industrial district where the recorded level for lead was 0.14 – that is, 200 times the legal limit of 0.0007. However, there are also major problems in several other towns. The effects can be particularly serious in mountain valleys where local meteorological conditions complicate dispersal of the 'white dust' from cement factories. The Bicaz-Taşca cement complex in the Bistriţa Valley is a good example (Mihăilescu *et al.* 1984). But no cases are more serious than the chemical plants which the wife of the late president was so intent on proliferating. At some plants processes which are dangerous enough for factory workers to wear gas masks present hazards to the local population inadequately protected from toxic emissions. Some of the worst cases relate to old factories which were built before the socialist period, like the carbon factory at Copşa Mică and the fertilizer plant at Valea Călugărească. In the latter case red dust from the tip pollutes surrounding farmland and forest; yet the material could be recycled and it is proposed that the waste should be removed to the Turnu Măgurele factory where facilities exist.

Yet the more modern units are not necessarily superior as regards pollution control. At Giurgiu the pollution problem assumes an international dimension since the chlorine factory makes its presence felt on the Bulgarian side of the Danube (Constantinescu 1990). International relations have been strained by this problem although the Romanians have been able to resist 'presiunea bulgară' with counterclaims relating to the Kozloduj atomic power-station, which has implications for the measurement of radiation and radioactivity (Vădineanu 1990). The pollution is not merely unpleasant. It constitutes a major health hazard, documented at Deva where ash and dust from the power-station is linked with a rapid increase in the incidence of cancer and in the number of complaints relating to the respiratory system (especially bronchitis). But there is no greater

139

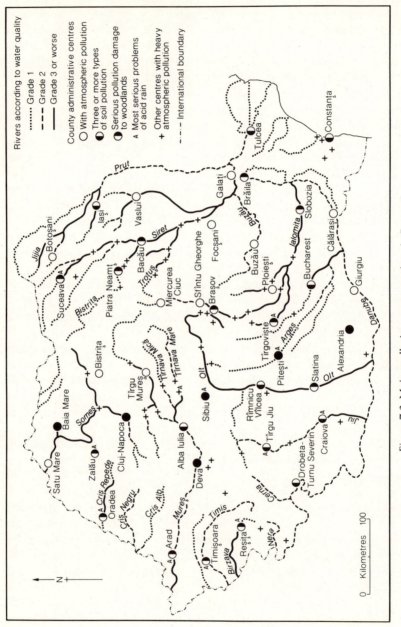

Figure 7.2 Major pollution concentrations in Romania
Source: Hâncu 1990b.

horror than the 'Suceava Syndrome' linked with air pollution originating from the artificial fibre plant at Suceava. The health hazard is seen most dramatically through malformed babies, and in response to public pressure the factory was closed down after the revolution for major overhaul (Mărăşanu 1990). If it is to reopen at all the public will require reassurance that the most effective pollution control equipment is installed, or alternatively that the most harmful processes are modernized and relocated. Meanwhile, the damage to agriculture is cumulatively substantial. Damage to cereal fields at Piteşti and Valea Călugărească; to orchards near Rîmnicu Vîlcea; and to vineyards at Copşa Mică and Mediaş involves losses of as much as a third of the harvest. Some pastureland is so badly contaminated as to be unusable (Vădineanu 1991: 115).

Water pollution

Water management has become increasingly difficult on account of pollution (Turnock 1979). Part of the problem arises when high phreatic water levels lead to the contamination of underground water supplies by fertilizers and pesticides. But the major problem concerns the discharge of pollutants into rivers. Sewage makes a considerable contribution because treatment facilities are limited. In 1945 only fifty-six towns (out of a total of 150) had sewerage, but only eight had purification systems. Now the great majority have sewerage (220 out of a total of 260) but purification systems are only available for 100 (30 per cent of which are not operating properly). But industry (and especially the chemical industry) is the major cause for there are many plants which generate both atmospheric and water pollution. The accumulation of waste means that pollution is diffused through each drainage system and problems are compounded by the absence or inadequacy of plant to treat processing water. Of some 3,500 polluting industrial units over half do not have purification plant at all or else lack sufficient capacity to treat all the effluent (Vădineanu 1991: 118). The principal river system (comprising some 20,050 km) breaks down into 7,800 km (38.9 per cent) of Category 1 rivers (relatively clean); 6,100 in Category 2 (30.4); and 6,150 km of Category 3 rivers (including 3,700 km of rivers which are completely polluted) (Hâncu 1990b). The 'dead' waterways include not only the Dîmboviţa below Bucharest but the lower reaches of rivers like the Bahlui, Bîrlad, Bistriţa, Olt and Trotuş associated with such provincial cities as Iaşi, Bîrlad, Piatra Neamţ, Slatina and Oneşti (formerly Gheorghe Gheorghiu-Dej), respectively (see Figure 7.2).

Extensive work was carried out on the Dîmboviţa in Bucharest during the 1980s so that clean water could flow on the surface while the sewage was diverted into a tunnel underneath the river bed and taken for treatment to a facility at Glina on the south-eastern edge of the city. But much remains to be done in other parts of the country where there are reports of technical difficulties, a lack of spare parts, and insufficient capacity for treating effluent. However, particularly serious problems arise where the pollution contaminates drinking water. This has been reported at the Roznov-Săvineşti complex in the Bistriţa Valley near Piatra Neamţ and, most seriously, at Arad where the huge water reserves associated with the alluvium of the Mureş Valley are threatened. The plant is discharging ammonia, nitrates and phosphates only 4 km away from a major water source for the city which scientists believe will be threatened within a decade as the pollution advances further into the groundwater reserves (Cîneţi 1990). Claims have been made about uranium mining at Crucea (in the Bistriţa Valley below Vatra Dornei) leading to radioactivity in the river water.

While much of the pollution arises from a gradual accumulation of harmful substances there are certain outrageous cases of dumping, among which the best example is the storage of toxic waste at the Free Port of Sulina in 1987. A contract with a chemical firm in Liechtenstein provided for the disposal of some 300,000 tonnes of waste material deriving from the production of dyes and paints. However, after the first consignment of some 1,150 tonnes had been accepted it became apparent that the material was highly dangerous and its diffusion by sea currents from leaking drums threatened the tourist industry along the Black Sea coast during 1988. The material was then shipped off to an unknown destination and legal action was taken against the port authorities (Vinicius 1990). But the coast has also been damaged by pollution originating from the oil refining and petrochemical complex at Midia-Năvodari: the seepage of polluted canal water threatens tourism at Mamaia and the drinking water supply for the city of Constanţa.

A rather unusual case of 'pollution' concerns the damage to salty waters through the unintentional seepage of fresh water! The salty waters of Lake Techirghiol are renowned for medical treatment but their value has declined through the reduction of the salt content (from some 80 g/litre to 40) through the infiltration of fresh water (Serban and Simiona 1990). This has arisen as a consequence of the development of water supplies for the coastal resorts and for local irrigation and stock-rearing complexes when the risk of damage to the lake was not taken into account. It has required investment in dyke construction on three sides of the lake to overcome this problem, and the salt content has now been restored to the level of 60 g/litre (Sărăţeanu 1990). A similar problem arose at Băile Herculane where the internationally famous spa was threatened by a sudden fall in the level of the River Cerna as a result of the diversion of water to the Motru and Tismana valleys in order to supply northern Oltenia's lignite-burning power-stations with cooling water. Additional investments then had to be made to bring water from the Bela Rece to feed into the Cerna immediately upstream of Băile Herculane and so remedy the situation.

SOILS AND VEGETATION

Agriculture has intensified during the post-war period and in recent years there has been a powerful drive to extend the area available with the result that a number of areas with natural and semi-natural vegetation have been lost. But agriculture also involves pollution in many different forms (Răuţă and Cârştea

1990). Figure 7.2 indicates those countries where serious soil pollution occurs in three or more forms (selected from a list including damage by fluorine, heavy metals, nitrates, nitrous oxide, pesticides, petrol, salt and sulphur dioxide, and by waste discharges from stock-rearing units and water purification plants). Nearly one-third (7,500,000 ha) of the agricultural land is subject to at least one form of pollution. Over 900,000 ha are chemically polluted (and 200,000 ha are rendered totally unproductive) through the use of fertilizers, insecticides and pesticides, consumption of which has risen from 5,900 tonnes in 1950 to 266,400 in 1966 and 1,199,200 in 1985. Around 200,000 ha are excessively affected in this way. A large fertilizer industry has been built but the application of the material still leaves a lot to be desired. The optimum application is not always properly established and there is also a good deal of carelessness with fertilizer dumped haphazardly when labour is short. Much better control is needed through checks on the harvested products (Baicu 1990). But there is also pollution from irrigation water which affects some 7.5 per cent of the total irrigated area of 2,710,000 ha. And further difficulties arise through road traffic, oil drilling, and a range of industries associated with chemistry: especially the fertilizer factories of Năvodari, Tîrgu Mureş, Turnu Măgurele and Valea Călugărească. Metallurgy industries should also be mentioned: the aluminium industry at Oradea, Slatina and Tulcea; other non-ferrous enterprises at Baia Mare, Bucharest, Căpuş, Cavnic, Copşa Mică, Săsar and Zlatna; and the iron and steel works of Hunedoara. Careless dumping of waste can have a damaging effect as in the case of shipments of rejected Italian tobacco which have arrived for burial on the 'Great Island' of Brăila (lying between the two main Danube river channels which has been dyked and drained for agriculture). The tobacco has polluted irrigation canals and caused unpleasantness through its foul odour, to say nothing of its dubious value as a compost.

Erosion is a major problem, for all land sloping by more than three degrees is liable to damage. On average 16.3 tonnes of material are lost from each hectare of land every year (a total of 126,000,000 tonnes for the country as a whole). Of course, the figure varies enormously with very high levels of erosion (exceeding 40 tonnes per hectare) in the sub-Carpathians where attempts have been made to restrict gullying by afforestation. Some 600,000 ha of land are affected by salt accumulation (a potential problem over a further 1,200,000 ha), while 2,300,000 ha need some correction for soil acidity. Some 7,600,000 ha of land have small or very small humus reserves (some 1,500,000 tonnes of humus are lost each year) and there are problems of soil structure over most of this area exacerbated by compaction through the use of heavy machinery. Irrigation has been linked with both salination and the loss of humus. In experiments at Belceşti in Moldavia it was found that irrigation has become counter-productive because only in dry seasons are the yields highest on the irrigated land. Ecological agriculture will now require specialization patterns appropriate to a restoration of humus and acceptable fertilizer applications (Ionescu 1982). A special regional problem arises from the changes of environment related to the drainage of the

Danube floodplain (balta) and the conversion of fishing grounds and seasonal grazings into cereal land (Iordan 1987). The works have been thoroughly unprofitable because the maize harvest (lower than expected) does not repay the investment, especially when the loss of fish (from the former lakes rather than the main river) is taken into account. Moreover, the microclimate has been affected: drought has allegedly become more frequent, while the vineyards on the lower terraces have suffered from the lack of reflected sunlight which gave the *podgoria* Greaca its traditional reputation in the Istanbul market (Preda 1990).

Forests

Soil erosion affects more than 4,000,000 ha where the soil losses range over 8–30 t/ha each year (Bally and Stănescu 1977). Here there must be greater afforestation (Traci and Costin 1966). However the existing woodland is by no means satisfactory despite the long-standing ecological concerns of Romanian silviculturists (Doniţa *et al.* 1978). The situation is very clearly laid out in the long-established silvicultural journal *Revista Pădurilor* (Turnock 1988). There are serious problems arising from air pollution: some 330,000 ha of woodlands are affected by acid rain (monitored since 1984). Much of the affected area lies in the vicinity of large industrial plants such as the Copşa Mică chemical and metallurgical works which has damaged some 50,000 ha of forest (Ianculescu 1977) (see Figure 7.2). This places great strain on the local silvicultural organization (*ocol silvic*) at Mediaş (Buzea 1989). Other major sources of acid rain are the sulphur dioxide emissions from chemical/metallurgical plants at Baia Mare and Zlatna. At a less serious level the list would include similar factories at Arad, Craiova, Făgăraş, Galaţi, Hunedoara, Iacobeni (near Vatra Dornei), Oradea, Pitesţi, Reşiţa, Rovinari, Suceava, Timişoara, Tîrgovişte, Valea Călugărească and Zalău (Negulescu 1990). Also, the cement work of Bicaz and the power-stations of Rovinari near Tîrgu Jiu and Turceni near Strehaia. The impact of acid rain is felt through the reduced growth rate.

But there is also a general weakness related to the persistence of woodcutting which exceeds the natural growth from year to year. While the area of the forest is fairly stable the amount of timber per unit area of woodland is falling. During the last 100 years the forest cover has declined by 29.2 per cent from 9,000,000 to 6,370,000 ha and the production potential has fallen by 32.7 per cent from 23,500,000 to 15,800,000 m³ per annum. This must be seen in the context of an annual consumption of some 30,000,000 million m³ (Giurgiu 1990). The forests have also become less varied with regard to species. Repeated calls have been made for better management of the forests with multiple use in mind (Giurgiu 1988). But there is no political will to reduce cutting to a level below the annual growth rate so that a programme of reconstruction can be undertaken, let alone increase the forested area from the present level of 28.6 per cent to at least one-third (Scutăreanu 1979).

DANUBE DELTA

A particularly good example of damage arising from inappropriate land-use policies is the Danube Delta (Banu *et al.* 1965) (see Figure 7.3). For this sensitive area has experienced a series of harmful development initiatives since the Second World War (Chiţu 1990). The late 1950s saw an attempt to produce reeds (with a high cellulose content) in areas of dyked marshland which eventually reached 30,000 ha. After a conference in 1956 experiments were conducted at Maliuc but large-scale exploitation went ahead before it was discovered that heavy cater-

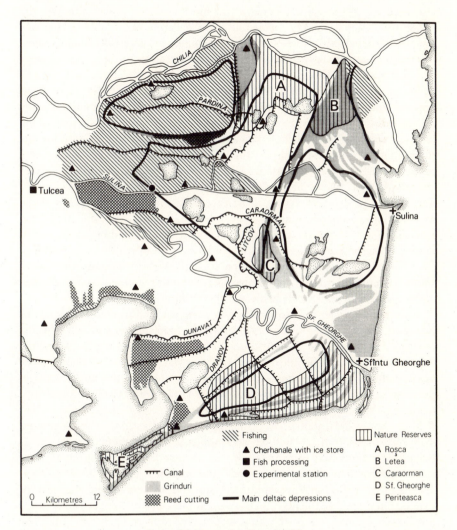

Figure 7.3 Romania's Danube delta
Source: A. C. Banu *et al.* 1965.

145

pillar tractors (supplied by the former GDR) damaged the root systems. The harvest was greatly reduced and much of the area became a wasteland, while the cellulose factory at Brăila had to find other sources of supply (including poplar trees from the plains and resinous timber from the mountain districts). Very light wide-tracked vehicles are now used but the reed harvest is now only 40,000 tonnes per annum compared with the 2–3 million tonnes once envisaged. There were environmental changes through the construction of alluvial platforms on which the harvested reed could be stored and the invasion of the reed beds by opportunistic hydrophilous species when reed growth was suppressed. There have also been adverse social consequences within the delta, because although the prison population drafted to remote outposts like Periprava for forced labour was gradually redeployed this left a civilian community at places like Maliuc where former reed workers now inhabit dirty, damp apartment housing with no work available apart from casual fishing and hunting.

New development strategies emerged as a result of the creation of an economic 'central' for the delta region. The 1970s brought a renewal of interest in the fishing industry (depressed by the changes to water circulation arising out of the reed exploitations). The delta's potential was valued as compensation for up-river fishing grounds lost by the dyking and draining of the flood plain (part of a massive programme of regulation work which has resulted in reduced water levels in the delta and damaged the traditional economy). But new species introduced into the delta have had an adverse effect on the native fish, and the organization of the industry in large basins (29 units totalling 50,000 ha) involves high development and management costs (to say nothing of the loss of natural marshland and woodlands) which have failed to realize even a sixth of the planned production. This situation has arisen partly because the fodder supply system broke down but problems have also arisen because of inadequate water circulation, particularly in areas with a substratum of peat.

During the 1980s a further şoc was experienced through the drive for increased agricultural production (to deliver the food surpluses essential for the paying-off of foreign debts). Agriculture is highly appropriate in 'gulfs' within the delta which have naturally silted up and have few lakes and swamps; but it is more questionable where elaborate drainage work is needed and this applies even in the western delta where the drainage of the land can be most easily accomplished (including the Chilia distributary where there is a succession of old deltas such as Babina, Ostrov and Tataru). The plan envisaged an expansion of the agricultural area from 11,000 to 90,000 ha and although the programme was never completed major works were undertaken at Pardina (27,000 ha), where three of the five proposed farms were made ready, and Sireasa (7,550 ha). There was also activity at Periprava but this was unsuccessful and the land is now managed for fishing. On the whole the investments were not justified by the results. With the levelling of ground inside the polders drainage was complicated (especially during the period of high water between March and June). The soils were not entirely suitable for arable farming and it is arguable that investments would have been more effective elsewhere. This is before consideration of the environmental impact which includes further losses of natural habitat essential for the survival of the delta's unique bird population. Areas set aside as reserves were threatened by this new programme where they lay adjacent to proposed reclamation sites: the juxtaposition of the extremely large Pardina project and the Roşca-Buhaiova reserve was particularly unfortunate. Moreover, grassland became over-grazed through increased numbers of cattle and sheep while fertilizers polluted the water, thereby encouraging the growth of algae which suffocated fish (Hopkins 1991). The consequences of eutrophication have been widespread affecting the Razim-Sinoie lacustrine complex which is managed so that the water circulation involves a throughput of enriched water from the water from the delta which is discharged into the Black Sea. There has been an irreversible drying-out of peats and luxuriant weed growth. Change also occurred in the woodlands through the planting of Canadian poplar on the river banks, despite the fact that the white willow is more adaptable to water-level fluctuations because of its well-developed root systems. The emphasis on Canadian poplar was also extended to the marine levees of Caraorman and Letea, partially destroying a natural forest ecosystem with particular aesthetic and floral value.

Further complications also arose through pollution by heavy metals originating at the Tulcea alumina plant and from dislocations associated with a bizarre project at Caraorman to quarry sand suitable for the recovery of rare metals. The navigation canal to Caraorman (never actually used to transport sand) had an adverse effect on water circulation and several lakes disappeared; throughout the delta the increase in navigation has had an adverse effect. However, it is the channels running transverse to the main rivers which should be eliminated in favour of a predominantly west–east water flow. Work to eliminate meanders on the Sfîntu Gheorghe channel (shortened from 109 to 70 km between 1985 and 1990) not only expedites navigation but increases the amount of sediment carried to the coast (significant in the context of a reduced silt load arising from the construction of dams at the Iron Gates and elsewhere). A system of concrete roads was envisaged for the agricultural zones and also as a transport link between Sulina and Sfîntu Gheorghe (associated with the programme to cut a trench parallel to the coast as a means of arresting coastal erosion in this section of the delta). But this programme has made only limited progress and it is unlikely that accessibility will be improved now that there is a desire to eliminate environmentally harmful projects and to manage the area as a national park with prime attention to environmental protection and the development of tourism (Roman 1990). Nature reserves have long existed (Letea forest in 1938; Periteasca-Gura Portiţei, Popina Island, Roşca-Buhaiova and Sfîntu Gheorghe–Perisor–Zătoane in 1961; Caraorman and Erenciuc forests in 1971) but protection must be afforded on a more comprehensive basis (Banu 1964).

NOISE AND OTHER FORMS OF POLLUTION

This is not a major problem in Romania outside the factories and building sites for traffic is still quite light by Western standards. But in rural areas where major mining and quarrying developments have occurred without an adequate infrastructure a succession of heavy lorries along country roads creates both discomfort and danger for the local population. Where the roads are unsurfaced then any vehicular traffic stirs up clouds of dust during dry periods (especially unpleasant under windy conditions). In many cases railways have eventually been provided but only after transitional periods during which road hauls of epic proportions have been organized, as with the transport of lignite from Berbeşti to Băbeni via a devious route through the villages of Milostea and Tomşani (Iacob 1985). The creation of large centralized wood processing complexes has led to a great increase in the number of timber-carrying lorries, replacing the forest railways which were previously used to transport timber down valley from the forests to local sawmills at the main line railheads.

Landscape

Local environments of ecological and historical interest have often been damaged through inadequate planning. The crucial importance of such matters has been emphasized for many years in the journal *Ocrotirea Naturii şi a Mediului Inconjurător* although it is only since the revolution and the appearance of newspapers like *Eco* that specific problems have been given prominence. The cement factory at Cîmpulung is proceeding to remove a mountain which has strong associations with military resistance during the First World War. The expansion of mining at Roşia Montană has overwhelmed the Cetaţii, a unique historic monument to the gold mining techniques of the Roman period (Bleahu 1990). The hydro-power project at Ştefăneşti in the Prut Valley made inadequate provision for the preservation of a unique element in the local vegetation, while both the Transfăgăraş Highway and the Danube–Black Sea Canal created avoidable damage to the glacial landscape of Bilea and the sand dunes of Agigea respectively. Inadequate controls over the quarrying of limestone have led to damage to the Furnica reservation near Sinaia. The work of listing and protecting monuments of nature has thus become an important preoccupation (Pop and Sălăgeanu 1965). But consideration should extend beyond a small number of outstanding cases such as Lacu Roşu/Bicaz Gorges and the famous Slănic salt mountain (Grasu and Turculeţ 1980; Sencu 1968).

More generally there is an increasing number of developments which create blots on the landscape which could be moderated in their impact by more careful planning. Damage in the Eastern Carpathians, an area of high tourist potential, has arisen from mining in the Căliman Mountains (reduced to a lunar landscape: *peisaj selenar*), although mining everywhere is creating problems of excessive tipping (to say nothing of the atmospheric and water pollution that arise from such nuisances) (Bold 1973). Some 19,000 ha of agricultural land have been lost by the creation of waste dumps and decanting ponds. The landscape around Anina in the Banat Mountains has been transformed by the decision to work the bituminous schists as a power-station fuel and a source of certain rare minerals. The countryside is now scarred by huge tips and open quarries, while substantial emissions of ash arise from poor planning of the power-station location and the height of the chimney. There is not even the consolation of economic benefits since the technically ingenious plan offers little prospect of viability.

Hydro-electric works, for long praised as one of the pillars of a green programme, are now being reappraised (Gruescu and Vasile 1985). There are potential benefits through flood control (reducing erosion), the retention of forests protecting sensitive catchments (when felling would increase sedimentation in lakes and accelerate run-off), the growth of fish stocks and the development of tourism in an ecologically acceptable manner. But in addition to the high economic and social costs arising from heavy investments there is considerable loss of land (previously used for agriculture and forest) and relocation of villages (Sandru *et al.* 1960). But then the reservoirs tend to silt up and cause scenic transformations which are not always desirable (Surdeanu 1984). Damage in the Retezat Mountains (following the Rîul Mare–Retezat hydro-power scheme) has been deplored and there is regret over the 'taming' of the Lotru Gorges through the creation of the vast water storage of Vidra which supplies Ciunget power-station through a tunnel bypassing the gorges altogether.

After the revolution, great concern was expressed over the proposed dam in the Mureş Valley at Răstoliţa, allegedly approved in 1989 without any local consultation (Dumitrescu 1990). Diversion of streams draining the southern flanks of the Căliman Mountains will cause rivers to dry up and there will be particular damage to the gorges on the Mureş in the Deda–Topliţa section, much appreciated by the rural population and by the townspeople of Reghin and Tîrgu Mureş. So the 1990s are already showing greater sympathy for a smaller scale of exploitation to minimize harmful changes such as destruction of scenic resources which have great importance for tourism in the future (Ujvári 1962). A sequence of small dams is under construction on the Zabala (a tributary of the Putna) in Vrancea. And in the case of the Olt (where work has been proceeding since 1970 on a total of twenty-four dams) the plan for a high dam in the gorge section near Racoviţa has been abandoned in favour of three small dams (at Racoviţa, Robeşti and Cîineni) which will allow a similar installed generating capacity but without flooding farmland and villages. Not that the tourist industry remains free from censure, because excessive pressure in some scenic areas rendered more accessible by cable cars and by the promotion of winter sports has caused some degeneration of a once rich flora. Attention is also being drawn to thoughtless behaviour by tourists in the way they park cars and dispose of rubbish, to say nothing of more outrageous vandalism like forest fire and wilful damage to traditional wooden buildings. Newly won political freedoms must be blended with a sense of responsibility towards the environment.

But the tourist infrastructure is inadequate and recreation facilities should be available on a more equitable basis. Information is rarely published, but data on green spaces in towns reveals some substantial inequalities (Muja 1984) (see Figure 7.4). When the information is generalized on a regional basis there are no major contrasts. The area of green space within the building perimeters, which averages 1.13 hectares per thousand persons across the country, varies only between 0.91 in the south-east and 1.52 in the north (see Table 7.1). Green space occupies 6.62 per cent of the building perimeter nationally and apart from the high figure of 8.39 in the north all the regional values fall between 6.12 and 6.98. But when individual towns are considered it transpires that twenty-two towns have more than 25 m² of green space per person (many of these being small towns with a population below 10,000) while twenty-three towns (distributed across all the population size groups) have less than 5 m² (see Table 7.2). Some of the suburban areas in large cities are badly served and the southern part of Bucharest is a case in point.

Table 7.1 Green spaces within urban building perimeters in Romania

Region	A	B	C	D		E		F
Centre	1.44	26.22	1.61	18.2	106	1.12	99	6.12
North	1.26	22.95	1.93	18.2	106	1.52	134	8.39
North-east	1.22	22.16	1.42	18.2	106	1.16	103	6.39
South-east	4.34	62.27	3.95	14.3	84	0.91	94	6.34
South-west	0.80	16.60	1.16	20.7	121	1.45	128	6.98
West	1.25	25.87	1.59	20.6	120	1.27	112	6.15
Romania	10.32	176.06	11.65	17.1	100	1.13	100	6.62

Source: Muja 1984, pp. 31–8.
Key: A Urban population (millions); B Building perimeter (thousand ha); C Green space (thousand ha); D Building perimeter: hectares per thousand of the population and relationship to the national average; E Green space: as D; F Green space as a percentage of the building perimeter.

Table 7.2 Romanian towns classified according to population size and provision of green space, 1980

Population ('000s)	Number of towns in each category of green space provision (square metres per person):					
	Below 5.0	5.0– 9.9	10.0– 14.9	15.0– 19.9	20.0– 24.9	25.0 and above	Total
50,000 and over	4	18	11	8	1	0	42
25,000–49,999	4	14	9	5	6	2	40
10,000–24,999	8	24	18	19	8	7	84
Below 10,000	7	16	18	9	7	13	70

Source: Muja 1984, pp. 31–8.

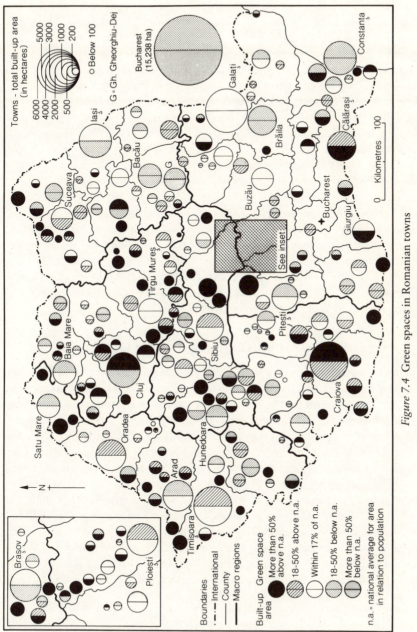

Figure 7.4 Green spaces in Romanian towns
Source: S. Muja 1984.

Culture

Pollution of Romanian culture by the former regime (and in particular the late president who is now condemned for a 'megalomania faraonică') has also attracted much comment (Maiorescu 1990). The manipulation of history and custom for ideological purposes has been seen in a most grotesque form in the programmes of *sistematizare* which used town and country planning to improve housing and services as a cover for the destruction of traditional buildings (many of great historic interest) and indeed a whole way of life. At the same time the new buildings offer little continuity with Romanian architectural traditions and the massive Republic Hall (Casa Republici) in the centre of Bucharest is a case in point (Modorcea 1990). But the villages have also been spoilt by a process of *depersonalizarea* involving apartment blocks which flagrantly disregard established architectural standards (Bleahu 1987). However a draconian plan to consolidate rural settlement and reduce the number of villages (currently exceeding 13,000) to between 5,000 and 6,000 was abandoned at the revolution with only a small number of complete clearances actually made (Turnock 1991). This destructive approach is all the more remarkable because it followed a period of strong support for the preservation of historic monuments, a policy fully outlined in such journals as *Arhitectura* and *Buletinul Monumentelor Istorice* (Giurescu 1989). There were certainly pressures for radical change but the outcome was still in the balance as late as 1975.

POLICIES FOR ENVIRONMENTAL PROTECTION

The first conservation measures were evident before the First World War and concerned the management of forestry and hunting. The first forest codes were followed by legislation for the protection of natural monuments and the setting up of nature reserves and national parks (Florea and Cătălin 1987). This followed a number of suggestions for nature reserves before 1914, including Slătioara forest in 1904. Then in 1923 came the bold proposal by the director of Cluj University's Botanic Garden that land in the Retezat Mountains (south of Hațeg) should be reserved for a national park, this being a period of land transfer from large estates to peasant proprietors. The call was repeated in 1928 during a congress of Romanian naturalists and the outcome was a law passed in 1930 for the protection of monuments of nature. A commission for natural monuments (Comisia Monumentelor Naturii) was set up in 1933 (surviving until 1944) and the first reserves were created at this time. The Retezat Park was inaugurated in 1935, with the aim of restricting pastoralism, woodcutting, hunting and fishing, but powers were inadequate until 1955 when a more effective authority was appointed to control an enlarged area of 13,000 hectares (including a scientific zone with limited access) (Popovici 1987). This should be seen in the context of genuine concern for conservation in the first years of socialism when a new commission was formed (following environmental protection legislation in 1950) and forest codes revised in 1954 and 1962 (Puşcariu 1975). A large number of reserves were designated: from 34 reserves in 1945 (15,000 ha) the total rose to some 420 by 1990 (220,000 ha) (see Figure 7.5). And there was a substantial publishing effort to emphasize their touristical and educational importance (Puşcariu 1972). Much of the literature is organized along systematic lines, singling out (say) forest or geological reserves (Bleahu *et al.* 1977, Nimigean *et al.* 1982, Toader and Nitu 1976). Other material deals with individual regions such as the historic provinces of Bucovina and Oltenia (Bobirnac *et al.* 1984, Seghedin 1983). But the local authorities are also keen to attract interest in their own reserves, as in Constanţa and Neamţ (Ciobanu *et al.* 1972, Sălăgeanu 1978).

Little was heard about pollution during the 1960s, although there was legislation in 1967 which laid down maximum levels of discharge for pollutants in water (Stugren 1965). But the former president became quite interested in conservation in the early 1970s following the United Nations environmental conference in Stockholm in 1971 (Vespremeanu 1981). He called for the elimination of all emissions and discharges of pollutants because damage was caused to the environment and materials recoverable by recycling were being lost (Ceauşescu 1973: 511). Legislation for environmental protection (*legea cu privire la protecţia mediului înconjurător*) was passed by the Grand National Assembly in 1973 and two years later a National Council for Environmental Protection (Consiliul Naţional pentry Protecţia Mediului Inconjurător – CNPMI) was established under the chairmanship of V. Ianovici. It was to play a co-ordinating role, complemented by district commissions at the local level, to provide integration with the wider regional planning process (Ianovici 1978). The expectation was that this machinery would 'gradually do away with the present pollution sources of water, air and environment' with special measures for the highly industrialized areas and that there would be 'adequate measures of environmental protection for any new investment ... so that initial ecological equilibrium should not be contaminated' (Ianovici and Popescu 1977: 13). But this organization could not adopt any effective sanctions and it was given no budget. Its volunteer members had environmental expertise but they tended to project narrow department interests (Vădineanu 1991: 110).

Particular importance would attach to the maintenance of a good environment in the Carpathians (especially in the forests and national parks) and also in the Danube Delta and along the Black Sea shore. With this in mind further laws were passed in 1976 to authorize programmes for the improvement of river basins (*amenajarea bazinelor hidrografice*) and the conservation of woodlands (*conservarea şi dezvoltarea fondului forestier*). The population as a whole (but especially youths who 'excel in such work') was to be mobilized to help in planting trees, laying out parks and playgrounds and in maintenance work (Ianovici and Popescu 1977: 20). There was certainly a powerful response from the scientists (Teodosie 1974). This was especially true of the geographers who rebuilt the unity of their discipline on the priority for research on the environment (Turnock 1982). There was a growth of research which went beyond the description

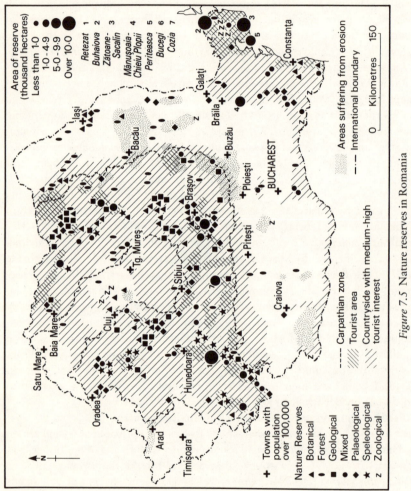

Figure 7.5 Nature reserves in Romania
Source: Atlas RSR 1979.

of environmental change and dealt specifically with pollution (Berindan 1975). There was a particularly strong lead from climatologists (Bălăuţă et al. 1973, Chiţu 1973, Erhan 1972, Gugiuman 1976, Trufaş and Trufaş 1978). The Romanian Geographical Society devoted much space to such issues in its *Buletin* when a single article on air and water pollution in 1972 was followed up two years later by case studies which filled an entire volume (Posea and Pisoţa 1972, Posea et al. 1974). Soil erosion featured prominently in a later edition (Buza et al. 1980). Several important textbooks were produced by geographers at this time, suitable for different levels of study (Roşu and Ungureanu 1977). There was particularly good coverage for schools, offering possibilities for a start in environmental education (Posea et al. 1978, Tufescu and Tufescu 1981, Tufescu et al. 1978, Zăvoianu et al. 1981). Other conservationists were also active in this way (Barnea and Papadopol 1975, Barnea and Ursu 1969, Ciplea and Ciplea 1978, Zamfir 1973). And an important contribution to ecological education was made by the natural science museums in the various counties of Romania.

Assessment

Some effective measures have been taken. The closure of the cement factory in Braşov has resulted from construction of new capacity in Hoghiz, while the abattoir and incinerator formerly operating near the centre of Bucharest have been relocated at Glina. Since the revolution the lignite-burning power-station at Braşov has been converted to burn methane gas and a programme is being drawn up to alter the profile of some factories near the centre of Bucharest and to transfer other units to the IMGB industrial zone on the southern edge. But there can be no doubt that good intentions have been totally inadequate. Although legal sanctions were provided for they were only occasionally applied and when action was taken the system of levying fines on the individuals responsible according to their ability to pay presented no problems for the enterprises which were only too happy to reimburse the employees concerned. Moreover, during the 1980s the Ceauşescu regime took decisions which contradicted the aims set out during the previous decade. For example the drive to increase agricultural production had negative implications for forestry despite the measures taken in 1976 to protect the woodlands. On the Năruja Valley of Vrancea extensive clearances were allowed so that a new state farm could be established for the summer grazing of livestock transported from lowland areas of the county. Pressure to expand production also encouraged corruption among foresters who might accept bribes to overlook unauthorized cutting or grazing.

Again, the commission established in 1975 took a strong interest in the Danube Delta and the biologists examined the problem of eutrophication, but it was no match for the powerful state enterprise (*centrala*) which was backed by the local authority, keen to exploit the delta in the way the president desired and to introduce its own team of specialists to provide the necessary climatic and limnological data. At the same time the government became increasingly sensitive over environmental issues and publication was discouraged. Factory workers were forbidden to talk about the emission of gases and the discharge of polluted waters and only occasional informal disclosures were made. The situation became particularly difficult from 1985 when academic and professional discussion of ecological problems (organized since 1970 by the university group 'Terra 2000') was discontinued. There was some criticism but it was 'hidden in long (preferably boring) treatises; sometimes the language was coded and sometimes publications were given misleading titles to divert the censor' (Fisher 1990: 2). A notable example was the work of 'Acţiunea Democratică' which operated from 1985 and produced a 'Green Report' in 1988 highlighting all the major forms of pollution related to the excessive emphasis on heavy industry. The report called for a change of policy and further environmental education to avert an ecological disaster (Ionescu 1988).

However, the greater freedom to publish and demonstrate since the revolution has indicated the full extent of public concern. Once the high costs of recycling had been appreciated the former regime revised its strategy of 1973 and 'neglected nearly all environmental problems' (Hâncu 1990b: 5). Advice directed along official channels was ignored, including the results of research on ecology and the construction industry 1987–9, and environmental protection agencies were marginalized through derisory funding: the CNPMI was seen as a phantom organization with just two specialists. Plans were not properly implemented, including measures for air filtering and water purification at twenty-six large industrial complexes (despite approval between 1987 and 1989). Evidently while there were installations at factories to control pollution many of these were of inadequate capacity when considered in relation to the *superconcentrarea* of industry with its gigantic units of production. Control equipment was often out of circulation because of maintenance problems or the drive to reduce the consumption of energy, and filtering equipment to control air pollution would sometimes be turned off at night.

The reasons for the failure to follow through the legislation of the 1970s are difficult to identify. There are studies which explain Ceauşescu's rather primitive approach to Communism, through his peasant background and persecution for subversive activity (Crowther 1988, Fischer 1989, Gilberg 1990). But they fail to deal with the strengthening of a revolutionary or 'extremist' approach during the 1980s (Pecsi 1989). Certainly economic crisis and pressure from creditors brought a maverick response by way of determination to liquidate foreign debts in the shortest possible time to avoid outside interference in economic planning. And this would have discouraged 'unproductive' investment in pollution control and conservation. But the late president also developed an obsession for socialist landscapes of his own making, stimulated by the necessity for limited redevelopment after the 1977 earthquake and models provided by cities like Pyongyang with which Ceauşescu was familiar through his close links with Asian Communist regimes. The scale of the Romanian construction industry also satisfied an important precondition. Thus a decade of reduced priority for industry which could have seen the deployment of resources into conservation in fact witnessed a surge in the opposite direction.

PROSPECTS FOR THE REVOLUTION

Government has shown appreciation of the problems, but of course it is easy to condemn the criminal follies of the 'Dictatura Ceauşistă' (Hâncu 1990b), it is less easy to act with determination. However, under the ruling National Salvation Front (Frontul Salvării Naţtionale – FSN) a new ministry has been created to combine general environmental concerns with the management of the forests and water resources (Ministerul Pădurilor Apelor şi Mediul Inconjurător). And there is to be an ecological commission (Comisia Ecologică) under A. Vădineanu (in charge of the environmental section of the new ministry) to undertake scientific research and provide legislators with the necessary information. A new quarterly journal 'Mediul Inconjurător' was launched to stimulate informed debate. Some observers are sceptical of Vădineanu's prospects in view of his failure to restrain harmful developments in the delta when he headed the inquiry team referred to above. There is similar doubt about the monitoring group (Agenţia de Ocrotirea Naturii) which is to be established in each region (Mihu 1990). This concern is deepened by methodological inadequacies regarding a comprehensive approach to pollution which should complement research specializing on different facets of the problem (Ungureanu 1982).

Yet some concrete measures have already been taken and Vădineanu's transfer to the post of adviser to the prime minister suggests that environmental matters command some priority. There will be an immediate effort to filter air and process water in the most polluted areas, recovering useful materials in the process. At Arad the polluted water will be pumped from the infected area and processed in an enlarged purification plant. Yet other improvements will take time. Pollution levels should be reduced by one-fifth by 1995 and European standards are to be reached soon after the year 2000 (Hâncu 1990b: 7–8). Legislation will be brought up to European standards and non-polluting technology will be encouraged. Work must be done to rebuild the infrastructure for dealing with the preservation of historic buildings. There must be a national strategy, and it will have to be applied most carefully in areas like Bucharest that have been badly treated in recent years (Giurescu 1989). Environmental education is to be strengthened and this will require work in the villages as well as presentations through the national media (Ionescu et al. 1989).

Special protection must be given to the most sensitive areas and the FSN government accepts that the Danube Delta is perhaps the most pressing case. Some additional nature reserves have been created (Hasmacul Mare, Istria and Sărăturile) with a combined area of 1,600 ha, supplementing the existing reserves with a total area in excess of 36,000 ha. There is to be a conservational authority with effective powers appropriate to a reserve of the biosphere (rezervaţie biosferă) declared in 1990 and endorsed by international organizations.

The reserve covers 591,200 ha (greatly enlarging the 18,145 ha Roşca-Letea biosphere reserve of 1979), with 103,000 ha consisting of a maritime buffer zone for 470,500 ha of land in the delta proper. This latter area divides into sixteen core areas (52,580 ha) along with buffer zones (230,000 ha including 103,000 ha of marine waters), ecological reconstruction zones (25,500 ha) and transition zones (162,120 ha). Only 103,000 ha are dyked for agriculture, forestry or fishing and of this area 28,250 ha are under water at present. A research programme is being put in hand by the reserve's scientific council and some elements of policy are being drawn up. There will have to be priority for the reduction of nutrients and the maintenance of a high water level. This has been complicated by the dyking of the Danube flood plain (restricting the volumes of sediment reaching the delta) and the low water level of recent years. The circulation of water needs careful co-ordination so as to follow a west–east alignment: some transport canals should be closed while others need to be dredged and widened. Woodland should be restored on some canal banks; industrial installations must be cleared away and small fish farms (of 50–100 ha) should be set up on ecological principles.

The mountains constitute another priority (Gîrlea 1977). Here again the developmental pressures, which extend to forestry, pastoralism and tourism as well as industry and mining, must be restrained (Boscaiu 1975). In the Bucegi Mountains the control on grazing by the silvicultural authorities has been in-effective since the revolution and there has been further damage caused by the armed forces (using heavy vehicles in areas with a fragile vegetation); also, tourists deviating from paths, damage trees and leave rubbish. A big educational effort is needed and a ranger service is being introduced. Damage caused by mining must be reduced and research on the Toroioaga Massif near Borşa has demonstrated the value of afforestation and containing walls for waste heaps in accommodating current levels of activity without harmful effects. Particular care is needed in the Căliman Mountains where the production of sulphur has been particularly harmful; also, in the Apuseni Mountains where the quarrying of bauxite and limestone is extensive, landscaping should be improved and quarries should be located away from the most scenic areas. Some areas of salt mining are affected by subsidence and there is scope at places like Coştiui, Ocna Şugatag and Ocnele Mări for recreation facilities on what is currently waste land. And there is a growing problem arising in connection with lignite quarrying in open pits, particularly on the edge of the hill country in Oltenia. Much land has been lost through the dumping of waste as well as the quarries themselves; while the failure to store topsoil in separate tips will complicate the restoration process.

The environmental difficulties are complicated by social problems through the uncertain future for the new industrial communities planted in those lignite fields which are highly uneconomic to exploit; and also for those members of the farming community whose ancestral lands (accessible through the demise of the cooperatives) cannot be reclaimed because of the quarrying. Further national parks are needed in such favoured areas as the Apuseni Mountains west of Cluj-

Napoca, Ceahlău north of Bicaz, the Rodna Mountains south of Borşa and Semenic-Cheile Caraşului east of Reşiţa. A programme was drawn up during the 1980s by specialists in forestry and natural monuments (Banarescu *et al.* 1979–80, Horeanu and Borcea 1982, Puşcariu 1981–2, Seghedin 1977). A dramatic gesture was made in 1990 with the designation of eleven new parks (total area 353,900 ha) in the mountains (supplementing the one existing park in the Retezat Mountains which dates back to 1935 and embraces 54,400 ha at present). More generally, a commission has been proposed to protect the entire Carpathian belt but although an authority has been set up (Comisia Zonei Montane – CZM, with offices in the twenty-two counties territorially involved) it is not clear how effective it can be as a force for conservation.

The ecological movement

After the revolution several ecological organizations have emerged. An ornithological society (Societatea Ornitologică Română) was formed in 1990 and there is a non-governmental association interested in the Danube Delta. However, the most significant development has been the formation of political parties as a result of initiatives in Bucharest and in provincial cities such as Arad and Turda. The two Bucharest-based parties, the Romanian Ecological Movement (Mişcarea Ecologistă din România – MER) and the Romanian Ecological Party (Partidul Ecologist Român – PER), did surprisingly well in the elections of May 1990, securing two senate places (one for each party) and a total of twenty seats in the Chamber of Deputies (twelve for MER and eight for PER). The two parties attracted 591,000 votes (the average for the two sets of returns) or 4.05 per cent of the total. Half of the votes (295,800) came from Bucharest and eight other industrial areas where the share of support was more than 20 per cent above the national average. In ascending order of support these were Prahova (6.15), Alba (6.20), Suceava (6.35), Timiş (6.75), Bacău (6.85), Hunedoara (7.00), Braşov (8.35) and Sibiu (11.25). Some 178,200 votes (30.2 per cent) came from fourteen counties where the level of support was within 20 per cent of the national average (that is, 3.25–4.85 per cent), and 117,000 (19.8 per cent) came from the other eighteen counties where the level of support was less than 20 per cent below the national average. The lowest values were scored in the predominantly Hungarian-speaking areas of Harghita (0.45), Mureş (1.0) and Covasna (1.1) where the voters supported the ethnic Hungarian party but there were also low shares in agricultural areas in North Transylvania (Sălaj and Satu Mare – both 1.95 per cent), and in the Wallachian Plain (1.7 per cent for Teleorman and 1.8 for Olt). Figure 7.6 indicates the particularly strong support which came from the more highly industrialized counties where the leaders of local authorities have often joined the ecological parties. It has been more difficult to stimulate interest in the agrarian counties where there is a measure of complacency, although interest increases when there is knowledge of the problems in other parts of the country.

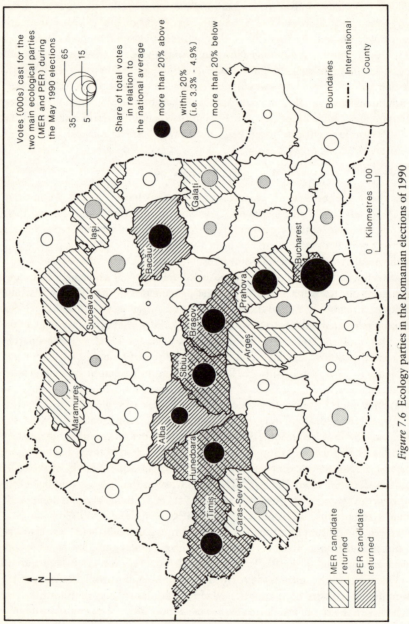

Figure 7.6 Ecology parties in the Romanian elections of 1990
Source: Mișcarea Ecologistă din România.

Preferences between the two ecological parties (which emerged from different groupings of specialists, with insufficient time available for consolidation before the elections) may have been influenced by the appeal of individual candidates and by the order in which the parties appeared on the ballot forms. The large number of parties contesting the elections meant that voters were faced with booklets rather than single sheets of paper with the parties listed according to draws carried out in each county. Many ecology supporters therefore voted for the first party to appear on the list. However, it would appear that MER is much the more active of the two ecology parties and its membership has been estimated at 100,000. It has a well-developed programme and has established close links with the green movement in Europe as a whole. The party champions human rights as a base for research into environmental problems and the formulation of appropriate economic policies (allowing foreign investment, private enterprise and ecologically acceptable production methods, including the eventual abandonment of chemical fertilizers, insecticides and pesticides). There is also a call for the close monitoring of pollution and for a new commitment to environmental protection, public health and respect for public opinion over the planning of towns and villages (the latter reference being made in connection with urban ecology, to restore individuality after years of mutilation in the interests of a dogmatic physical planning programme). And, finally, MER campaigns for a non-political education system to increase public awareness of major issues not only of ecology but of Romanian culture and international relations as well. MER organization, which exists not only at national level but through more than 100 branches in the various counties and also in the larger towns, emphasizes scientific work and its use in politics.

There is also a strong accent on youth. A total of twenty-nine commissions work on different issues such as agriculture, ecological mapping, education/mass media, new forms of energy, nuclear power/radioactivity, and the protection of buildings. There is also an ambitious publishing programme focusing on the weekly newspaper *Eco*. Sales of some 200,000 have been achieved, but the government controls the paper supply and the printing presses so that a number of print runs have been restricted to below 40,000 when economic viability requires 100,000 sales. MER's activities have inevitably been given a mixed reception by people who are basically sympathetic to the ecological movement. There are some misgivings about direct involvement in politics but these reservations arise primarily from the realities of pre-revolutionary public life when the Communist Party constrained the activities of scientists and the expression of opinion. Moreover, past experiences of PCR membership to secure advancement led many people to associate politics with crude opportunism. But some are beginning to adjust to a new democracy and appreciate that MER exists only as a pressure group to provide constructive opposition in parliament. But it is also significant that the law currently restricts organizations to professional associations and political parties! Registration of MER in the latter category brought the advantage of premises allocated by the interim government anxious to fragment the opposition as much as possible (only 250 signatures were required to achieve recognition). So co-operation with the manoeuvrings of FSN has inevitably tarnished the party in the eyes of critics. Equally, MER's association with the interim government in the early days of the revolution created some misunderstandings abroad after FSN decided to set itself up as a party to contest the elections. This, however, is likely to be no more than a transitional difficulty.

WESTERN TECHNOLOGY

The international dimension is certainly prominent as evidenced by the agreements over the quality of water in the Danube which has led to a secretariat in Bucharest. But, in particular, Romanian academics have kept a close eye on Western literature, despite foreign currency shortages which have greatly complicated the supply of books. There are several valuable texts which draw heavily on non-Romanian sources (Roşu 1987). So there is widespread appreciation that Western technology is essential for the solution of pollution problems. In the past Western assistance has been geared largely to industrial production, and the organization responsible for pollution monitoring and control equipment (Intreprinderea de Aparate şi Utilaje pentru Cercetare, set up in 1978 in Bucharest, with branches in Cluj-Napoca, Galaţi, Ploieşti, Suceava and Timişoara), was thrown back on to domestic resources and forced to use outdated technology to produce some thirty different types of detection equipment. Since the revolution, arrangements have been made for foreign collaboration and there have been discussions with the German firm Lurgi regarding the most serious problems at chemical factories.

This is very much in line with public opinion since demonstrations at Suceava demanded international co-operation to modernize the artificial fibre factory. It is appropriate here to refer to the political links between MER and the green parties represented in the European Parliament; also the remarkable MER initiative over an ecological university in the capital (Universitatea Ecologică Bucureşti – UEB). The rector, Dr D. Drimer, enjoys close links with North America and obtained considerable financial and technical support, making it possible for UEB to admit some 1,800 students in autumn, 1990. This private institution charges higher fees than the state-controlled universities and five years of study are required for graduation (six in the case of Medicine). Moreover there is, as yet, no law which sanctions private enterprise in the field of education. Nevertheless there has been a very encouraging level of interest. The four faculties (Engineering, Law, Medicine and Natural Sciences) are all related to the ecological programme through the objective of turning out graduates with skills in the monitoring and control of pollution (Drimer 1990).

Romania does not adhere to all international arrangements (for example, the Ramsar Convention for the protection of wetlands) but there is now a greater readiness to embrace international standards, and foreign assistance over conservation programmes is gaining momentum. There has been involvement with

UNESCO programmes (linked with the three biosphere reserves established in 1979) and with the World Heritage Convention. There has also been re-entry into the World Conservation Union (IUCN) following a lapse through most of the 1980s at a time when pressures for great commitment to conservation in the Danube Delta were being resisted. Now IUCN's East European Programme is lending strong support to the Romanian initiative in setting up the Danube Delta Biosphere Reserve (DDBR). This involves professional assistance in a new development plan for the delta (for example in tourism) and should develop into a permanent presence through an office in Tulcea. The DDBR management has also set up a facility at Uzlina on the Sfîntu Gheorghe channel to be used as an international centre for education and training related to the delta's ecological problems.

CONCLUSION

The lack of statistical data frustrates an entirely balanced assessment of pollution problems in Romania. However, it is clear that development under Communism was very harmful and a political will to improve matters in the 1970s lost much of its resolve under the changed circumstances of the following decade. The partial freedom for the press since the revolution has brought new evidence to light, emphasizing the particularly serious hazards arising out of the chemical industry and schemes of land and water management (hydroelectric, land drainage and irrigation projects) previously portrayed as beneficial. There is genuine indignation over the mismanagement of the economy which has caused serious environmental damage without any benefits to the nation's wealth. Yet it is important that this mood should not lead to polemics; rather, it should be harnessed to promote progressive policies for the future, including an effective programme of environmental education. A start has been made, but important decisions have still to be taken and progress will inevitably be slow in view of the painful economic transition which has to be undertaken in the switch from central planning to a market economy.

8

FORMER YUGOSLAVIA

Professor B. Jancar-Webster

All references to Yugoslavia are to be taken as meaning former Yugoslavia

INTRODUCTION

At the beginning of the 1990s, former Yugoslavia is in deep trouble. It is hardly the moment to talk about environmental protection. For those of us who have followed the country's progress since the Second World War, the dissolution of the federal republic is a sobering experience. Environmental voices are hard if not impossible to hear in the alarums of war, and even if heard, the urgency of the battle leaves them unheeded. In one day of fighting, greater environmental destruction occurs than in several years of ordinary living. The damage from the war already is over $20 billion and still rising. During 1990 and 1991, there was not one ecological protest throughout Yugoslavia, a far cry from the anti-nuclear demonstrations that swept the country four years earlier. In an earlier article, the author expressed hope that the conflict between republics would develop techniques for the common management of pollution control 'while there was still time' (Jancar 1985: 242). In the 1990s that hope has given way to the realization that time may already have run out. The Yugoslav tragedy is an important lesson in understanding the close connections between technology, institutional structures, and ideology in promoting environmental degradation in the course of economic development. As the dissolution of the Communist states indicates without a doubt, technology is not neutral: it is closely tied to economic and political organization. The kind of infrastructure realized by technology has a strong influence on the organization of institutions, and ideology informs both institutions and the development of technology.

Environmental protection in Yugoslavia began to be taken seriously in the 1970s. Before the Second World War Yugoslavia was an underdeveloped, primarily agricultural country. The Second World War devastated both population and land, and brought industrial development to a standstill. Most of the country's industrial and urban growth has been forced to occur in the time-span since the war. The rate of growth of both industrialization and urbanization in Yugoslavia during this period was among the highest in the world, and little thought was given to their impact on the environment. Yugoslavia is faced today with increasingly serious environmental problems which only yesterday seemed

164

Figure 8.1 Pollution in former Yugoslavia

inconsequential. The chronic economic crisis in progress since the mid-1980s has not helped the problem of importing technology with scarcer and scarcer hard currency resources. But the real impediment to progress is the rupture between the six constituent republics (Bosnia-Hercegovina, Croatia, Macedonia, Montenegro, Serbia and Slovenia) brought on by the inadequacy of the ideological and institutional solutions developed during the forty-year rule of the Yugoslav League of Communists (YCL).

POST-WAR ECONOMIC DEVELOPMENT

In 1948, Yugoslavia had a population of 15,901,000. In 1979, the population had grown to 22,160,000, a 39 per cent increase. Growth slowed considerably during the next decade. In 1990, the population stood at 23,800,000, an increase of 7.5 per cent over the 1979 figure (Federal Executive Committee 1980: 31, World Resources Institute 1990: 253). During the first time-period, interrepublican migration was heavy, with thousands of peasants leaving the land in search of jobs in the industrializing cities. According to the 1971 census, more than 1 million people had changed their place of residence since the previous census ten years earlier. Of these, 460,000 had migrated from Bosnia-Hercegovina and 568,000 had migrated to Serbia, primarily to Belgrade. The unidirectional thrust of the migration is illustrated by urbanization figures. In 1948 there were 83 urban centres with populations above 10,000. In 1971 there were 148. Between 1931 and 1971 Ljubljana grew 2.9 times; Belgrade 3.1 times; Zagreb 3.5 times; Skopje 4.6 times; Niš 14.6 times; and Split 16.5 times. In 1946 approximately one-fifth of the population lived in cities; in 1985 about half. As can be seen from the figures, urban growth was not the same in every republic. For example, in Macedonia in 1971 over 30 per cent of the non-agricultural population was living in the capital, Skopje. By contrast, only 12.6 per cent of the same population in Slovenia was living in the Slovenian capital of Ljubljana. Yugoslavs are quick to point out that the process of urbanization in Yugoslavia had been one of the most accelerated in the history of urban development, generating a whole set of pollution problems that were virtually non-existent in the 1960s.

Industrialization has also been rapid. Between 1947 and 1975, the average rate of economic growth was 6.2 per cent, and between 1953 and 1964 an impressive 8.6 per cent. By contrast, the economy grew at 1.2 per cent in 1987, and declined in 1990. In 1947 industry accounted for approximately 19 per cent of the work-force. In 1977 industry accounted for 40 per cent. In 1951 there were 6,825 cars and 1,150 buses. In 1976 there were over 17,300,000 cars and 21,000 buses. Between 1970 and 1987 the total number of motor vehicles in Yugoslavia increased almost four times and the number of passenger cars 4.18 times (OECD 1989: 251–3). During the period of most rapid economic growth the production of energy increased over 18 times, with coal being the most important domestic energy source. The number of paved roads grew from one paved road out of Belgrade before the Second World War to a network of highways criss-crossing the country and covering over 20,000 kilometres by 1979 (Federal Executive Committee 1980: 95). Between 1975 and 1989 the road network has increased by 20 per cent, the highest rate of increase of any of the OECD countries (OECD 1989: 247).

Impact on the environment

These developments upset traditional land-use practices and wrought havoc with the environment. From a total of 15,000,000 acres of arable land, the amount of agricultural land under cultivation fell by 4 per cent from 10,300,000 ha in 1961 to 9,900,000 in 1976. In 1987 the amount of land under cultivation fell a full 21 per cent to 7,800,000 ha or 0.33 per capita (World Resources Institute 1990: 281). Since 1961 then, 24 per cent of the cultivated land has been lost to farming. About 20,000 ha continue to be lost to production yearly and the situation has reached crisis proportions. After a 16 per cent increase between 1937 and 1961 the forested area then decreased by a substantial 12 per cent from 10,300,000 ha in 1961 to 9,100,000 in 1976. The loss of forests was a primary factor in the spread of erosion which has now become critical as well. An intensive reforestation programme undertaken with the help of volunteers, Young Pioneers, and schoolchildren, and in progress since the 1960s, has ameliorated the situation somewhat. Official data given to the OECD indicate that annual forest depletion from lumbering and environmental causes appears to be offset by annual growth (and these figures also contradict those published internally by indicating an increase in forest cover since 1950) (OECD 1989: 115–19). Total forest cover in Yugoslavia in 1987 was 9,500,000 ha, a solid 4 per cent increase over 1976. But environmentalists are not encouraged by this performance and they believe that thousands more trees will have to be planted each year to stop erosion and restore Yugoslavia's forests.

Key

⫶: ▬ (National status only)

1 Triglav
2 Risnjak
3 Brioni
4 Plitvicka Jezera
5 Paklenica
6 Kozara
7 Krka
8 Kornati
9 Mljet
10 Sutjeska
11 Fruska Gora
12 Tara
13 Durmitor
14 Lovcen
15 Skadarsko Jezero
16 Biogradska Gora
17 Djerdap
18 Kopaonik
19 Sara
20 Mavrovo
21 Galicia
22 Pelister

X (Biosphere reserves)

1 Velebit
2 Reka Tara

◉ (World heritage — natural and cultural)

1 Skocjanske Jame
2 Plitvicka Jezera
3 Split
4 Dubrovnik
5 Kotor
6 Durmitor I Reka Tara
7 Studenica
8 Staria Ras - Sopocani
9 Ohrid sa Ohridskim Jezerom

Figure 8.2 Water quality of former Yugoslavia's major rivers

Source: Environmental Status Reports: 1990, Vol 2, *Albania, Bulgaria, Romania and Yugoslavia*

The quality of water (see Figure 8.2) has steadily deteriorated so that, according to the four-class system adopted in law in 1968 and amended in 1978, virtually every river has fallen from a high water classification to a lower, sometimes by as much as two grades. In Kosovo, where rivers were relatively unpolluted in the 1960s, analyses showed that by 1979 not a single river in the province was of a water quality which could be designated first class (*Privredni pregled* (Belgrade), 29 March 1979). Water quality indicators of selected rivers in Yugoslavia published by OECD indicate a slight decrease since the 1970s in biological oxygen demand for the Danube and an increase for the Drava during that same time period. Even more significant have been the increases in concentrations of nitrates, ammonium, cadmium, chromium and copper (*Politika*, 20 June 1984; OECD 1989: 61–7).

The Adriatic Sea along the Yugoslav coast has fallen victim to increasing pollution despite Yugoslavia's signing of international and regional agreements, membership in the UNEP Regional Seas Programme and the formation of the Alps–Adria bioregion. In the summer of 1989 algae blooms, similar to those that occurred on the Italian coast and clogged the canals of Venice, appeared intermittently along the Yugoslav coast as well. The water quality of the two rivers of Yugoslavia protected by regional compacts, the Neretva and the Sava, are a national disgrace. The Sava is Yugoslavia's main water artery. Along its shores stretches the majority of the country's most advanced industry and its water is used for both industrial and household consumption by Belgrade. Pollution caused by oil and other pollutants seeping into the river earned the Sava a name reminiscent of The Silent Spring, 'a river without life' (*Politika*, 25 December 1984). Although over half the population is now supplied by public water systems and 74 per cent are listed as having access to safe water, the surface water designated for urban settlements is increasingly threatened by contamination from industrial and agricultural wastes. In the mid-1970s one-third of the urban water systems lacked adequate treatment facilities, and many local water systems required substantial reconstruction because they had been built by laymen, not experts. A big problem in the 1980s was the construction of water treatment plants. By 1984, sewage disposal had also become a major problem. Raw sewage continues to be dumped untreated into rivers, lakes and the sea (*Vjesnik* (Zagreb), 5 June 1982).

Of all the forms of pollution, air pollution is perhaps the most serious and most visible in the larger cities but it remains highly localized. Newspapers in the republican capitals and major urban centres publish daily data on air quality. Almost without exception during the winter months, the ambient air quality in one of Yugoslavia's major cities exceeds the permissible legal level for the major pollutants. The main culprit is sulphur dioxide. In Belgrade, particles of coal dust the size of ash descend from the sky in the winter, covering buildings and streets with black film, while the air is saturated with the acrid smell of coal fires. Newspapers have reported that in some parts of Skopje, air pollution is so high that it impedes 'normal breathing' (*Vjesnik-Sedam dana*, 8 December 1979: 15). In

1990 Belgrade earned the dubious distinction of having the highest concentration of sulphur dioxide and NOs per cubic metre of air of any capital in the world, including Lisbon, Prague, Tokyo and Brussels (*Dnevik* (Belgrade), 3 September 1990: 1).

Fortunately for Yugoslavia air pollution remains primarily an isolated urban problem and average population density is the ninth lowest out of the twenty-six European countries. However, it must be remembered that half of that population now lives in cities (World Resources Institute 1990: 269–71). Although the country is a signer of the European Transboundary Air Pollution Agreement, the long-range transport of air pollution has not been a significant domestic issue even after the disaster of Chernobyl. The official study, prepared in 1976 for the Federal Executive Council and the executive councils of the republics and provinces, identified sixty-eight areas where permanent or critical air pollution had been monitored. Of these twenty-five were in Serbia, with severe contamination registered in Belgrade, Kraljevo and Niš. On the territory of Serbia alone experts have cited as many as 2,500 industrial installations with inadmissible levels of emissions. Happily, Yugoslavia benefits from the Kosava, a strong westerly wind which periodically blows across the central plain, sweeping away the coal specks, dust particles and other pollutants which blacken Belgrade's skies, leaving the air cold and clear. Health officials count on this wind to prevent the development of killer smogs or similar phenomena dangerous to human health.

The growth of heavy industry has brought a rapid increase in solid waste. The Adriatic port city of Rijeka has been cited as 'probably the most threatened area in the North Adriatic' as a result of the continued discharge of waste water directly into the sea. The Neretva river polluted by wastes discharged into the whole length of its stream, also pours its polluted contents into the Adriatic. Waste disposal has reached a point where recycling strategies or utilizing secondary by-products in the production of new goods are finally being seriously discussed.

Urban development

Perhaps the most salient environmentally degrading feature of the post-war period has been the chaotic growth of towns at the expense of the countryside. Far exceeding the planned forecasts, the expansion of the cities has demanded the creation of totally new infrastructures or the enlargement of existing facilities of limited capacity to handle the environmental problems generated by the influx of migrants to the urban areas. Among the worst polluted cities is the metallurgical town of Zenica in Bosnia-Hercegovina where experts say the cost of clean-up exceeds $220 million (Cerović 1991: 8). The newcomers to the cities left their farms to be cared for, more often than not, by their ageing parents, or abandoned them altogether. No longer able to undertake the physical work necessary to make the small 5 ha parcel yield its modest livelihood, those who have remained on the land have preferred to sell their plots to the newly prosperous in the towns for summer vacation homes. Erosion, soil deterioration, and the permanent loss of arable land have been the results.

For their part the cities have expanded with very little planning, or with un-anticipated rapid growth outstripping planning capabilities (*Politika*, 10 June 1984: 63). Although investment in housing and utilities represents around 10 per cent of realized investment by the social sector, over 80 per cent of all Yugoslav housing is built privately. Much of it is built according to whatever regulations have been drawn up for designated areas, such as for an urban residence or vacation home. Construction costs are extremely high for most Yugoslavs and the purchase of a ready-built house is beyond their pocketbook. The housing problem in the cities is acute. People have to wait years for a cheap rent, city or enterprise-owned apartment. Connections pay: like elsewhere in the former Communist countries, the best municipal housing is awarded to those with the best party and government connections. Unable to find housing and with no money to purchase it, the immigrants to the cities stop at the city limits and put up shanty towns typical of urban development in the Third World. City inspectors are reluctant to evict or convict the inhabitants of these primitive

Figure 8.3 Protected areas in former Yugoslavia

Source: Environmental Status Reports: 1990, Vol 2, *Albania, Bulgaria, Romania and Yugoslavia*.

dwellings because of possible public outcry and what would appear to be discrimination against the poor. So the unplanned sprawl expands around the city's perimeter, upsetting the orderly development of urban sewage, water, and transportation systems.

Green areas

Last, but not least, there is the problem of preserving the country's green areas, natural landmarks and unspoiled regions (see Figure 8.3). Here the record is only slightly better. In the immediate post-war period the then centralized government voted to set aside restricted areas for nature reservations and natural landmarks following the example of the former Soviet Union and other East European countries. In 1951, there were three national parks and three regional parks; in 1976 there were fifteen national parks with a total of 301,704 ha and seventeen regional parks. But in 1991 only eight national parks had survived, two of which are classified biosphere reserves under the United Nations 'Man and the Biosphere Programme' (MAB), with another two included among the wetlands of international importance (OECD 1989: 145–9). The two biosphere reserves are the geological wonders formed from the karstic soil of southern Croatia, known as the Plitvice Lakes, and Durmitor National Park in Montenegro.

Economic development needs have made heavy demands on some of Yugoslavia's most beautiful scenery. The Adriatic is governed by three sets of regulations, but industrial and tourist inroads have spoiled much of the coast's natural beauty. An attempt was made in the mid-1980s to build a dam on the Tara River in Montenegro at a point where it would flood the extraordinary Tara river canyon. Organized pressure by experts and scientists succeeded in transferring the dam site upriver from the canyon, but did not succeed in stopping the project altogether.

The self-management system which came into being in the 1960s required each enterprise, whether a national park or a factory, to finance its own budget. This requirement has posed additional problems for the preservation and maintenance of the protected areas. In 1976 only twelve of the fifteen national parks had secure incomes. By 1989 this number had dwindled to eight. The successful were those, like Plitvice, that were able to combine conservation profitably with tourism. Today the Croatian firm Rade Končar of Zagreb is promoting the train it designed for transporting tourists around the Plitvice Park. Run by a non-polluting electric motor, it has a maximum speed of 25 km/hr and can transport 72 passengers in its three cars. The batteries will power the train for seven hours at a time, and are expected to last for ten years (*Tanjug*, 17 November 1990). The record for regional parks is even less impressive. Only two of the seventeen have assured financial means. The scarcity of financial resources has meant that many of these parks are parks in name only with no park facilities. They are constant prey to encroachments from the surrounding forest or agricultural areas.

THE TECHNOLOGY–SOCIAL SYSTEM– ENVIRONMENTAL CONNECTION

Since Yugoslavia's main economic and urban growth occurred in the post-war period, at a time when the consequences of pollution were beginning to be known, the question may justifiably be asked why no environmental precautions were taken. Why does Yugoslavia, which prided itself on its unique self-management system, exhibit the same dreary pattern of pollution which has characterized the economic growth of all the industrialized states? True, Yugoslavia's pollution does not yet match the environmental catastrophes in Poland, former East Germany or Czechoslovakia. The geography of the country precludes a disaster like the Aral Sea. Yet pollution from its chemical plants rivals that from plants in Romania whose toxic fumes were the catalyst for the first Bulgarian environmental protest in the Bulgarian city of Ruse across the Danube River. If environmental degradation is not as severe in Yugoslavia as in other East European countries, credit goes to geography and an economic crisis in progress since the 1970s, not to self-management or the far-sightedness of the Yugoslav leaders.

The answer lies in the interconnection between technology, institutions and ideology. The Marxist-Leninist ideology which informed the world's Communist parties proved to be the last expression of the industrial paradigm, not a new interpretation of reality (Pirages and Ehrlich 1974: Ch. 1). The industrial paradigm celebrates man's dominance of nature, views economic and material development as progress, conceives organizations in terms of hierarchy, and praises hard work for economic self-betterment. The Marxist-Leninist additions were Lenin's three concepts of the dictatorship of the proletariat, democratic centralism and the leading role of the party, and Stalin's use of revolution from above (forced collectivization and industrialization) to effect economic growth. These four concepts enabled the Communist parties to assume the monopoly of all aspects of the societies they ruled, particularly the economy. Most important to this chapter, they locked those societies into a political/economic vice which permitted no alternatives for fear the Communist hierarchy (*nomenklatura*) would lose power.

The Yugoslav variant of this ideology – socialist self-management – compounded the political rigidity of the one-party monopolistic state by creating regional equivalents: six republican one-party monopolies and local industry-government–party organizations. The result was that the technological rigidity imposed by the nature of industrial technology was further constrained by the politics of political fragmentation. The country may have been called Yugoslavia – but that country never had a common market, common economic development, or a common culture. The National Liberation War, which was the ideological foundation of the post-war state, meant different things to different nationalities. After the adoption of a fully-fledged system of self-management in 1974, a common social system and a common government gradually disappeared. The elections of 1990 demonstrated that there was consensus over virtually nothing in Yugoslavia and no love of common country to build a common future. Self-management had become one more variant on the old theme of ethnic rivalry made more severe by Communist one-party rigidity.

If the ideology had operated in isolation, then Yugoslavia and the other former Communist countries would have been able to resolve their economic problems and to have taken care of their environmental problems by the simple reformation of legislation and the immediate changeover to democracy and a market economy. But the interlock of technology and infrastructure stood in the way. The relationship between technology, the social system, and the environment may be characterized by four main elements. The first is that throughout history the type of technology prevailing at any time has been closely linked with the political, social, and economic systems. Like all systems, the social system is an ever-changing mix of stability and resilience, to use ecologists' terms. In the past systemic tendencies to stability have prevailed except in the face of a violent overthrow of the government or technological innovation. Today the pace of change has quickened. In the West the microchip revolution facilitated not only the accurate monitoring of our economy but the precise documentation of the impact of our technology on the environment. The problem is that the technological system which produces our pollution has shown itself to be particularly intractable and inflexible. The rigidity of the industrial system has its origins in the energy infrastructure where power plants may last as long as forty years, where long lead times are necessary to develop new energy projects and where there are entrenched public perceptions of cost, need and environmental acceptability (Davis 1990). When we talk about changing an energy technology we are talking about changing an entire economic and social system.

The second feature of technology is that civilization as we know it has evolved from the exploitation of the most readily available and least costly (in terms of energy needed to access it) energy use linked to specific technological inventions and to increasingly costly energy systems linked to new technological inventions. The first stage of the industrial revolution was based upon coal. The exploitation of coal was made possible by the invention of the steam engine, which brought automation to the mines. Coal in turn powered the steam engine, and together they made mass production economic by reducing the cost of land and sea transportation. The concentration of resources at the mines and in the new factories, necessitated by the powering of the steam engine with coal, made rapid industrialization possible. Resource concentration also contributed to the rise of the modern polluted and polluting city.

The second phase of the industrial revolution began at the end of the nineteenth century with the developments of electric power, petroleum, and natural gas as energy sources. The key inventions were the internal-combustion engine, and the chemical and metallurgical industries. The environmental impact of their widespread application was the increased pollution of the inner city by chemical and metallurgical plants, and the growth of suburbs as the better-off sought to escape urban blight, first along the railroads and then along the newly paved highways. In a short time, a network of highways criss-crossed the industrialized world where only wagon paths had been before. Western Europe and the east and west coasts of the US were transformed into one vast suburbia. The third stage is the age of the microchip, biotechnology, and advanced materials. The information and automation age has globalized national economies, enabling us to transmit data around the world in seconds. Companies can order parts made in one part of the world, have them assembled in another part, and sold in yet a third part. Equally important we have gone to the moon and seen 'spaceship earth' in its entirety from afar. Technology has forged a global interdependence that has undermined our former concepts of national sovereignty and international behaviour and heightened our vulnerability as passengers aboard a fragile and unique planet (Rosenau 1990).

The third element in the relationship between technology, the socio-political system, and the environment is that the benefits of technology increase linearly in proportion to its scale of application, whereas the environmental and social disjunctures resulting from the application of technology increase nonlinearly as the scale of application increases. By the time the negative environmental effect registers in public consciousness, society is already accustomed culturally and institutionally to the technological benefit. As a technology matures it tends to become more homogeneous, less innovative and adaptive. Success freezes the corporate enterprise into a mould dictated by fear lest departure from a successful formula jeopardizes aggregate capital investment, marketing structure and supporting bureaucracies. After an initial phase of competition, one technological variation begins to win. Economies of scale, marketing, and production give it a competitive edge and research tends to be directed only at marginal improvements. The technology has produced what Brooks terms a technological monoculture (Brooks 1986). The new technology and its supporting systems (the marriage of the automotive and petrochemical industry) constitute a more and more self-contained social system, unable to adapt to the changes necessitated by its success.

A final factor is that all human action on the environment tends to reduce variability. The domestication of animals, for example, was a selection of the most tractable of the wild creatures. A lawn is composed of only a few of the many varieties of grasses. Technology is a highly selective process. By contrast, natural ecosystems, among which human society may be included, are marked by high variability. Nature is not neutral or simply acted upon. Human management in its effort to maintain the steady state of an ecosystem impacted by human activity provokes a response from the ecosystem as it tries to adapt its patterns of behaviour in the face of disturbance. Because of the delay between our selective management and nature's response, we become aware of nature's reaction through the element of 'surprise' (Holling 1986). This surprise may be an unexpected discrete event (Chernobyl), discontinuities in long-term trends (industrialization), or the sudden emergence of new information into the political consciousness (global warming). Discontinuities may not manifest themselves in a single generation. The negative by-products of our rough predictions may not be known until they appear as a surprise to future generations (Orr and Soroos 1979).

Our ability to respond to surprise depends on the resilience of our human institutions. The tendency for technical monoculture to develop huge supporting bureaucracies to minimize risk and enhance survival increases the stability of the system at the expense of resilience. Thus the logic of technological development is for conservatism in institutions and ideology in all socio-economic systems after the initial period of innovation and exploration. What the East European experience demonstrates is that given similar technologies, difference in socio-political system contributes to changing the speed with which these technologies' environmental disbenefits manifest themselves by surprise and possibly the scope of the surprise itself. More particularly in those countries that experienced Communist rule, system conservatism shortened the time period within which the technological monoculture evolved towards paralysis and the system collapsed. While political and economic pluralism has produced environmental degradation, none of the countries exhibiting this kind of system has yet approached the ecological barrier, as is the case in parts of Czechoslovakia and former East Germany.

THE YUGOSLAV JUGGERNAUT

Even Yugoslavia's most economically developed republics are not yet at the threshold of the third phase of the industrial revolution. More importantly there are profound differences in the level of development between the north and south. While Croatia and Slovenia may be at the brink of the information revolution, Macedonia, Montenegro and Bosnia-Hercegovina are more likely moving to the end of the second phase. Because the third phase is not yet stabilized anywhere in the world in a permanent institutional form, the introduction of the new technology can only be achieved by innovation and competition.

Self-management

To hasten passage through the second stage of the industrial revolution, Stalinism created a technological monoculture upon the introduction of a technology. Self-management was a reduction in scale, not a change in form, from the Stalinist model, focusing all the rigidity at the local level which was least capable of adaptation without causing enormous social dislocation. By the beginning of the 1980s self-management found itself in the same situation as the Stalinist centralized economic system, unable to provide the necessary climate for innovation and the system collapsed. Yugoslavia is now in a phase of economic and political adaptation to social surprise. The more advanced republics wish to accelerate the movement towards democracy, privatization and a market

175

economy to gain access to the new high technology. The less developed republics, with Serbia at their head, cling to the old forms of economic and political organization invigorated by frantic appeals to nationalist symbols, quarrels, and dreams going back almost a century.

Environmental surprise did not exert as catastrophic an effect on Yugoslavia as it did on the other East European countries. Nevertheless, it became increasingly visible during the 1980s, and was catapulted into prominence by Chernobyl. The author has analysed elsewhere the structural reasons for Yugoslavia's environmental mismanagement and they need only be summarized here (Jancar 1987). Foremost among these should be mentioned the monopoly of the republican Communist leaderships of the institutions of state, law, the economy, and information. If the Soviet Union could formerly be viewed as a centipede with a thousand lower administrative units reporting to the central ministries in Moscow, then the metaphor for Yugoslavia was a hydra, the many-headed monster of Greek mythology. According to the 1974 Constitution, whatever laws or economic decisions were made at the federal level could only be made with the consensus of the republican leaderships. The Chamber of Republics and Provinces of the bi-cameral federal legislature did not reach a decision by majority vote, but each republican and provincial delegation acted as a bloc. Every decision a delegation made had to secure the approval of the corresponding republican or provincial legislature.

Self-management enabled the republics to decide their own internal affairs, assigning the federal government a very limited set of powers when compared with the American system (see Article 8 of the Constitution of the SFRY). In the environmental area the federal government had jurisdiction only in matters concerning two or more republics and in international transboundary pollution, as for example the 1974 law on sea coasts or the 1977 law on maritime and internal shipping (*Sluzebni list*, SFRJ, 10 January 1974). The republican and provincial assemblies in agreement with the Chamber of Republics and Provinces decided the most crucial questions: the size of the federal budget and economic issues (such as monetary regulations, price control, credits for the underdeveloped areas of Yugoslavia, and taxation). In addition, the republican assemblies decided independently of the Federal Chamber questions relating to the supervision of federal administrative agencies and the formulation of enforcement policies for the implementation of federal statutes. These provisions gave the republics jurisdiction over both federal and republican environmental enforcement organs (Article 286 of the Constitution). Limited federal jurisdiction was thus juxtaposed against substantial republican power.

Self-management also gave considerable scope to the districts (*opštine*) for the protection of the local environment through the passing of regulations, the planning and implementation of pollution control measures, and the organization of their own environmental inspection systems. However, every local government unit was required by law to be self-supporting: in other words it had to raise the money to fund local social welfare programmes (pensions, medical care, education, public transport, water systems) from the revenues generated by local industry. At the same time, localities had little say regarding the projects they had to fund. The required social welfare programmes and the share of money to be given to the corresponding social welfare organization formed to administer them were determined by the Federal Assembly.

Environmental protection was not a mandated public service. As a consequence environmental programmes could not benefit from an assurance of permanent funding from public monies. Each programme had to seek its own funding. Since local government and local industry had substantial portions of their budget already allocated to mandated fields, there was little incentive to undertake costly environmental improvements – especially in the less-advantaged communities. The pay-your-own-way system resulted in the richer republics and districts being able to provide better services and environmental protection than the poorer. It also contributed to a special relationship between local industry and local government, with the latter guaranteeing the former's survival through dubious bank loans and other inefficient practices in order that the former might guarantee the latter's ability to meet its mandated budget. In a situation redolent with the company/town mentality there was no space for environmental considerations. The mining town of Zenica is an example. Heralded as an example of self-management in the 1960s, by the 1990s it was known as the most polluted town in Yugoslavia.

Finally, self-management resulted in a parcelization of environmental programmes, as complete as the fragmentation of the economic enterprise. If a project was to be undertaken that extended beyond the boundaries of one enterprise or locality, the project partners had to draw up a self-managing 'compact'. This agreement specified the contracting parties, the work to be done, and what each party was to contribute materially and financially to the project. Negotiations of this kind are highly sensitive. Organizations were reluctant to sign a compact unless there was some impending ecological disaster or economic benefit to be derived. The two major inter-republican environmental agreements – the 1980 Neretva and Sava river compacts – remained largely on paper, unimplementable because the vague wording committed the contracting parties in principle only (*Vjesnik*, 5 June 1982). Self-management entrenched local interest at the expense of the common interest. Agreements did not become operative because of conflict between contracting parties over environmentally irrelevant issues. These ranged from disputes over the location of a regional park's headquarters to fear of economic disadvantagement, which delayed the signing of the Una river 'Forever Wild' compact, and led to deliberate procrastination in fulfilling the compact after signature to make a political point, as was the case in the Sava River Compact (*Vjesnik*, 26 January 1983). There have been two success stories: the cleaning up of Lake Palić in the Vojvodina near the old Hungarian resort town of Subotica, and the signing of a compact on a sewage system for Lake Ohrid in Macedonia. The first project, however, was totally local and required no outside monies or materials. The second succeeded in the first stage of construction but ran into financial difficulties and skyrocketing costs because of rampant inflation of the dinar. The latter's initial success may be attributed to the co-operation extended by international organizations, such as OECD, UNESCO, and the World Bank which agreed to put up 40 per cent of the capital.

Yugoslavia today is a prisoner of the self-management system. The social and environmental disadvantages accruing to the successful introduction of a technology in a free society were both intensified and hastened by a prior decision to freeze the whole system into a mould informed by ideological principles. Far from remedying the defects of the Soviet Communist system, self-management rendered them even more difficult to control. Because the economic system provided the security and control function which maintained local, republican and federal rulers in power, Communists at all levels of government found themselves in a situation where they could not innovate without bringing the whole system down. At the same time that Yugoslavia needs economic reform to compete on the world market and improve living standards at home, regime stability depended on the maintenance of the local and republican monopolies engendered by the self-management system. Self-management forced reduction in variability by insisting on identical forms of economic and political institutions in all the republics held together by the confederal thread of the party. The deliberate interruption of the process of competitive technological development characteristic of the early stages of industrialization consolidated a plethora of technological monocultures and destroyed civil society. Administrative paralysis set in.

Energy

Nowhere is the connection between technological and political rigidity so evident as in the field of energy. The country has only limited options regarding choice of fuel, and the huge indebtedness which has accumulated during the 1980s ($20 billion of foreign debt and unknown sums of internal debt) has not improved them over the years. Yugoslav efficiency in energy use, while better than that of the other former Communist countries, is bad in comparison to that of the countries of Western Europe. Between 1975 and 1987 total energy requirements per capita rose 54 per cent, the highest of all the OECD countries. In 1987 the energy required to produce one dollar of GNP in constant 1980 US dollars was 1.3 times more than that needed in Greece, the least efficient energy user in Western Europe (World Resources Institute 1990: 317). To meet its energy requirements Yugoslavia has tried to increase domestic production. In 1987 total Yugoslav production of fuel was 1,046 petajoules. By contrast total consumption was 1,663 petajoules (OECD 1989: 229, World Resources Institute 1990: 247, 319). Between 1977 and 1987 overall consumption increased by 38 per cent and production by 39 per cent. The largest increase in domestic production was in that of coal which increased 59 per cent over the 1977 figure. During this period the nuclear power plant at Krško on the border between Slovenia and Croatia was also brought on line. With virtually no oil reserves and very modest gas reserves Yugoslavia has been forced to develop the highly polluting lignite and brown coal reserves or else to import.

In the 1960s Yugoslavia contracted to build a pipeline from the Hungarian border to Belgrade in order to heat the city with imported Soviet oil rather than polluting coal. Construction of the pipeline was halted when the government ran out of funds. Polish coal was an alternative until its import was stopped – once again due to lack of money. The collapse of the Council for Mutual Economic Assistance (CMEA) and the economy of the former Soviet Union meant the virtual end of a trade relationship which accounted for 45 per cent of Yugoslavia's trade with the outside world in 1987. Like other East European countries Yugoslavia can no longer count on Soviet oil unless it finds the hard currency to pay for it. Despite these setbacks the country remains dependent upon oil imports, with over 45 per cent of its fuel consumption in oil. Since 1980 oil imports have decreased only slightly. In 1987 they met almost a quarter of the country's energy requirements (OECD 1989: 221–3). Experts argue that replacement of imported oil with domestically produced fuel is a costly business. In 1982, it was estimated that it would cost twice as much to replace a million tonnes of imported oil with domestic fuel (*Economska politika*, 25 January 1982: 19–20). One of the reasons for the high domestic development cost is that much of Yugoslavia's imported oil is paid for by reciprocal purchase of goods and services. One energy expert told the author that less than 20 per cent of the imports required payments in dollars.

To reduce foreign dependency the federal energy plan up to the year 2000 called for the accelerated development of coal and hydropower, as well as conservation measures. The major reserves of coal lie in Kosovo, the autonomous province whose Albanian majority has been locked in struggle with Serbia since the beginning of the 1980s. In the mid-1980s, given the instability in the area, hydropower appeared to be a better alternative. The need to develop domestic sources of energy was directly responsible for the willingness on the part of three republican leaderships and the federal government to flood the Tara river canyon. During the winter of 1983 a cold snap made such a draw on available energy supplies that Bosnia-Hercegovina and Montenegro, already energy-poor, were forced to cut off electricity for consumers during certain periods of the day. Neither republic had significant coal reserves. Studies showed that the production of hydroelectricity had not yet been exploited to capacity on the Drina river system and on the Morača, a smaller river flowing west down the Montenegrin mountains into the Adriatic. The two republics, together with Serbia, were determined to explore the possibilities for hydropower development. In January/February 1984, the Serbian power company allegedly cut off electricity to Serbian consumers to show solidarity with its counterparts in Bosnia-Hercegovina and Montenegro. Public concern was enhanced by reports from the mass media. In May 1984 the Federal Chamber of Republics and Provinces approved the agreement for the exploitation of the hydroenergetic potential of the Drina and Morača rivers.

Everything had happened relatively quickly. Opposition to the agreement did not get into action until after the bill was passed. For the first time in the history of Yugoslavia a federal agency, the Federal Council for the Protection and Improvement of the Human Environment, co-ordinated an opposition effort concentrated in the officially recognized expert and professional organizations grouped under the Socialist Alliance. The effort was successful in getting the government to reconsider. A compromise was reached whereby a dam would not be built at the high bridge over the canyon but up-river where it would flood an area that was outside the MAB reserve. Given the available technology, the commitment for economic development and the industrial ideology of socialism, the government could not forgo construction but it could change location. Five years later massive popular demonstrations forced the Hungarian government to abandon its plans for the Nagymaros dam on the Danube. This seeming victory for the environmentalists may be illusory. Since its election in 1990 the Slovak government has repeatedly confirmed its determination to begin electricity production from the already completed dam their side had agreed to build upstream at Gabčíkovo. The argument is the same one used earlier by the three Yugoslav republics. With no hard currency to import foreign oil what other source of energy is there in a land with virtually no other energy alternatives?

Nuclear power

The need for energy led the Yugoslavs to nuclear power. Nuclear energy, however, has been problematic from the very beginning. The initial proposal to locate the first nuclear power-station on Vir Island in the northern Adriatic had to be abandoned because of local popular protest and the refusal of local government to go along with the plant. There was protest, also, about Krško – not from the town but from scientists in nearby Zagreb. The plant was built and went on line in the early 1980s. In December 1985 the Federal Executive Council approved the purchase of four new nuclear reactors costing billions of dollars. Immediately, the Union of Engineers and Technicians of Yugoslavia organized a round-table discussion of the decision to which it invited foreign as well as domestic experts. The round table revealed deep divisions between those experts for and against nuclear power, generating a wave of controversy about the wisdom of going even further into debt for nuclear power. In March 1986 there were elections to the Federal Parliament during which the nuclear issue attracted considerable public interest. However, when the new government assumed office in early April one of its first decisions was the purchase of one of the reactors (Jancar 1987: 300–1, 331–5).

Then came Chernobyl. Typically, the official media were silent on the amount of radioactivity that fell on Yugoslavia. One Slovenian official told the author he reported the amounts to the IAEO but not to Slovenians. The public responded immediately and massively. In a high school on the outskirts of Belgrade a boy requested permission to circulate a petition asking for the end of nuclear power. The local Communist Youth League office had just changed leadership and the new head was reluctant to rock the boat so early in his tenure. But the boy persisted. The anti-nuclear movement swept through Yugoslavia like a storm. At its head were the informal peace and ecology movements in Slovenia which had the closest contacts with the green movement in Europe and with ecological protest. From Slovenia the movement went to Croatia and thence to Serbia. Eventually 120,000 signed the petition. The nuclear lobby was forced to defend itself on the mass media and in public meetings. In Slovenia the government held a series of town meetings to explain the importance of nuclear power to the republic and to reassure anxious citizens about its safety. The movement did not accept the lobby's explanations. During 1987 and 1988 the anti-nuclear protest exploded, with massive demonstrations all over the country crossing republican lines. In the summer of 1988 the Communist Youth League, the umbrella organization for the youth leagues of the six republics, presented a bill to the Federal Parliament. In December 1988 the Federal Parliament voted a moratorium on nuclear power until the year 2000.

Again these protests cannot be called a victory for the environmentalists. The rigidity of energy needs tied to an inefficient infrastructure remains. The ecology group within the Socialist Alliance in Slovenia went on to form an independent green movement and in the summer of 1989 formed a green party (Greens of Slovenia) to compete in the first free elections held in the spring of 1990. They won 8.8 per cent of the popular vote, campaigning on a platform of closing Krško. Their victory earned them three ministries, seventeen seats in the 240-member parliament, one vice-presidency of parliament, one vice-presidency of the Executive Council of Slovenia, and one place in the presidency of Slovenia. In an interview with the author in the summer of 1990 the Minister of Energy described the building of Krško as a policy of the time of irresponsible giganto-mania, made possible by the political monopoly of the Communist Party. But, he continued, the truth is that the Slovenians cannot afford to close the plant: they need the energy, and plans have removed the closing down of operations to the farther future. Given the huge investment, the energy the plant generates, and Slovenia's present economic straits, it would be irresponsible for the government to shut off nuclear power. Communist or democratic, Slovenia needs energy and Communist or democratic, Slovenia needs economic development.

The Yugoslav experience is a lesson in environmental/economic trade-offs. The hydra of decentralization in a one-party monopoly provides no more basis for competition or innovation than does a more centralized system of the same type. The much vaunted 'socialist market' turned out to be no market at all. Even the wealthier republics such as Slovenia have found out that costly investment in outmoded polluting technologies cannot readily be replaced by new, even cost-lier, less-polluting technologies. Neither infrastructure nor capital is there to exchange the smoke-stack industries of the industrializing period for the high-tech industries of post-industrialism. If technological change is a challenge to

Slovenia, how much more so is it for the far less developed republics of Bosnia-Hercegovina and Macedonia?

THE ENVIRONMENTAL FUTURE OF 'THE YUGOSLAV SPACE'

Yugoslavia is at a crossroads. The odds are that it will disappear as a country to be replaced by new countries formed within the old territory. The new democracies of Slovenia and Croatia have said they cannot work with Communist governments like that of Serbia and they have declared independence. Serbia's nationalism so far has remained rooted in former Communist practice. Macedonia has also declared independence under a Communist government. Only parts of Bosnia-Hercegovina and Montenegro with Serbia appear still willing to reconstitute Yugoslavia. When peace marchers from Western Europe marched through Ljubljana, Zagreb, Belgrade and Sarajevo in the autumn of 1991, only thirty people turned out to watch them pass in Zagreb, then under siege by Serbian units, and only in Sarajevo did the thousands of young people join in singing 'Yugoslavija' (National Public Radio News, 29 September 1991).

What, then, are the chances for improvement of the environment in the changed conditions? Under the Communist federal system, every republic developed some form of environmental administration and regulations. There are over 150 environmental statutes now on the books in Yugoslavia with amendments to existing laws modifying that number each year. The 1980s saw modifications of laws on waste disposal and forest practice, and changes in land laws regarding agricultural practice. Croatia, Serbia and Slovenia passed the most comprehensive legislation, but all the republics have laws covering the major pollution areas. Since the mid-1980s the Five-Year Plans have incorporated environmental procedures into the planning system, requiring enterprises to report environmental conditions and making the granting of construction loans contingent upon the inclusion of appropriate pollution controls in plant plans for all new industrial development.

The problem is that, as in the other former Communist countries, the laws were not obeyed. Enterprise managers chose to import 'dirty' but less expensive foreign machinery rather than pay for the more expensive equipment even though they pay no import tax on pollution-control devices. Short-term gain here, as everywhere in the industrial world, prevailed over long-term benefits. Second, there was little standardization of environmental norms and practices. Slovenia approved the highest and more complete set of environmental standards, but efforts to integrate standards and practices ran aground on the shoals of nationalism. Starting with 1974, each republic has had control of environmental management within its territory. Despite such developments as the interrepublican 'self-managing association of interest' (or SIZ as they were called in the 1970s) and the creation of a Federal Council on the Environment and Territorial Planning, republican control of republican environmental affairs proved a

formidable impediment to the formation of a coherent Yugoslav federal environmental policy. The interrepublican agencies served as forums for the exchange of
interrepublican environmental concerns, but they could not impose environmental obligations upon unwilling republics. As a consequence the more
advanced republics of Croatia and Slovenia reached independence with a more
developed and better enforced organizational and legal environmental management structure than did the less developed republics. Even in Slovenia, however,
the observance of environmental regulations was patchy and uneven. Third,
comprehensive data on environmental pollution in Yugoslavia has been very
poor and Yugoslav entries in international compendia of environmental information continue to be erratic.

Since the end of the 1980s environmental management has increasingly
centred on developing independent mechanisms to address the probability of
disassociation from the federal republic. Slovenia undertook a review of environmental conditions, which it published in 1990 in issues of the official legislative
gazette *Porocevalec* (2 February 1990 and 3 March 1990). The review is
intended to serve as the information base for the eventual formulation and
passage of a complete set of environmental laws which will be geared to market
economic conditions and Slovenian independence. That same year Croatia drew
up a law on solid waste disposal, modelled on European Community guidelines
(*Bilten* (Zagreb), June 1990). The law was to be followed by similar legislation in
other areas. In Serbia a comprehensive draft law on environmental protection has
been before the legislature. In May 1991 the newly formed Ecological Council of
the ruling Socialist Party of Serbia, the renamed Serbian Communist Party,
recommended the establishment of an independent Serbian Ministry for Ecology
with full power to monitor and co-ordinate environmental regulations resulting
from the passage of the new law (*Politika*, 28 May 1991: 6).

At the end of the 1980s Croatia and Slovenia began to reach out and establish
regional environmental ties. Each joined the Alps–Adria regional environmental
programme. Both prepared environmental assessments for the programme and
now look to the programme for financial and material assistance in improving
environmental conditions. Yugoslavia as a whole was accepted as a participant in
the EC FAR programme for Eastern Europe and in January 1991 experts identified three groups of projects as priorities for the united country. These included
the standardization of environmental norms throughout the republics to bring
them up to European Community standards based on a new study of environmental conditions throughout the republics. The FAR programme offered some
$23 million to Yugoslavia after the federal government had determined its
priorities among some forty-eight different important environmental projects.
Among the projects to receive funding were air pollution control in Belgrade,
Zagreb and Ljubljana, then the Vardar valley and the Sava river (Cerović, 19
January 1991. But Croatia and Slovenia's declarations of independence coupled
with the outbreak of fighting put all projects on hold; the US Environmental
Protection Agency also halted its projects. Until the question of independence is

resolved there will be little foreign support or investment in environmental projects.

Environmental awareness

Within each republic environmental awareness varies considerably. Like all the Communist countries, information about the environment was censored for a long time and there was no transrepublican official environmental organization. Self-management enabled local residents to protest local issues such as the public referendum against the nuclear power plant on Vir Island, the public referendum for the sewage system on the Yugoslav side of Lake Ohrid, and the co-operation given by the local community to the cleaning-up of Lake Palić. But the public in general only began to be sensitized to environmental issues after Chernobyl – and in different degrees in the different republics.

Environmentalism has a long history in Slovenia, dating back to the 1970s when university students braved loss of career to demonstrate on environmental issues. The Slovenian movement developed its philosophy from the German Greens, but its organization was rooted in its own experience. In the first years, when all grass-roots expression was considered suspect by the leadership, students gained recruits to the movement by word of mouth only. This practice was followed all the way up to the end of the Communist system. Word of planned demonstrations was also spread by word of mouth, while the mode of demonstration was left up to the individuals in the different local groups. Gradually the network expanded. In the mid-1980s the movement followed the peace and feminist movements in accepting the umbrella of the Slovenian Communist Youth Movement (CYM). The movement broke with the CYM in 1989 to form an independent organization. The new organization continued its method of personal recruitment, information dissemination, and suggestions for action. It was one of the major participants in the mass demonstration in the main square of Ljubljana protesting the prosecution of a Slovenian student who refused to do compulsory military service and demanding that his trial take place in Slovenia. In the summer of 1989 the Greens decided that conditions were ripe for them to form a party even though they had no legal authority to do so. The party was based on the democratic and networking principles of the informal organization with a great deal of latitude left to the local units to decide what actions to undertake. When the Communist Party called for elections the Greens decided to run candidates. In early 1990 they joined the umbrella opposition group DEMOS which stood for democracy and independence.

The Greens' electoral programme set forth the party's political ideals: democracy, private property, economic competition, and a moderate nationalism. Behind these values lay the post-materialist values of the European Greens. As a representative of the Greens told the author, the Greens promoted their values as being global, not related to the interests of specific groups as were the values of the old and reformed political parties. In addition, the Green focus was on the quality not quantity of life. Quality of life was best improved by opposing all forms of bureaucratization including technology. Like Greens elsewhere the Slovenian Greens called for a shift to alternative energy sources, a less hierarchically structured mass society, and a return to the valuation of the individual. The Greens saw the end of the monopoly of the Communist ideology as the definitive break with bureaucratization in Slovenia. This ideology had celebrated, in the opinion of P. Jamničar (Secretary-General of the Slovenian Greens, interviewed in 1990), the destructive aspects of modern society: bureaucracy, technocracy, and political hierarchical power. For them the West would never adequately resolve its environmental problems until the economic monopolies were brought down.

The fact that the Greens were among the winners in the 1990 election would seem to augur well for the Slovenian environment. But the Greens, like the other political parties, could not escape political reality. When the author interviewed representatives of the party in 1990 they were against disassociation with Yugoslavia on the grounds that independence would mean the militarization of Slovenia and the perpetuation of the technocratic-political hierarchy. However, after the December 1990 referendum went overwhelmingly in favour of independence, all but the most radical parliamentary Greens voted for independence as well. Assuredly the Greens have not forgotten their values, or their dedication to forging a new society in Slovenia, but the fact is that the public is a product of the old industrial paradigm reinforced by forty years of Communist socialization and the historic dreams of nationalism. If the Greens are to remain in power and be re-elected they have to bend before the national-industrialist wind.

In Croatia the environmentalists only began to organize seriously in the late 1980s, after Chernobyl. Before the 1990 elections there were groups formed in virtually every large town in Croatia. After a heated controversy between forming a party or retaining an NGO network, these groups agreed to come together in an umbrella group (the Green Alliance for Croatia) just prior to the elections. Running for office was made voluntary. Groups from Zagreb and Split did run candidates, but only a candidate from Split was elected, primarily with the support of the city's still vigorous Communist Party machine. After the election the Alliance was dragged down into petty squabbles and intrigues between its members. Like all other groups in Croatia its activities have almost entirely been subsumed by the struggle for national independence.

In Bosnia-Hercegovina environmental awareness has been on the increase since the late 1980s. A group calling itself the Una Emeralds Society of Bihač, seat of the Communist government of the 'first liberated area' during the Second World War, won the 1990 United Nations' Environmental Program Award for Ecology. The Society has 24,000 members and since 1987 has managed a school for talented young artists between the ages of 10 and 15 who are invited to spend ten days along the Una River painting nature. The award was given for Una Emeralds' achievements in successfully promoting environmental protection among young people and in urging the introduction of environmental protection as a subject in schools (*Tanjug* (Belgrade), 4 June 1990).

Environmental awareness in Serbia also came to maturity in the late 1980s. In 1989 Vukasin Petrović founded the Ecological Centre in Belgrade for the co-ordination of environmental NGOs in Serbia and throughout Yugoslavia. In 1991, along with others from ECO-Centre (as it came to be called), he organized the Ecological Forum. Asserting that the worst pollution in Yugoslavia was political and nationalistic, Petrović offered the Forum as an organization which would reach out to the most committed people in the environmental movement throughout the Yugoslav 'space' (*prostor*) regardless of political orientation and facilitate communication between them (Stanković 1991). In an interview with the Press Petrović stated that his chief concern was with the ecological future of this 'space', now that Yugoslavia was breaking up. No matter what happened politically, he argued that the ecological problems would persist and demand common, co-ordinated action. The ideas of the Ecological Forum are not in tune with the nationalist slogans heard in Serbia in 1991. The organization did receive funding from the East Central European Environmental Centre in Budapest to hold a conference of regional environmental NGOs in the fall of 1991, but how long can it continue its transrepublican and international outlook before it also is consumed by Serbian nationalism? Like the Slovenian and Croatian public, the Serbian public is not yet ready for post-industrial values. Moreover, the very fact that the Forum is located in Belgrade and organized by a Serb makes it suspect to the green movements in Croatia and Slovenia. The political reality of the 'Yugoslav space' is against Serbian co-ordination of interrepublican communication.

Regional co-operation

If regional co-operation between environmental groups and personalities is frustrated by nationalism, explosive nationalist passions make co-operation between republican governments impossible. Co-operation was difficult under a unified Yugoslavia. It will take years for co-operation to resume and it is doubtful that it ever will resume within a South Slav framework. Croatian and Slovene environmental practitioners see their best hope in republican membership in the European Community. Like many in official positions in the new East Central Europe, they tend to see the European Community as the solution to every problem, be it economic, political, or environmental. Under the confederal umbrella of the Community, they reason, all nations in Europe will find a European home.

Environmentalists in Croatia and Slovenia may expect disillusionment. Western Europe is not about to incorporate the weak economies of Eastern Europe into the Community until each country has made the transition to a viable market economy and until each has brought its economic and environmental standards up to those of the Community. How Croatia and Slovenia will do that without substantial Community or US assistance is problematic. Pushed by Germany, the Community only reluctantly recognized the republics in January 1992; the United States did not. The alternative is to meld the ideas of the Slovenian Greens and the Serbian Ecological Forum, where environmental activists would unite to promote post-industrial values, technologies, and organizations. This option remains bleak as long as each republic stays in the institutional strait-jacket of Communist industrial organization, from which it cannot escape without substantial Western technological and financial aid. And Western aid cannot go forward in the unstable climate of civil war. The outbreak of violence, first between Slovenia and Serbia and then between Croatia and Serbia, can be interpreted as the ultimate social surprise of a failed ecosystem. The man-in-the-street knows violence is destructive. Croatian and Serbian youth are hiding to escape the draft. But this society is caught in the vicious web of early industrial technology imposed and maintained by a monopolistic economic system controlled by persons educated in nineteenth-century industrial values: the hierarchy of race and religion, power determined by military control, and progress measured by quantities of units of output.

CONCLUSION

As stated at the beginning, the Yugoslav story is a lesson for the West. The recession in the United States is indicative of the persistence of the industry–government–technocratic union which has characterized socialism and capitalism alike. Countries no longer develop in isolation but are part of the global system of interdependence. In past centuries the great powers imposed their political solutions on the small powers. The West is continuing to try to do that in Yugoslavia. Nationalism is no guarantee of democracy. And we have seen how the Greens, like other groups and parties in Croatia and Slovenia, have succumbed to the nationalist line. Nevertheless, to insist that Yugoslavia should remain united is to endorse a return to coercion and political monopoly. Under that system the lands of Yugoslavia were rapidly and systematically degraded. Experience in the West demonstrates that environmental protection and pollution prevention is most effectively implemented in countries with strong stable democracies. By this criteria the West would appear to have a greater interest in fostering democracy in Yugoslavia rather than unity. Democracy, we suggest, requires a new alliance between consumer, producer, and environmental groups in the area interested in constructing the future of the planet, not in settling old nationalist scores. If the new alliance is forged in all the republics of Yugoslavia, Dubrovnik need no longer fear the guns offshore and sustainable development in a clean environment may become a reality rather than an unattainable dream.

9

ENVIRONMENTAL ISSUES IN THE NITRA VALLEY OF SLOVAKIA

D. Turnock

EDITORIAL NOTE

This is the first of two additional contributions which have been made on the occasion of the launch of the paperback edition. The aim of this chapter is to document, through a specific regional study, the approach to environmental issues under the transition which is now much further advanced than it was when the book was first written. The region is a classic pollution blackspot on account of lignite mining, electricity generation and associated heavy industry. Although some protective measures were taken under Communism, the priority for conservation is now enhanced, and the current approach is highlighted through action to reduce pollution at individual plants, to modify the economic structure, to protect fragile ecosystems and to introduce a stronger environmental dimension into planning for all areas (Andrews 1993). Changes in the planning of settlements are also dealt with. The same issues are then dealt with more broadly across Eastern Europe as a whole in the final chapter.

INTRODUCTION

Like the better known Hron and Váh valleys the Nitra Valley started to assume a pronounced industrial character in the nineteenth century. The study area concerns the middle and upper sections of the valley, embracing the present counties of Nitra, Prievidza and Topolčany (Table 9.1). They have a combined population of just over half a million (513.5 thousand in 1990), with five towns of more than 10,000 population (Nitra 90.7 thousand at the beginning of 1990, Prievidza 52.4; Topolčany 37.6, Partizánske 26.7 and Zlaté Moravce 16.3). All three counties support substantial industrial activity but the importance of agriculture becomes progressively stronger when attention shifts from Prievidza (the uppermost section) through Topolčany to Nitra which has the best physical resources. Unemployment rates have risen steeply through the transition years though the situation is close to the average for Slovakia. Prievidza came seventh in the list of thirty-eight counties as regards the unemployment rate in 1990 (i.e., thirty-one counties had higher rates) while Nitra came eighth and Topolčany

Table 9.1 Study area profile, 1990 (except where otherwise stated)

Indicator	Nitra	Prievidza	Topolčany
Population (000s)	212.1	139.7	161.7
Employment (000s)	86.5	61.5	67.1
Employment in Industry (000s)	26.1	31.5	30.4
Employment in Agriculture (000s)	16.4	5.7	9.7
Industrial Establishments	78.0	44.0	42.0
Industrial Production (billion crowns)	6.47	6.79	6.98
Ditto (thousand crowns per capita)	30.7	55.9	61.9
Agricultural Production (billion crowns)	2.42	0.74	1.81
Crop Production (billion crowns)	1.09	0.27	0.77
Unemployment (%)	1.14	0.75	0.94
Ditto (1993)	14.70	13.64	15.87
Foreign Investment (% share to 6/1993)	5.47	0.06	0.03

Source: Statistická Ročenka Slovenska 1991.

thirteenth. In 1993 the positions showed some slippage with Prievidza fourteenth, Nitra sixteenth and Topolčany nineteenth. It is suggested that delayed privatization of state-owned mining and manufacturing facilities in these areas may have affected the situation. Nitra has attracted considerable foreign investment (5.47 per cent of the total for Slovakia) but Prievidza and Topolčany have done badly. Investment has been directed not so much at large enterprises heavily dependent on orders from the FSU, but at smaller firms in Nitra county (for example Samsung of South Korea have invested in Calex at Zlaté Moravce). In Prievidza only the food industry has attracted investment (from Nestlé). However, reduced industrial activity (Smith 1994) provides a window of opportunity for a revamping of environmental policies and it is evident that there is now a momentum for change, both nationally and locally.

INDUSTRIAL DEVELOPMENT

The industrial growth of the region may be related to the arrival of the railway in the late nineteenth century. Taking off from Palarikovo (near Nové Zamky) on the Budapest–Bratislava main line of 1851, the railway reached Nitra in 1876, Topolčany in 1881, Chynorany in 1884, Priedvidza in 1896 and Handlová in 1913. An extension from Handlová through to the Ostrava–Žilina–Lučenec–Miskolc main line of 1872 was completed only in 1931. However, long before this, manufacturing traditions had been established in the northernmost part of the valley at Prievidza, Nitrianske Pravno and Valaská Belá. Particularly significant was the glass industry, comparable with the long-established crystal glass industry of the Czech lands. This industry still continues, making use of the silica found at Zliechovsky Gapel, and the pure limestone that is still worked in

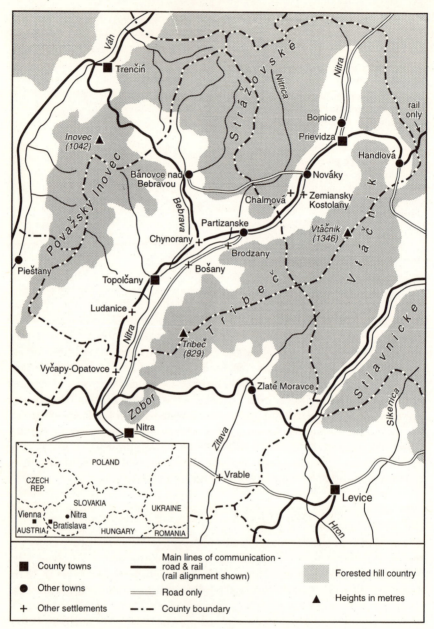

Figure 9.1 The Upper Nitra Valley

areas such as Bystričany. The limestones of the area were also useful for other purposes including agriculture and the carbide and sugar refining industries. Mention should also be made of gold-washing activity near Nitrianske Pravno, which created the first problems of dereliction through the accumulation of waste material.

Under capitalism the local forests supported processing capacity at Dolné Vestenice and Velke Uherce (furniture), while forest fruits were canned at Prievidza and the availability of tanning products in the forests paved the way for the tannery at Bošany. The local production of potatoes gave rise to starch factories at Brodzany and Chynorany, and the area also boasts a range of useful building materials. But, more significantly, the upper section of the valley was transformed by lignite mines and quarries in the areas of Cigel, Handlová and Nováky (Szöllös 1993). After the formation of the Czechoslovak state some of these industries were unable to compete with more efficient producers in the Czech lands and several temporary closures took place. However, the lignite was much in demand and this sector of the local economy continued to grow. Output fluctuated according to geological conditions, with production from Cigel being particularly steady (Szöllös 1994). The fuel was initially taken away by rail for use elsewhere and in 1957 there were still substantial railway flows heading southwestwards towards Nitra and Bratislava as well as eastwards to Košice via Ružomberok and southeastwards to Zvolen and Lučenec (Ivanička 1973, Map 11). But major industrial development started in the inter-war years with the establishment of a shoe factory at the confluence of the Nitra and Nitrica rivers, a nodal point that was named Batovany (after the name of the company: Bata). It provides a striking example of a strategy that was subsequently adopted under Socialism in other parts of Slovakia (as at Žiar nad Hronom in 1953) to introduce new industrial towns in areas with a scattered rural settlement (of the 'Kopanitsa' or 'Lazy' type; originating in the migration of pastoralists from other parts of the Carpathians) lying at a distance from established urban centres. After World War II, Batovany was renamed Partizánske to honour the local 'Vtáčnik' resistance organization and to reflect the ethos of the new socialist regime. The footwear industry was greatly extended to supply the FSU, which offered a huge and steady market.

The power station in the town burned lignite as did the larger power station of Zemianske Kostolany built in the 1950s, close to the chemical factory at Nováky, which dates back to a project started during World War II to produce war materials in the then-independent Slovakia (Ivanička 1961). There was already a small power station at Handlová where a carbide and ferrosilica plant was also built in the inter-war period. The Nováky chemical plant was subsequently converted to a hydrocarbon base for the manufacture of PVC, plastic and detergents in addition to carbide and soda. The power station, originally of 178.8 MW capacity (made up of seven blocks of 22.4 MW and one of 32 MW) was enlarged by an additional 440 MW (4 × 110 MW blocks) between 1964 and 1975 (though the original capacity was subsequently reduced, after restructuring, to 108.8 MW: 2 × 22. 4 MW plus 2 × 32 MW). Most of the lignite supply – 1.5 million tonnes per annum – comes from Nováky (also Cigel and Handlová, along with Dolina near Velky Krtíš), although Cigel was the main source in the past (Szöllös 1993). This is easily the largest power station in the area and it has been a major supplier to the electricity grid in Western Slovakia (with major supplies to the nearby Nováky chemical plant and the Žiar nad Hronom aluminium smelter) (Szöllös 1994). However, there are smaller thermal stations in the area at Handlová, Levice, Topolčany, Žiar nad Hronom and Zlaté Moravce (not to the mention the nuclear projects of Jaslovske Bohunice and Mochovce which, along with the hydro project of Gabčikovo on the Danube, will meet the deficit in the western part of the country).

Growth in the Upper Nitra valley during the 1950s resulted in considerable immigration. This ceased by the 1960s and growth is now the result of natural increase. Industrial development lower down the Nitra Valley has continued with particular emphasis on the chemical industry at both Nitra (plastics) and Topolčany (synthetic fibres); although the linkages extend still further south to the Sala complex close to the Danube. Dolné Vestenice (a village in the Nitrica Valley close to Partizánske) acquired a factory producing synthetic rubber, using premises previously occupied by the timber industry. However, the Upper Nitra area is still the industrial heartland dominated by the energy-chemical complex focusing on the lignite fields, the power station and chemical complex (diversified by the production of building materials at Zemianske Kostolany; 'Poróbetón' cinder concrete referred to below, p. 197), using female labour for the most part to complement the male employment at the chemical plant. There is also a range of local industries in small settlements such as Kamenica pod Vtáčnikom, Lehota pod Vtáčnikom and Lehotka; in the former case a small factory originally opened to process textile waste is now active in producing textile materials for the furniture industry. Then there is also the footwear–rubber complex of Partizánske (with the Bošany tannery and the Dolné Vestenice rubber plant) diversified by small food and glass industries in Partizánske and a range of local industries in the surrounding villages of Janova Ves, Klátova Nova Ves, Skačany, Velky Klíž, Velké Krštenany and Žabokreky nad Nitrou.

POLLUTION AND OTHER ENVIRONMENTAL PROBLEMS

The relatively modern industries have a better pollution record than those in the upper valley, which despite the modest scale of development compared with such principal industrial regions of Eastern Europe as Northern Moravia or Upper Silesia, became a typical pollution blackspot through the dependence on lignite without adequate control of atmospheric emissions and waste water discharges. Prievidza county produces the greatest weight of sulphur dioxide emissions in Slovakia: 123.1 thousand tonnes in 1990 (22.6 per cent of the national total), a situation that will only be eased with the completion of the programme to install filters. As regards the deposition per unit of area, Prievidza comes second to Košice City with 128.2 tonnes per square kilometre. Rates in the lower parts of the Nitra Valley are much lower; falling into the 7.1–14.0 group at Topolčany and 3.6–7.1 in Nitra. Prievidza is again prominent for the quantity of nitrogen oxide emissions: 32.7 thousand tonnes; equivalent to 34.1 tonnes per square kilometre (exceeded only by Košice City according to both criteria) (Drgoňa *et al.* 1993). On this basis Prievidza is one of the areas with an extremely disturbed environment mapped by the Slovak Environment Ministry in 1993 (see Figure 9.2).

The power station consumes 1.5 million tonnes of lignite annually (Drdoš *et al.* 1994) and, before improved methods of treatment were introduced, was mainly responsible for the 229,000 tonnes of emissions discharged each year; consisting of sulphur dioxide and arsenic as well as ash and dust. The lignite has a sulphur content of 1.36–3.56 per cent and an ash content of between 20 and 40 per cent. The emissions have given rise to health problems, with chronic cases of arsenic poisoning. Each year the chemical plant of Nováky released 6,000 tonnes of waste calcium carbide and generated 1.2 million tonnes of waste material. Again, this affects public health, since about one hundred people in Nováky were found to be affected by diseases related to toxicity (Krajčír 1993). Studies in medical geography are continuing, although links with pollution are not entirely clear in many cases (Drdoš *et al.* 1994: 142). The chemical plant also released into the Nitra River some 11.0 million cubic metres of waste water containing chlorides, chlorinated carbohydrates and insoluable substances. But water pollution also arose from the emissions which affected the ground water (along with inorganic fertilizer and pesticide residuals). The situation is most serious in karstic areas with abundant underground water sources where pollution has arisen from tipping; notably the storage of ash at Chalmová. Where dumps occur large quantites of sludge tends to seep into the ground, particularly in limestone areas.

Water pollution has resulted in great damage to fishing and agriculture while some industries such as food processing have been forced to close because of pollution effects. At the same time the soil was contaminated with heavy metals (copper, lead, zinc etc.) and arsenic, even in the northern part of the basin where there is very little industry. Writing on the pollution problems in the 1960s K. Ivanička anticipated the extension to the power station and the diversification of the chemical complex into PVC, but he also appreciated the threat for light industry and expected the starch industry (compromised by the decline in potato growing) to be closed down when the plant became obsolete and to be relocated at either Orava or Turiec (Ivanička 1961: 138). Apparently the factory still operates, but with a reduced production level. However, it should not be overlooked that large-scale industrial development has brought some benefits in terms of water management. There are a number of water storages, which are a positive element in ecological terms through water conservation as well as the flora and fauna sustained: note, for example, the reservoir of Nitrianske Rudno

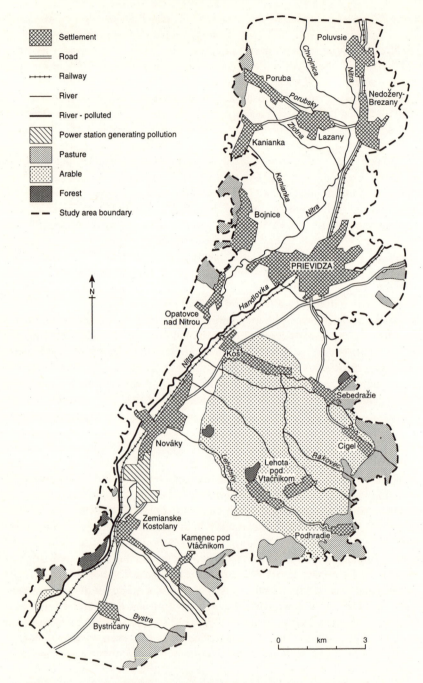

Settlement
Road
Railway
River
River - polluted
Power station generating pollution
Pasture
Arable
Forest
Study area boundary

Poluvsie
Chvojnica
Nitra
Poruba
Porubsky
Nedožery-
Brezany
Zlotná
Lazany
Kanianka
Kanianka
Nitra
Bojnice
PRIEVIDZA
Handlovka
N
Opatovce
nad Nitrou
Koš
Sebedražie
Nitra
Cigel
Nováky
Rakovec
Lehota
pod
Vtáčnikom
Lehotsky
Zemianske
Kostolany
Kamenec pod
Vtáčnikom
Podhradie
Bystra
Bystričany

0   km   3

Figure 9.2 Aspects of land use, settlement and pollution in the Prievidza area (after
Drdos and Szekely 1994)

194

Table 9.2 Pollution in Slovakia: the counties of the study area and other counties badly affected in 1990

County	Emissions in tonnes (000s)					Deposition in tonnes/km² of county territory				
	A	B	C	D	E	A	B	C	D	E
Bratislava	3.8	34.0	7.8	3.2	1.1	10.3	92.4	21.2	8.7	3.0
Galanta	7.9	15.7	8.6	16.6	1.4	8.0	15.9	8.7	16.8	1.4
Košice	30.2	40.4	36.6	111.0	4.3	123.8	165.8	150.0	454.9	17.8
Liptovsky Mikuláš	18.6	14.0	2.6	6.6	1.6	9.5	7.1	1.3	3.4	0.8
Martin	9.8	10.8	3.3	4.0	0.9	8.7	9.6	2.9	3.5	0.8
Nitra	5.3	7.7	2.0	6.5	1.6	3.7	5.3	1.4	4.5	1.1
Nove Zámky	5.4	18.9	3.7	6.4	1.6	4.0	14.0	2.7	4.8	1.2
Povazská Bystrica	23.2	11.1	2.9	5.6	0.4	19.4	9.3	2.4	4.7	0.3
Prievidza	44.4	123.1	32.7	6.1	1.5	46.3	128.2	34.1	6.4	1.6
Spišska Nova Ves	7.5	20.1	1.1	5.7	1.3	4.9	13.1	0.7	3.7	0.9
Topolčany	6.6	11.0	1.8	7.3	1.7	4.9	8.1	1.3	5.4	1.3
Trebišov	31.8	68.3	27.1	6.5	1.5	24.1	51.7	20.5	4.9	1.1
Vranov nad Toplou	2.3	13.2	2.6	4.5	1.2	2.7	15.8	3.1	5.3	1.4
Žilina	7.9	14.8	5.7	6.7	1.6	7.2	13.5	5.2	6.1	1.5

A Solid Emissions; B Sulphur Dioxide; C Nitrogen Oxides; D Carbon Monoxide; E Hydro-Carbons
Source: Statistická Ročenka Slovenska 1991: 67–8.

in the upper Nitrica Valley (Dubcová and Kramáreková 1992). Another important development has been the construction of small hydro-electric projects (for example at Jelsovce and Nitra), which have their origins in the inter-war years.

Lignite exploitation from open pits caused much disruption at Lehota pod Vtáčnikom between 1979 and 1988. Some 9.5 million m³ of overburden were moved to allow 1.6 million tonnes of coal to be extracted. Three large rockfalls occurred in the process and there was great disruption through the need to divert streams in order to protect the lignite pits. This required the laying of some lengths of concrete piping, with consequent adjustments to the water table which was depressed by as much as twenty-five metres in some cases. This surface extraction of lignite consumed some one hundred hectares of agricultural land, most of which was reclaimed as grazing land during 1991–2. Most of the coal is obtained by underground mining, which is less disruptive on the whole, but there have been landslides at the foot of the Vtáčnik Mountains; also many cracks in the soil and subsidence depressions as well. Subsidence has been a particular problem for agriculture (depending on the depth of the workings and the stability of the basement geology): 502 hectares have been badly affected and about half this area has been devastated.

However, modern farming methods have themselves exerted a negative influence. Mention should be made of the pollution (of both air, soil and ground water) resulting from the accumulation of organic waste by large stock-rearing units. Also large-scale vegetable production has created a so-called 'evacuated landscape' with contaminated waters and a strong concentration of biotic elements: also an increased concentration of lead, zinc and other elements. Some 19,000 hectares of arable land have been contaminated and need grassing over. Grassland has also been affected by modern farming since the natural meadows were ploughed and 'improved' by the sowing of new grasses; while the piedmont grasslands at the foot of the Vtáčnik Mountains were drained, with consequent disturbance of the natural drainage system (exacerbated by mining subsidence) and soil forming process.

In the Vtáčnik and Strážovské Vrchy mountains a total of some 7,000 hectares of woodland have been damaged through toxic emissions. Beech, oak and spruce trees have all been affected. Woodlands in the Vtáčnik Mountains are also affected by landslides and by mining subsidence; the latter has resulted in the widespread emergence of a secondary, man-modified, surface. However, the greatest changes to the land are not specific to the Nitra Valley but have occurred throughout the areas of arable farming as a result of creating very large fields. This facilitates large-scale crop production but is extremely harmful as far as wildlife is concerned: hares, partridges and rabbits are no longer so common because habitats (comprising many small patches of woodland and scrub) have been destroyed. As the natural vegetation has been impoverished some plants have become extinct and several species of bird are threatened: hawk, sparrow-hawk and various waterbirds. In addition, marshes were drained and this had a similar damaging effect on wildlife.

Reference should also be made to the buildings of the area. Much of the accommodation is quite new but many of the flats in Prievidza are cramped, while the quality of housing in parts of Nováky is unsatisfactory. The rural areas offer good conditions where they are close to the towns and have attracted a commuter population, often accommodated in small apartment blocks; further away from the urban cores the quality of the accommodation decreases. In Prievidza county all the communities with a below average living environment are rural: Bystričany, Cigel, Kamenec pod Vtáčnikom, Lazany, Nedožery-Brezany, Opatovce nad Nitrou, Podhradie, Poluvsie and Poruba. On the other hand, living space per person is relatively high because of depopulation.

ACTION TO IMPROVE THE ENVIRONMENT

A major area of concern has been the Zemianske Kostolany power station where sulphur dioxide emissions have been reduced by the adoption of the Austrian 'Voitsberg' system, which is now being applied throughout the station. The environment in the immediate vicinity has been much improved and the rising ground on the western side of the valley, overlooking the lake created by the damming of the Nitra River to create a reservoir of cooling and process water, is an attractive place for the construction of chalets which provide weekend recreation for Nováky people. A particular problem has been the storage of ash, which was initially dumped at Chalmová close to the power station. There was a disaster in 1963 when the oldest tip collapsed and large quantities of waste material slumped into the Nitra River, polluting the whole downstream section accompanied by 'complete liquidation of the river ecosystem' (Drdoš *et al.* 1994: 139). The accumulation of ash also affected the thermal spa of Chalmová through the changes in water temperature (a decrease of four degrees centigrade in two springs and an increase of two degrees in another) and displacement of the springs.

Alternative arrangements for the storage of ash have been made in a karst sinkhole (Drienok) some three kilometres away, and this will accommodate the waste for several more years (Jakál 1993). A total amount of 1.2 million tonnes of slag and ash is being deposited each year (Drdoš *et al.* 1994). As a result of the opening of the new sludge beds in 1990 the Chalmova thermal water springs have started to attract renewed local interest; nestling close to the huge ash tip which has now been grassed over, with the benefit of 20–30 cm of top soil obtained from lignite workings and from building sites in Prievidza. However, the new arrangements are not entirely satisfactory because the water that transports the ash in the pipeline from the power station is penetrating the sludge bed and is entering the karst formations. This then impacts on the alluvial plain of the Nitra where former lakes have reappeared and where new swamps have been created. The rise of the groundwater table has launched hydrodynamic processes leading to the collapse of caverns in the phreatic zone (accelerated by the pressure of road traffic) linked with subsidence depressions on the surface. In 1991 a series of depressions was formed on arable land. A longer-term solution to the ash problem (short of the abandonment of lignite burning at the power station) lies in the use of the waste for building materials. As already noted above, this is already happening on a limited scale through the factory for 'Poróbetón' (i.e., porous concrete) in Zemianske Kostolany, which dates to the opening of the power station in 1957.

Lignite mining would appear to have a limited future and some writers have envisaged the termination of all exploitation before the end of the twentieth century (Drdoš and Szekely 1994: 229). The future would seem to lie in conversion of the power station from its present dependence on lignite to fluid combustion, supporting a local manufacturing establishment concerned increasingly with small light industries 'as potential carriers of innovative behaviour' (ibid.). However, the 'smokestack' industries persist none the less; although some changes to the production profile of chemical plants have reduced pollution, while filters installed at the power station in 1991–2 had such an effect that bees returned to the area after a forty year absence. But the Upper Nitra is Slovakia's leading coalfield (currently producing four-fifths of total national output). And such is the continuing political pressure to retain jobs in the coal industry that development is still proceeding in a bid the slow the present fall in production. The village of Koš has been evacuated to make way for an extension of mining in the Nováky area. The villagers are, however, being rehoused in a good situation at Kanianka above Prievidza with scope for individual family homes as well as apartments.

Dumping will have to be regulated and this requires research to establish the extent of the problem. The Upper Nitra Valley contains numerous cases where materials have been dumped in tips or else discharged into landfills and slurry beds. In Prievidza district 213 dumps were enumerated: one hundred and seven for domestic waste (and known to the local authority) plus sixteen spoil heaps, twelve industrial waste dumps and seventy-eight with a mixed profile: a total of 213 dumps in an area of 958.6 km^2 with fifty-two settlements and total population of 138.5 thousand. The Chalmová ash dump has already been referred to; another major dump is at Podubie, used by the Nováky chemical plant. Only thirty-one dumps are actually legal; meaning that one hundred and eighty-two are illegal. Moreover in only twenty-six cases was dumping controlled: in the other one hundred and eighty-seven cases it was indiscriminate. Some of these tips constitute hazards, especially those that are not recorded and about which there is uncertainty as to their content. 'Solid municipal waste dumps, agricultural waste dumps or dumping sites of industrial waste are potential sources of water and soil contamination and they can cause hygienic problems' (Kramáreková and Dubcová 1993: 127). Much tighter control is needed and many dumps should be closed and landscaped. The proximity of most dumps to settlements simplifies the problems of treatment, although in the short term at least the dumps pose a hazard to public health and threaten both agricultural land and watercourses. Of one hundred and sixty-four dumps actually surveyed contamination was noted in twenty-one cases and was considered probable in another eighty-six cases. Contamination can occur through leakage into soil forming material and also erosion and landsliding. Water resources (including the thermal and curative springs) need to be endowed by protection zones while nature reserves and recreational areas also require special attention. Some improvements have been effected at dumps: for example, sewage sludges are dehydrated and subsequently deposited but the handling of agricultural and toxic waste, however, remains unsatisfactory.

Improving water quality in the Nitra River remains an issue. Studies have been carried out to assess purity levels through concentrations of dissolved oxygen and assessments have been made to establish the cost of making improvements. Achieving the lowest possible discharges by using the best available technology would require an annual cost of US $14.4 million (Paulsen 1993: 28). On the other hand, an improvement to the level of 6.0 mg/l would cost only US $6.6 million (and US $2.8 million for a 4.0 mg/l level).

THE BUILT ENVIRONMENT

The built environment will benefit from a higher priority for conservation and architectural quality. Improvements have been effected in housing conditions through the concentration of building for industrial workers in the city of Prievidza rather than the more polluted environment around Nováky. Prievidza has grown very rapidly as both a district town (with status similar to Nitra and Topolčany) and as a dormitory for factory workers who commute daily down the valley. Also within easy commuting range is the village of Kanianka which gained 1,100 new inhabitants between 1987 and 1990 as a result of resettlement from Koš to make way for the expansion of lignite mining. Prievidza's population was 53,424 in 1991 compared with 6,096 in 1950. The cleaner air of Prievidza makes the town a convenient location for light industry, which includes food processing and the manufacture of furniture. Building materials and textiles are also produced. The textile industry is in evidence in the smaller settlements of Nitrianske Pravno and Nitrianske Rudno, while engineering and glass-making are established at Valaská Belá along with the furniture industry at Pravenec. Indeed, the latter installation is the centre of the industry in the district, with the capacities at Kamenec pod Vtáčnikom and Prievidza as affiliated units. There have been improvements in food distribution. In 1961 Ivanička emphasized the inadequacy of local supplies of dairy products which had to be supplied from Bánovce nad Bebravou and Topolčany. He recommended new facilities in the area at Partizánske.

Prievidza has the benefit of proximity to the spa of Bojnice (with a population of just over 5,000) and many of the townspeople (as well as Slovaks from further afield) visit the resort at weekends (while of course there are commuters from Bojnice travelling daily to Prievidza by bus on weekdays). The thermal springs here have been known since the sixteenth century and they are among Slovakia's 2,300 mineral and geothermal springs (1,200 of which have been examined by government institutions and administered by the Ministry of Health). The Carpathian mineral water resources of the Nitra Basin arise where a longitudinal fault creates inlets for waters at Bojnice (and also at Chalmová and Male Bielice in the southern part of the Magura and Strážov Highlands). The akrathothermal waters result from rapid infiltration through limestone, as the water descends through the limestone and dolomite into deeper strata, becoming warm and mineralized in the process. At Bojnice, which has a warm, humid climate, akrathothermal waters form a system of shallow mineral spring channels, welling up from 1,200–1,500 metres below ground (Kris et al. 1995). The thermal lake (30 metres in diameter and 6.3 metres deep, with water at 37° centigrade) gives the greatest water yield. The Cajka baths are still appreciated, and balneal treatment is still used to ameliorate neurone diseases and some mental disorders. The resort is dominated architecturally by the medieval castle, which has many architectural details in the French style as a result of nineteenth-century rebuilding by the Palffy family. The pleasant aspect of walking in the wooded surroundings of the resort has been enhanced by the planting of trees on ground affected by subsidence.

However, the growth of Prievidza reflects a powerful centralizing of both population and industry (milling for example) which has resulted in

Table 9.3 Planning for the Nitra Valley settlements

	Variant I				Variant II			
	1980	*1995*	*2010*	*2030*	*1980*	*1995*	*2010*	*2030*
Population (000s)								
TOWNS	160.6	209.1	258.5	298.0	160.6	197.3	232.0	259.0
Bánovce nad Bebravou	15.3	21.3	25.5	30.0	15.3	17.5	25.0	30.0
Nitra	76.6	102.0	132.0	150.0	76.6	94.0	107.0	117.0
Partizánske	23.3	28.3	31.0	34.0	23.3	28.3	34.0	40.0
Topolčany	31.3	39.0	48.0	60.0	31.3	39.0	44.0	48.0
Zlaté Moravce	14.1	18.5	22.0	24.0	14.1	18.5	22.0	24.0
GROWTH VILLAGES	11.7	14.4	16.0	17.6	18.4	23.7	34.8	58.1
Bošany	4.1	5.1	5.5	5.6	4.1	5.1	5.8	6.6
Ludanice	–	–	–	–	2.7	3.7	6.0	11.0
Preselany	–	–	–	–	1.6	2.5	5.0	12.5
Vrable	7.6	9.3	10.5	12.0	7.6	9.3	11.5	15.0
Vyčapy-Opatovce	–	–	–	–	2.4	3.1	6.5	13.0
OTHER RURAL AREAS	189.3	163.3	146.9	139.1	182.6	165.8	154.6	137.6
TOTAL	361.6	386.8	421.4	454.7	361.6	386.8	421.4	454.7
Housing Units (000s)								
TOWNS	47.6	71.6	94.7	119.4	47.6	68.9	85.2	103.8
Bánovce nad Bebravou	4.5	7.2	9.4	12.0	4.5	7.2	9.4	12.0
Nitra	23.1	35.5	48.4	60.2	23.1	32.8	39.2	47.0
Partizánske	6.6	9.5	11.3	13.6	6.6	9.5	12.4	16.0
Topolčany	9.2	13.1	17.5	24.0	9.2	13.1	16.1	19.2
Zlaté Moravce	4.2	6.3	8.1	9.6	4.2	6.3	8.1	9.6
GROWTH VILLAGES	3.5	4.8	5.8	7.0	5.3	7.8	12.7	23.2
Bošany	1.2	1.7	2.0	2.2	1.2	1.7	2.1	2.6
Ludanice	–	–	–	–	0.7	1.2	2.2	4.4
Preseleny	–	–	–	–	0.5	0.8	1.8	5.0
Vrable	2.3	3.1	3.8	4.8	2.3	3.1	4.2	6.0
Vyčapy-Opatovce	–	–	–	–	0.6	1.0	2.4	5.2
OTHER RURAL AREAS	52.2	53.4	53.3	55.6	52.2	53.1	55.9	55.0
TOTAL	103.3	129.8	153.8	182.0	103.3	129.8	153.8	182.0
Commuting (000s)								
TOWNS	+30.6	+23.8	n.a.	+14.5	+30.6	+25.6	n.a.	+21.0
Bánovce nad Bebravou	+4.0	+4.0	n.a.	+1.0	+4.0	+4.0	n.a.	+1.0
Nitra	+12.1	+7.7	n.a.	+3.7	+12.1	+9.5	n.a.	+9.4
Partizánske	+5.3	+4.7	n.a.	+5.0	+5.3	+4.7	n.a.	+3.7
Topolčany	+4.7	+2.8	n.a.	+1.3	+4.7	+2.8	n.a.	+3.5
Zlaté Moravce	+4.5	+4.6	n.a.	+3.5	+4.5	+4.6	n.a.	+3.5

	Variant I				Variant II			
	1980	1995	2010	2030	1980	1995	2010	2030
GROWTH VILLAGES	+2.1	+3.5	n.a.	+2.8	+0.9	+1.1	n.a.	−6.6
Bošany	+0.5	+0.1	n.a.	−0.3	+0.5	+0.1	n.a.	−0.6
Ludanice	–	–	–	–	−0.6	−1.1	n.a.	−2.6
Preselany	–	–	–	–	−0.2	−0.6	n.a.	−3.0
Vrable	+1.6	+3.4	n.a.	+3.1	+1.6	+3.4	n.a.	+2.6
Vyčapy-Opatovce	–	–	–	–	−0.4	−0.7	n.a.	−3.0
OTHER RURAL AREAS	−32.7	−27.3	n.a.	−17.3	−31.5	−26.7	n.a.	−14.4
TOTAL	0.0	0.0	0.0	0.0	0.0	0.0	0.0	0.0

Source: Statny Institut Urbanizmu Uzemneho Planovania 1984, *Uzemna prognoza velkého územneho celku Hornonitrianska Sidelno-Regionalna Aglomeracia*, Bratislava, SIUUP.

depopulation and a relative decline in services and housing quality in the remoter rural areas. Planning under socialism encouraged the growth of the towns which was achieved through an extension of boundaries as well as an increase in the housing stock and its availability to migrants from the countryside at low rents. In the case of Nitra, Olas (1991) has documented the growth of the 'Nitra Villages' with extensions to the city boundary in 1961 (bringing in Hor. Krskany and Kynek); 1974 (Chrenova and Hrnciarovce); 1975 (Dražovce and Janikovce); and 1976 (Ivanka and Lužianky). In 1991 there were eighty-six official administrative units in Nitra county, three with a population exceeding 5,000 and forty-one with a population below 1,000. In 1921 there were one hundred and seventeen units, only one with a population larger than 5,000 but eighty-three below 1,000 (Olas 1992). The services in the smaller places are very limited, for a study in the Levice county (southeast of Nitra) shows that forty-three of the eighty-three units were in the Grade D Category for services; indicating a level of provision below that for Grade C (nineteen units), which involved a kindergarten, four-year primary school, culture centre, library and sports ground (Baráth 1992). Many Grade D villages claimed none of these facilities, possessing only a church which did not feature in the official classification scheme.

Planning in the 1980s for Nitra and Topolčany counties, along with the adjacent Bánovce nad Bebravou area of Trenčin county, anticipated further centralization especially in the county towns (Table 9.3). But a second variant gave more emphasis to Partizánske as well as to the rural areas where five growth villages were recognized compared with only two in the first. It is interesting to note that while commuting levels were eventually expected to fall the additional growth villages in the second variant were seen largely as dormitories with a daily export of labour. Until 1990 population trends appear to have been roughly in line with the forecasts for the towns, except in the case of Topolčany where the

1990 population of 32.6 thousand was below the 1985 estimate of 34.8 thousand. However, the plans of the Communist period have now been set aside and each community is free to take initiatives. Thus some of the Nitra villages have opted out in the hope that they secure a better deal through funding directly from the state's environment fund and that the improvement in services (perhaps with a cultural/community centre and/or sports facilities) will offset such disadvantages as the loss of subsidized public transport. The villages are also free from the Communist controls on settlement densities which aimed at twenty-five persons per hectare and housing plots of 600–800 m². Where larger plots on the edge of a village were reduced down to the official level by transfer to cooperative farms, such land can now be retrieved; but where plot boundaries were redrawn so that additional houses could be built in the centre of a village there will be no compensation since the 'surplus' land was officially purchased.

There is considerable potential for tourism in the rural areas, especially in the main valleys where the damming of the rivers has created opportunities for water-based recreation (Jordan 1992). Such is the demand for water that further dams will be needed in the future. There are also further opportunities in villages on the edge of the mountains, which are convenient places for excursions to higher ground. Brodzany, near Partizánske, is a case in point. As well as offering access to the high ground, the village is attractively laid out along a linear axis following a narrow tributary valley on the edge of the hills. Recent settlement has encroached on the Nitra floodplain and is complemented by the construction of second homes for Partizánske people on a steep slope overlooking the village proper. Brodzany also boasts a château set in landscaped grounds (where the river has been transformed into a cascade of waterfalls). With the château now a museum open to the public, the grounds provide an attractive recreation ground. The village is also of interest for its recent landscape history, with the phase of private estate management followed by the Socialist period, which led to the construction of new buildings for livestock (served by a monorail system for the supply of fodder); a complex recently superseded by further purpose-built accommodation. The recent modernization of the village can also be traced through improvements to local services (notably a small power station built since World War II) and local industries, which include a box factory and two units using former estate farm premises: a workshop producing components for the Partizánske shoe-making complex and a motor vehicle repair shop (the latter occupying accommodation formerly used by the estate farm).

LANDSCAPE PROTECTION

Protected landscape area status was accorded to Ponitrie in 1985. A protected area of 37.7 thousand hectares covers the central part of the Tribeč and Vtáčnik Mountains: a forested interfluvial axis, approximately fifty-six kilometres by fifteen kilometres, with Vtáčnik Mountain (1,346 m) as the highest point. The area includes part of Žiar nad Hronom county as well as sections of the Nitra,

Prievidza and Topolčany counties with which this review is primarily concerned. Within the protected area sixteen separate reserves have been established (Olas 1993). These are known variously as state nature reserves, protected national formations, protected sites, protected natural environments and study plots. Tribeč consists of sedimentary rocks, including much limestone; oak, oak-hornbeam and beechwood are characteristic, with plant communities such as Adonis vernalis, Dictamnus albus and Stipa joannis. By contrast, the volcanic Vtáčnik are composed of andesites. The typical woodland species are beech and mixtures of beech and fir; and there are some rare montane plant species such as Adenostyles alliariae, Doronicum austriacum and Luzula sylvatica. The wildlife includes native predators such as the lynx as well as fallow deer introduced in 1867 and protected fauna species, which include Cochlodina remota, Saga pedo and Tibicen haemalodes.

Policy sets out to promote awareness of this heritage through appropriate cultural and economic policies including programmes for water management, health and recreation. Access to the protected area is facilitated by a network of waymarked paths, including a route along the watershed from Nitra to Vtáčnik. This gives access to some of the reserves among which those close to the city of Nitra are particularly well-known. Zoborska Iesostep (wood-steppe) comprises a reserve of 24.4 hectares dating back to 1952. On the limestone surface there are cliffs and clint fields (limestone pavements) and a diverse xerophilous vegetation grading from steppe and forest-steppe communities to oak groves and oak-hornbeam stands. In the latter there are small islands of Quercus pubescens, with Q. petraea on acid subsoils. Diversity must be maintained, but there is a threat from heavy recreational pressure by the inhabitants of the city. There is a growing population on the Zobor foothills where weekend cottages established amidst vineyards have been enlarged and transformed into permanent residences, with increasing subdivision of plots and an improved infrastructure involving public transport and other basic services. The smaller area of Lupka (9.8 hectares) was also declared a reserve in 1952, and it offers fauna including rare species of insects. The reserve is adversely affected by a refuse dump that already existed when the reserve was set up. It does not cater for toxic waste, however, and will eventually be closed. Finally (as regards the Nitra area) Žibrica was declared a reserve in 1954 over an area of 28.1 hectares, which includes a Halstatt fortified settlement. Once again there is a combination of steppe, forest-steppe and forest communities in an area once subject to heavy exploitation for the extraction of high-quality limestone.

THE CONTEMPORARY PLANNING CONTEXT

The wider context to the issue concerns a research priority to establish the vegetation history of sensitive areas as a basis for estimates of ecological stability and recommendations for conservation in the future (Hornik 1993). This work takes place in the context of wider investigations dealing with ecological themes in historical geography, which highlight the changes in the mountain environment over centuries of rural colonization by pastoralists and later by tourists, as a basis for revised plans for the expansion of tourism to concentrate on the less sensitive areas and contribute to sustainable rural development (Lehotsky *et al.* 1993; Seko 1988). Studies in biogeography also highlight the long-term ecological effects of the settlement process (Plesník 1988). This work may be combined with an environmental approach to landscape mapping (Otahel *et al.* 1993). Indeed, under legislation passed in 1992, a new element has been introduced into physical planning, with maps of ecological stability that will start with the land-use pattern and the drainage system. The surveys will highlight biocorridors and recommend appropriate management (including reductions in fertiliser applications or a switch from arable to pastoral farming); also biolines, such as belts of woodland to subdivide the vast open fields which were created in order to enhance the efficiency of the state and cooperative farms. Such exercises will also highlight conflicts (for example, mining subsidence, effluent from intensive stock rearing units and traffic congestion) and the incidence of acid rain. This is now an integral part of the territorial planning documentation (Uzemnoplanovacia dokumentacia) for each 'major territorial area' (Velkehoúzemyćho Celku). As before, the planning process involves mapping (at a variety of scales) to show house plots, building lines, infrastructure and the official limits to each settlement. But the environmental recommendations introduce a new element, and there is now discussion as to how far these recommendations should be mandatory. This work is being coordinated by the Slovak Environment Agency (Slovenská Agentúra Zivetného Prostredia) based at Banská Bystrica with a network of seven regional offices (one of which is in Nitra). This organization cooperates with the agriculture and water authorities – which make their own environmental assessments – and is also responsible for the identification of conservation areas and species preservation.

The ultimate spatial context for the planning will depend on decisions concerning the administrative regions. The counties inherited from Communism are to be grouped into larger units. This could result in a set of six large regions focusing on the cities of Banská Bystrica, Košice, Nitra, Prešov, Trnava and Žilina, with Bratislava as a separate region. However, a number of variants have been considered, which would give the status of regional administrative centre to a number of other towns: Levice, Lučenec, Martin, Michalovce, Prievidza, Rimavská Sobota, Spišská Nova Ves and Trenčin. This would increase the number of regions to as many as fifteen. The seven-region formula would unify the Nitra Valley with a single region extending from the Danube at Komárno to cover the districts of Levice and Nové Zamky as well as the three covered by this study; whereas the adoption of a separate region for Prievidza would divide the valley into two. Much more difficult, however, is the problem of reconciling a system based on city regions with the aspirations of the Hungarian community for a system of small administrative units where Hungarians would always be influential, and in many cases dominant. For the area along the southern border, where they are overwhelmingly concentrated, a total of twenty-two units have been proposed for an area with a total population of 824.9 thousand people, 61.5 per cent of whom are Hungarian. The average population for a unit would be just 37.5 thousand (but the figures would actually vary between 71.3 thousand for Nové Zamky to 20.0 thousand for Filakovo). There would be a Hungarian majority in fifteen of the twenty-two units; with the highest proportion in Dunajska Streda (90.1 per cent) and the lowest in Levice (21.3 per cent). The advocacy of a scheme that is at such total variance to the city region formula preferred by the mainstream community serves to highlight the intractability of the ethnic issue.

CONCLUSION

The Nitra Valley has experienced a sustained programme of industrialization that has damaged the environment in a number of ways. However, the most severe problems arise from the dependence on lignite (instead of hard coal), which yields very large quantities of waste, and in the particular case of the local Nitra Valley lignite contains such toxic materials as arsenic. The prominence of the inversion phenomenon also adds to the pollution problems through slowing down the dispersal of emissions into the atmosphere. State policies towards agriculture in the Socialist period have also had negative effects on the Nitra Valley and not all the damage can be connected with industrialization. Deficiencies in the built environment can also be related to national policies, which have impacted on the settlements of the Nitra Valley along with other parts of the country.

There is much public concern about the environment and some improvements have been made. But the long-term solution will require major economic restructuring to emphasize small industries and a growth of tourism (with the eventual rehabilitation of the Chalmová spa as one important objective). Better environmental management will depend not merely on specific measures to solve local problems (which may necessitate the gradual run-down of the coal industry and a reprofiling of the Nováky chemical plant). It will also require national planning policies to emphasize ecological corridors, restore biolines across the countryside and protect vulnerable landscapes. Moreover, a revised system of planning control, like that promoted by the Slovak Environmental Agency, will have to give greater priority to the rural areas and the environment generally. Finally the local administrative regions will need some revision to provide the necessary coordination across critical areas and provide an effective spatial framework to mobilise community action.

10

A REVIEW OF ENVIRONMENTAL ISSUES IN THE LIGHT OF THE TRANSITION

F. W. Carter and D. Turnock

INTRODUCTION

Eastern Europe's environmental crisis has not been solved since the demise of Communism and it continues to play a significant role in the transition process (Russell 1991a; 1991b). The founding of a new economic system and a move towards fresh property relations during the privatization process are very reliant on improved ecological circumstances. The search for solutions goes on, but increasing economic problems have complicated possible methods of solving the issue, and the population of the region continues to experience a declining quality of life as a consequence of appalling pollution levels. Although the average citizen is now more aware of threats from increased environmental degradation, many people are anxious that the new system should produce improved environmental conditions rapidly. Therefore, protecting environmental quality while pursuing economic development poses a particularly difficult challenge for the governments of East–Central Europe (DeBardeleben and Hannigan 1994; Enache 1994; Bassa 1994; Špes 1994).

Heavy investment in economic restructuring is still necessary in the region, although the severity or omnipresence of some air pollution problems are not as bad as first feared. Nevertheless, the post-Communist societies of these countries must ensure that efforts and resources used for pollution control give the best results. This may be particularly acute given their restricted financial means and the most cost-effective ways for solving environmental problems must be found (Edeburn 1993; Paulsen 1993). For example, accountability for existing environmental damage and environmental reconstruction costs encroach on the cost-benefit considerations of potential investors both at home and abroad.

Some form of balance sheet should be drawn up on the progress achieved in this part of Europe where, it has been argued, the continent's most severe environmental problems exist (Nefedova 1994). Immediately, certain questions arise about the chances of potential environmental success in East–Central Europe. For example, in the short term will it be possible to obtain desirable globally-imposed environmental levels in this region, given the sluggish rate at which its institutions are evolving? How can the legacy of four decades of centralized Communist planning in the fields of technology, economic structure, social organization, and complete disregard for the environment be overcome under new market-oriented economies? For how many more years will old Communist capital stock be used, and with what negative results on the contemporary environment? To what extent will environmentally sensitive changes in production methods have on reducing emission rates? What effect will the spread of new technologies, market-oriented commerce and consumerism, private land-holding, new government environmental legislation and growth of environmentally conscious political parties/institutions, have on reversing earlier environmental degradation? While precise answers to such questions cannot be attempted here, it may be possible to indicate how far East–Central Europe has gone down the road to surmounting some of these major environmental difficulties.

MAJOR FORMS OF POLLUTION

The recent publication of maps based on late 1980s environmental data for East–Central Europe provide a clear benchmark for the inheritance from Communism (Nefedova 1992). Clearly, there is a marked zone of environmental degradation stretching from Leipzig through north-west Bohemia and Prague to Cracow. Farther south there are high risk zones around Budapest and to a lesser extent in south-central Romania and southern Bulgaria. These areas result from intense industrialization and weakly developed environmental consciousness; they are also locations possessing a complex mixture of heavy pollution and very delicate ecosystems.

Air pollution

The World Health Organization has recently published a report on air quality over Europe (Nilsson 1995). It reveals that SO_2 short-term values are being widely exceeded, none more so than in the Black Triangle covering southern Poland, south-east Germany and the northern part of the Czech Republic. Here excessive concentrations are found both in urban areas and the surrounding countryside. The heavily industrialized Black Triangle is a region with a long history of ecological degradation (Pape 1993a). External organizations are slow to come to its aid. Neither the World Bank nor the European Bank for Reconstruction and Development (EBRD) has taken much interest in solving its ecological problems, whilst the EC PHARE project has been sluggish in implementing its 'Black Triangle Programme' (Pape 1994). Sofia and Ruse in Bulgaria are amongst the worst places outside this area to suffer environmentally; both are subjected to more than two hundred days annually when concentrations of total suspended particulates (TSP) rise above the accepted limit of one hundred and twenty micrograms/m³ per twenty-four hours. Nitrogen oxides also exceed safety limits (i.e., more than sixty micrograms/m³) in many urban centres and are known to cause damage to the respiratory system; Belgrade has the highest levels of NO_2 with an annual average exceeding one hundred micrograms/m³ (Nilsson 1995: 3).

However, Eastern Europe is not unique. Calculations for sulphur dioxide emissions over the period 1980–92 focus on a core area with more than 500 g per hectare extending from the Black Triangle of Eastern Europe westwards through Germany and the Benelux countries to embrace the whole of England apart from the West Country. A concentric zone with values exceeding 100 kg would cover the northern half of Eastern Europe including northern parts of Romania and former Yugoslavia, northern Italy, north-east France, central Scotland, Denmark, southern Sweden and western parts of Belarus and Ukraine. Table 10.1 profiles the situation across Europe with regard to sulphur and nitrogen oxides. For sulphur the level of emissions in relation to population was 38.6 tonnes per thousand population in 1992; rather more than double the rate in both Western Europe (18.6) and the FSU (17.6). Within Eastern Europe rates in the Czech Republic were 74.5 per thousand population (almost double the East European average), while they were less than half the average in Albania (17.8) and Croatia (16.7), and were negligible in Macedonia (2.4). For nitrogen oxides the East European rate is slightly lower than the West European, while there are still major variations within the region. However, there is an encouraging sign that pollution levels are falling, although more accurate data is needed and many people in Eastern Europe would like to see much more improvement in air quality, given the scale of the economic recession, with rising unemployment and a decline in industrial output.

A recent study has helped pinpoint the main culprits – thermal power stations. Four of the twenty-five largest SO_2 emitters on our continent are located in East–Central Europe, and they were responsible for 4.04 per cent of the total emissions: Maritsa Istok (Bulgaria), Belchatów (Poland), Prunéřov and Tušimice (Czech Republic). A wider examination covering the hundred largest polluters shows that twenty-seven are in Eastern Europe where the three countries of Bulgaria, the Czech Republic and former Yugoslavia have the highest values in Europe for emissions from these major sources related to population (Table 10.2).

In North Bohemia coal is king (Jeřabek 1994). Over the last four and a half decades, one hundred and twelve villages have been razed to the ground in order to acquire this source of energy (Piňos 1993; Pavlínek et al. 1994). The enormous concentration of power plants burning washed brown coal and lignite without desulphurization, together with adverse climatic conditions for pollution dispersion, lead to poor air quality in the North Bohemian Basin and also in the Labe (Elbe) Valley, which is Bohemia's most fertile agricultural area (Kopačka 1994). The situation is especially serious during the winter when temperature inversion produces smogs (Quitt 1989). Whilst air pollution levels have declined recently

Table 10.1 Emissions of sulphur and nitrogen oxides[a]

| | Sulphur | | | Nitrogen Oxides | | |
| | 1980 | 1992 | | 1980 | 1992 | |
	A	A	B[b]	A	A	B[b]
Albania	60*	60*	17.8	30*	30*	8.9
Bosnia-Hercegovina	102*	100*	22.9+	35*	52*	11.9+
Bulgaria	1025	560	62.5	416+	240	26.8
Croatia	75	80+	16.7+	100*	124*	25.9+
Czech Republic	1128	769	74.5−	937	698	67.6−
Hungary	816	451+	43.7	273	211+	20.4
Macedonia	5*	5*	2.4	2*	2*	1.0
Poland	2050	1410	36.8	1500+	1130	29.5
Romania	900	702+	30.9	680*	680*	29.9
Slovakia	350	187+	35.3	227+	224	42.3
Slovenia	118	94	47.1	43	46	23.0
Yugoslavia	350*	335*	32.2+	160*	196*	18.9
EASTERN EUROPE	6979	4753	38.6	4413	3633	29.5
WESTERN EUROPE[c]	14068	8177	18.6	14039	14210	32.4
FORMER SOVIET UNION[d]	6172	3931	17.6	3248	3628	16.3
EUROPE	27219	16861	21.5	21694	21471	27.3

A Total emissions '000t B Emissions t. per thousand of population.
[a] Official data received at the ECE Secretariat except where otherwise stated.
[b] Belarus 1990; Bosnia-Hercegovina, Croatia and Russia 1991; Slovenia 1993.
[c] Austria, +Belgium, Denmark, +Finland, +France, +Germany, +Greece, +Iceland, Ireland, +Italy, +Luxembourg, Netherlands, Norway, +Portugal, +Spain, Sweden, Switzerland, +*Turkey, +United Kingdom, Yugoslavia
[d] +Belarus, *Estonia, *Latvia, *Lithuania, *Moldova, Russia, Ukraine
* estimates by MSC–WCCC
+ (interpolated)
Source: Acid News 5 (1994), p. 14.

this is due largely to a drop in electricity demand during the industrial recession and closure of several major energy devouring plants. Moreover, restructuring has led to a drop in industrial output, and the weather over the last few years has been milder; most important, however, have been proposals adopted to eradicate not only large but also modest pollution sources locally (Plamínková 1995; Crockford 1993).

The most significant factor now in North Bohemia, as in other parts of the Czech Republic, is pollution from domestic heating appliances (Vavroušek 1992, 1994). Even in areas of heavy air pollution, related to the dense location of power plants and industrial concerns, domestic heating units bestow a disproportion-ately high percentage of pollution in relation to fuel consumption. Unfortunately, small amounts of smoke and gas emission near ground level from individual outlets do not make pollution control equipment installation worthwhile. In

Table 10.2 Sulphur emissions from the 100 biggest sources[a]

Region/Country	Number of sources	Sulphur emissions '000t t/'000 pop.		Largest source of emissions and '000t	
EASTERN EUROPE					
Bulgaria	4	484	54.0	Maritsa East	325
Czech Republic	5	385	37.3	Prunéřov	137
Poland	6	395	10.3	Belchatów	168
Romania	6	289	12.7	Isalnita	79
Yugoslavia	4	202	19.4	Nikola Tesla	78
WESTERN EUROPE					
Germany	12	780	9.7	Janschwalde	157
Greece	2	148	14.4	Megalopolis	76
Italy	6	234	4.1	Brindisi Sud	57
Spain	4	586	15.0	Puentes	271
Turkey	6	616	10.3	Afsin-Elbistan	288
United Kingdom	13	896	15.5	Drax	132
FORMER SOVIET UNION					
Russia	10	924	6.2	Montsegorsk	212
Ukraine	13	793	15.2	Krivorozhskaya	95

[a] The table deals with all countries with two or more polluters in the top 100. Countries with one polluter only are: Estonia, France, Hungary, Ireland, Macedonia, Moldova, Portugal, Slovenia and Slovakia
Source: C. Ågren 1994, 'Emissions: the worst hundred sources', *Acid News* (3) p. 3; M. Barrett and R. Protheroe, *Sulphur emission from large point sources in Europe*, (Swedish NGO Secretariat on Acid Rain), Gothenburg, 1994.

North Bohemia, the government earmarked 1,000 million Czech crowns in 1993 for domestic heating unit conversions. Gas and electric central heating has replaced domestic boilers and coal-burning stoves in many places. It is hoped that this conversion process will be completed by late 1995. Furthermore, in North Bohemia, the Czech Electricity Company (České energetické závody: CEZ) is planning to close some power plants, whilst the rest will be subject to desulphurization. Other, less efficient, plants in this area will probably go when the Temelín nuclear power plant is completed. Overall in the Czech Republic, it is hoped that SO_2 emissions will have fallen to one-tenth of their 1994 levels by 1998; financial help will come mainly from government and Czech business sources and is not dependent on foreign aid.

Other regions remain heavily polluted. One of these is the Upper Silesian Industrial Region, with its coal mining, iron, steel, metallurgy and machine manufacturing industries. It remains Poland's major industrial area with a network of twenty-five urban centres and 2. 6 million inhabitants. Environmental protection is among proposals made to foreign investors for the regional development of the area. Local authorities would like to attract investment for improving public utilities such as waste water treatment and incineration plants (Anon [a] 1994). Air pollution remains Upper Silesia's chief problem and the energy sector is its leading polluter. Of the eighty antiquated plants listed for closure in 1990 only three had actually done so by the summer of 1993. Admittedly, a few cases of restructuring were being planned, but emissions from seventy-seven of these plants continued as before (Poborski 1994). More recently dust emissions in the region have declined considerably; two-fifths of them due to economic recession and three-fifths to better environmental measures (Hlawiczka 1995). In the short-term, inexpensive measures related to pay-off time and cost-benefit could be primary factors when considering other possible solutions for reducing air pollution (Bjorklund 1993).

Whilst air pollution from coal-fired power plants and heavy industry is fairly well documented in East–Central Europe, contributions from motor vehicle use are less well-known. Lack of precise data helps to confuse the issue, but a recent independent study has revealed that car pollution levels are lower than in the West. Car ownership is only about two-fifths that of western Europe and high motoring costs often reduce distances driven. Thus, both total and per capita car emissions (CO, HC_S, NO_X) are lower than elsewhere on the Continent. Estimated figures for the period up to 2010 based on car totals and annual distances driven suggest that fundamentally per capita levels will not alter relative to total population. Be this as it may, their increasing concentration in certain large urban areas will not be prevented – for example, between 1991 and 1992 car totals in Warsaw doubled. Further, maintaining per capita levels does not disguise the fact that contributions to air pollution from cars in East–Central Europe are greater than their western European counterparts, in spite of a significant increase in imported West European second-hand cars. Many East European varieties are now relatively old, use highly polluting fuels and lack up-to-date pollution control equipment (Boyle 1995).

The fact that cars are relatively old (fifteen years on average, which is double the US figure) means that relatively few cars have modern pollution control equipment. Some cars made in Eastern Europe are notorious polluters: for example, the East German-made Trabant with its two-stroke engine, which emits high levels of hydrocarbons. But all motor vehicles run on highly polluting fuels, with a high sulphur content in the diesel oil. The lead content in petrol is high (0.2 g per litre in the former Czechoslovakia, 0.3g in Hungary and up to 0.6g in Poland) compared with the level of 0.15 in Western Europe. So far, very little use is made of lead-free petrol. Indeed, the difficulty of finding lead-free petrol in many areas is a problem for the tourist industry when many western cars cannot run on any other fuel. Although an increase in the immediate future may be prevented by lower average incomes (arising from economic restructuring) and rising energy costs, pollution is likely to increase with an estimated 13 million cars by 2010, each averaging some 8,000 kilometres per year (Walls 1993: 5).

Future plans suggest a rash of new motorways through East–Central Europe. Through its influence on land use and travel patterns, transport planning helps determine pollution levels, and all seems set for a great expansion of the motorway network in the area. During the next couple of decades, combined total motorway length in Poland, Slovakia, the Czech Republic and Hungary will experience a three-fold increase. Over the next decade-and-a-half around 2,000 km will be added in Poland, a further 800 km in the Czech Republic by 2005; Hungary hopes to complete about 500 km by 2000, and Slovakia some 300 km during the next couple of years (Andersson 1994). In addition, there are plans for a major new international routeway (via Baltica) from Estonia through Latvia–Lithuania and Kaliningrad to Warsaw where it will link up with other already existing main routes. Quite apart from the additional traffic there is the important matter of the impact of the motorways themselves. Impact assessments were made into some of the first motorways to be built in Eastern Europe (Lehotský et al. 1989) and such work will be more important in future.

Water

Besides air, water ranks as a major recipient of environmental pollution. During years of Communist rule water quality suffered serious decline; it has now been recognized as a serious problem and schemes are afoot to redress the balance (Jancar-Webster 1995). Waste accumulations pollute water supplies: in the Katowice area of Poland underground water is polluted because it seeps through tip heaps. Some two-thirds of Hungary's drinking water supply has no protection from sewage contamination, which amounts to some two million cubic metres of sewage leaking into the soil each year (Reeves 1993). Water quality is particularly bad in the northern hill region (Varga et al. 1990). Heavy metals, associated with mining activities, are often present in the soil (Persanyi 1990).

The Vistula below Cracow is threatened by heavy metals: chrome, iron, lead and, especially, mercury. Moreover, fly ash from Warsaw power stations like Żerań is being dumped on the banks of the Vistula and causes acidic pollution of the river. The Labe (Elbe) river has long been used by the Czechs and countries downstream to dump waste. Water quality through biological pollution is becoming increasingly a matter for concern. For example, the River Jezera is significant as one of three rivers supplying the Prague agglomeration with fresh drinking water. Unfortunately the river is subject to three major pollution forms, namely, domestic effluent, waste material from industrial plants and seepage from agricultural land. BOD measurements have been taken and considerable amounts of nitrates, ammonium, phosphates and various other undissolved substances have been recorded. Contaminated run-off has been measured to estimate its effects on water quality farther downstream (Šulc 1994).

Waste water from an area of some 817,000 km^2 drains into the Danube River basin, an area populated by about 67 million people. Spatially, the basin encompasses most of Austria, Hungary, Romania, the Slovak Republic and former Yugoslavia; it contains about one third of Bulgaria (G. Gergov 1994) and the Czech Republic, as well as notable parts of Germany and Moldova and small portions of the Ukraine. The greatest influence on water quality comes from urban and industrial effluent; this is particularly serious as only some major urban centres in the Danube basin have satisfactory waste water treatment plants. Further pollution is added from agricultural by-products, provincial housing, solid waste and sludge. In comparison with pollution discharged by all European rivers flowing into the North Sea, the Danube releases into the Black Sea higher total loads of organic materials, phosphates, heavy metals and oil. The most urgent contamination problems arise in the Danube's tributaries, where the addition of municipal and industrial waste water has a bigger effect on the comparably weaker river/stream flows. High concentrations of certain pollutants (e.g., organic and inorganic micro-pollutants), create health risks and endanger ground water resources. For example, several urban centres in southern Hungary were discovered to have undrinkable polluted groundwater. As a result, potable water for local residents had to be conveyed from alternative safer sources in plastic bags. In Romania, micro-pollutants and nitrates are affecting major aquifers on the lower stretches of the River Siret, endangering domestic water supplies. Once the main problems have been delimitated, discussions can then begin with government representatives to define priority ventures and assist in plan implementation. For example, total investment requirements for the Siret River Basin alone could exceed 250 million ECUs (Anon [b] 1993).

Great changes have taken place on the rivers through hydro-electric schemes, which have combined power generation with water storage, flood control and navigation facilities. However, there have been considerable effects on the environment as well as disruption for the human settlements affected by dam and reservoir construction. Major projects are now seen as a hallmark of authoritarian Communist planning and are increasingly controversial even where there are substantial benefits, an image now reserved for the World Bank. Recent problems on the Labe have been linked to a 22-megawatt hydro-electric dam project being built at Dolní Žleb in northwest Bohemia (Steinhauserová 1994). The ecological balance of the river upstream will occur through flooding on the dam's completion. Two sensitive ecological areas, one under regional protection and the other designated as a national park, will be effected. This has attracted the attention of protest groups like 'Dětí Země' (Children of the Earth) and 'Ekotrans' who believe the project will ruin unique wildlife in the Labe river canyons (Paul 1994). The issues surrounding the massive Gabčikovo-Nagymáros project on the Hungarian–Slovak border are discussed fully below.

Dumping of waste

If waste is not emitted into the atmosphere or discharged into rivers it tends to be dumped rather than incinerated or recycled, although these more environmentally-friendly options are being considered much more seriously at present (Bialic 1994). About 8 million tonnes of hazardous waste is produced annually in Poland. The leading offenders are Szczecin province (1.6 million tonnes) on the

Baltic coast, followed by Legnica, Bydgoszcz, Katowice and Gdańsk provinces (Wysienska 1993). Slime and particles from the steel industry's gas treatment plants, sewage treatment residues, slaughterhouse waste, and debris from galvanizing plants, refinery processes, used cooling agents, oils and lubricants are the main ingredients. Unfortunately, two-fifths of Poland's hazardous dumps operate without any legal safety measures whilst there is no data on illegal dumping. Evidence suggests, however, that this is on the increase (Mering 1994; Andrzejewski and Baranowski 1993).

The central western part of former GDR (Hettstedt and Gera) has been affected by lignite quarries (Merseburg/Mucheln); and the relics of mining for potash salts (Stassfurt) and uranium (Ronneburg). Around Eisleben, dumping associated with the Mansfeld metals complex, has created a series of tall conical dumps, which are not only prominent landscape features but hazardous on account of the concentrations of arsenic, cadmium and zinc. Although these operations have largely disappeared since 1989 due to the closure of the last copper mine (Sangerhausen), the winding up of the joint GDR–Soviet Wismut uranium company, and competition from fertiliser producers in West Germany, the tip heaps remain.

Problems arise elsewhere through some large dumps of toxic waste which may easily become unstable. The gravel and shale dam at Mojkovac in Montenegro was threatened by collapse after flash floods in 1992. The dump arose through the reopening of a medieval mine by the GDR Briskovo company and construction of a processing plant to produce some 4,500 tonnes per year of lead and zinc prior to the collapse of the Soviet market, which led to closure in 1991. There was a risk that poisonous sludge (accumulating at a rate of 35,000 tonnes per year between 1976 and 1991) would be carried into the Tara (a UNESCO-protected river and a major tourist attraction separating two national parks) and on to the Drina, Sava and Danube. Fortunately, the problem was contained, but the dump will clearly become a major hazard if it pollutes the river system.

Dumping remains a serious issue throughout Eastern Europe because in addition to the tips of the Socialist period which were not properly sanctioned and regulated, there are further dangers from toxic waste imported from the West where much tighter regulations are in force. Soviet troops left large amounts of fuel and oil behind in places like the Ralsko airbase near Milovice which has subsequently drained into the ground. In some cases there has been enough fuel available for private individuals to dig oilwells! However, the costs for total restoration are heavy and there are many former Red Army installations that require attention (285,000 ha in the former GDR alone) along with industrial areas like the Greifswald nuclear power station site (Fisher 1992).

NUCLEAR POWER

The nuclear debate continues in an atmosphere of controversy, especially concerning East–Central Europe where the industry faces an uncertain future

(Kramer 1995). The subject is laden with conflicting and contested beliefs; major international banks have adopted a rather wary strategy, but the issue still aggravates public opinion. Meanwhile, the nuclear industries of North America and Western Europe persevere in their hopeful gaze towards the former Soviet bloc for continuing business (Anon [c] 1994; Anon [d] 1991). The safety angle in East–Central Europe's civil nuclear power sector continues to cause apprehension amongst potential Western cooperators. Assistance programmes have been frustrated by three main problems: first, an inability to understand which parts of the nuclear power industry can be kept and extended; second, how to cure present institutional defects; and third, to appreciate the enormity of resources necessary for plant modernization. Conversely, two inherent internal factors have to be taken into consideration regarding the dearth of sufficient finances for solving the region's nuclear safety problems. First, the former Soviet Union's influence in operating East–Central Europe's nuclear power programmes and second, the persistent contemporary economic crisis. It is clear that if Western assistance is going to effectively contribute to the region's civil nuclear power needs, some critical evaluation of the above factors is necessary (Mussapi and Thomas 1994).

In the Czech Republic the nuclear debate continues to be a contentious issue (Wedmore 1994). Much centres around the Temelín nuclear plant 100 km south of Prague in southern Bohemia and near the Austrian border (*Mladá fronta Dnes* 1995). The Soviet-designed power plant, a legacy from the Communist era, suffered early teething troubles and construction was suspended in 1990. Plans were reactivated in 1993 and Westinghouse Corporation, an American Pittsburg-based electrical equipment manufacturer, won a contract worth more than $400 million to complete plant construction and raise it to internationally accepted standards. This involved certain modifications, including the supply of new instrument devices and control systems. However, in the event of a nuclear accident, Westinghouse were not be held liable for damages. Czech state funds would be used to compensate any citizens, or neighbouring states, affected by such a catastrophe. Further, geological research has shown that although Temelín is situated on a region of faulting, no active tectonic zones have so far been located there (Kalvoda 1992).

Temelín's critics are worried about the high costs of compensation should an accident occur at the plant. Moreover, any damage in nearby Austria from such a disaster would have to be covered by Westinghouse according to Austrian law. Critics believe that future plans by the Czech Republic to augment its reliance on nuclear power are not only hazardous, but also unproductive and expendable. Such critics maintain that money would be better spent on conserving energy, the purchase of pollution-abatement technology and a change over to gas-fired power stations. Some of their worst fears have been underscored by a decision from the country's power monopoly (CEZ) to publish a list of ten possible 'temporary' nuclear waste storage sites from the year 2004 onwards. Waste from the Temelín and Dukovany (near Brno) plants will be stored at Dolní Ves, Horní

215

Cerekev, Nová Ves, Karlovka, Kozcin, Pacejov, Pastuchovice, Skalka, Srby and Setřichov nad Bystrici for fifty years. It will then be transferred to an as yet unchosen permanent location somewhere in the Czech Republic, pending a final decision made in 1996. Obviously, strong reactions have been registered by the ten towns involved, which in some cases could lose tourist potential by the decision. This has prompted the CEZ to promise the local authorities of the ten towns up to 15 million Czech crowns in government subsidies annually, in an attempt to persuade them to accept nuclear waste dumps on their doorsteps (Drda 1994).

Similarly, a nuclear controversy is facing the Slovak Republic over completion of its Soviet-designed Mochovce power plant, 120 km east of Bratislava. In January 1995, the government chose Electricité de France as Slovakia's western partner to help modernize the plant and cover a quarter of the costs. The EBRD has been considering whether to loan DM 412.5 million to complete the project, but the Mochovce site has provoked intense opposition locally, in the Czech Republic and from nearby Austria. In turn, this has led to intervention by the European Parliament and various international environmental groups and prompted the Bank to delay its decision. The EBRD has always stipulated a precondition to its loan, namely closure of the nearby old and extremely dangerous Jaslovské Bohunice nuclear plant by the year 2000 (*Pravda* 1995). Meanwhile, Slovaks argue that the Mochovce project is needed for supplying the country's energy needs. They intend to press ahead with the plant irrespective of the EBRD decision, possibly with Russian assistance (LeBor 1995; Anon [e] 1994), for the Russian government is prepared to give a $450 million loan for the completion of the two reactors. Should Russia win the contract (i.e, provision of technology and manpower) Slovakia would not be required to close the two units at Jaslovské Bohunice. Moreover, the Russians might accept quantities of radioactive waste for which Slovakia has no long term storage capacity. All this is seen as part of Russia's aim to rekindle closer ties with former Soviet satellites. In this case technology transfer (e.g., the offer of aircraft to Hungary and Slovakia) could advance Russia's geopolitical interests by discouraging former allies from seeking NATO membership (Lawson 1994).

The Paks nuclear power station in southern Hungary provides nearly half of the country's electricity demands. There is a need to increase energy capacity but the country faces a dearth of power resources combined with a poor energy efficiency record (i.e., about half of the Western European average in relation to its GDP). The official policy for boosting domestic energy supplies can be achieved in one of the following ways: first, construct a new lignite burning power plant using poor quality domestic sources and producing high SO_2 emissions; second, erect a new hard coal power plant based on expensive imported supplies; third, enlarge the Paks plant, which by East–Central European standards has an enviable safety record. It has already attracted several French and Canadian companies interested in selling nuclear technology there (Andersson 1993). However, the 'Energy Club', an NGO umbrella organization for green groups opposed to nuclear power, is very disparaging of any policy based on this energy form, particularly regarding the probability of potential accidents and problem of nuclear waste disposal. The 'Energy Club' would prefer to see improved energy efficiency, priority given to less energy-intensive industries and greater use of alternative, more renewable fuels (e.g., solar and geothermal). Its members maintain that no final decision should be made without first informing both the general public and those local communities involved (Amon 1994).

Alternative views on Hungary's energy policy suggest that greater emphasis should be placed on renewable resources (i.e., solar and geothermal), whilst planning priority should be given to less energy-intensive industries. Improvements should also be made regarding energy efficiency; efforts should be made to replace outmoded, heavy power-consuming electrical and other equipment, commonly found in many Hungarian households (Anon [f] 1994). Western help has been forthcoming: in 1994, Hungary's power sector received a loan of $100 million from the World Bank to develop and fulfil suitable environmental policies, finance generation, and improve its energy management system. Whichever way this money is spent, all the signs suggest that Hungary's nuclear power industry will preserve its *status quo* position.

Bulgaria has a large nuclear power station at Kozloduy that has been radically overhauled (*Trud* 1995). There are also plans for a second project at Belene, which are encountering mixed feelings among the public. Only Slovenia in the former Yugoslavia, possesses a nuclear power plant, located at Krško on the border with Croatia. Originally, the plant was planned under the political monopoly of a Communist government in which its energy was to be shared between Slovenia and Croatia. It remains in operation despite post-1989 attempts to close it down. Krško generates considerable low and intermediate radioactive waste, which creates disposal problems. Studies have explored possible radioactive dumping sites for the Croatian-owned plant and potential locations have been earmarked (Šaler 1993). Austria requested that the Slovenians terminate production at the Krško plant as it posed a possible source of danger; the Slovenian government stated in response that Krško is not the only nuclear power plant in central Europe that poses a potential threat to Austria. In reality, Slovenia cannot afford to close Krško down, because of its significant energy role for the Slovenian economy (Anon [g] 1994).

Both Romania and Poland have less immediate fears from domestic nuclear pollution. Lack of funds could further delay completion of Romania's only nuclear power plant at Cernavoda, 150 km east of Bucharest. Original plans for it to become operational in 1985 were abandoned because of financial problems. In 1991, its prospects were revived when a contract was signed with a Canadian partner. Now a further $230 million are still needed to complete the first reactor, which has been constructed under Canadian supervision; a second unit will come on stream in 1998, if enough financial resources are forthcoming (*Romanian Press Service* 1993; Anon [h] 1994). In Poland, four reactors were under construction at the Zarnowiec site near Gdańsk, but work on them was abandoned in 1990 after the break-up of the Soviet bloc. Poland remains the only East–Central European country to have no plans for commercially produced nuclear energy by the end of this century. According to official policy, Poland will only start generating a small amount of nuclear energy in 2010, based on an installed capacity of 6,000 megawatts (Polska *Agencja Prasowa* 1990).

TRANSBOUNDARY ISSUES

Pollution problems in Eastern Europe continue to be aggravated by effluent and emissions emanating from other countries. Poland's problems in the Black Triangle are exacerbated by pollution from the Czech Republic and Germany carried by the prevailing wind. Poland is also affected by the poor condition of the Baltic Sea, which continues to be a reception area for polluting substances from the surrounding countries. These include nutrients drained into the sea by rivers (particularly phosphorus and nitrogen compounds) and sewerage, especially from urban areas, which indirectly affects the BOD loading (oxygen balance) of the sea (Hiltunen, 1994). The Baltic is also threatened by a rise in sea level linked with the phenomenon of global warming. A recent meeting in Berlin (March–April 1995) on Climate Change (FCCC) reported on a General Circulation Model (GCM) showing that greenhouse gases would induce a global rise in sea level of about 4 cm per decade on average (Bennetts 1995). Polish views of this impact vary; the most negative impact envisages violent land encroachment by sea water obliterating shorelines and destroying agricultural areas. Flooding would ruin low-lying areas like Żuławy, the Słowiński-Kaszubski Lake District and the Hel Peninsula. A less pessimistic view suggests that Baltic Sea encroachment would only lead to coastal erosion (Wałdoch 1992).

Transboundary pollution also contributes to Slovakia's problems. For example, repeated oil leaks into eastern Slovakia have occurred from a faulty Ukrainian pipeline. Oil from the sub-Carpathian pipeline, 80 km from the Slovak border, has found its way into the Latorica and Uh rivers and wells in the area have become affected. No long-term repair programme has been adopted by the Ukrainian authorities, although each leak is mended as it happens. The absence of a bilateral water management agreement between the Ukrainians and Slovaks has meant the prevention and treatment of spillages remains a complicated issue. In the meantime, the Slovaks have built retaining walls along the rivers to collect oil, but they are not adapted to resist rapidly rising spring floods. As a result, drinking water remains under direct threat and alternative potable supplies have to be quickly available (Anon [l] 1994).

In the former Yugoslavia, war conditions have rather deflected interest from pollution problems. Until the situation reaches some kind of successful conclusion it will be difficult to estimate how much environmental damage has been achieved. The destruction of dams to impede water and electric energy into enemy territory has scarred the landscape, while uncontrolled wastes from defective industrial plants, military occupation and local livestock farms helps pollute groundwater (Jancar-Webster 1995: 64). Even areas outside the present conflict can be affected. For example, the earthwork Peruca Dam on the Cetina River near Split provides water and electricity for the Dalmatian coast. Fears of hidden explosives and threats from autumn rains could flood an area inhabited by about 50,000 people should it collapse. It also helps to regulate water flow for two other dams further downstream. In turn, such thoughts have complicated strategic water supply problems along the Adriatic coast dependent on the Cetina River catchment area. Much of Croatia's south Adriatic region already possesses zones where water supply is becoming critical; both industrial development and tourism could in future be affected by lack of fresh water (Ridjanović 1993).

Unfortunately, the military situation in Croatia during 1991–4 curtailed activities (e.g., in the Plitvice and Paklenica National Parks; Eidsvik and Deželič 1993). Conversely, in Slovenia, the more peaceful situation has enabled the Institute for Conservation of Natural and Cultural Heritage to undertake similar responsibilities to those designated its Croatian counterpart, as, for example, work carried out in Kočevsko Natural Park (Anko 1994). Other serious implications of the war in former Yugoslavia include the shelling of Dubrovnik by Serb and Montenegran artillery (severely damaging one of most important historic townscapes in Europe) and the pillaging of art treasures from repositories in the war zone. International assistance will be needed to trace valuables taken from places like Vukovar on Croatia's eastern frontier.

IMPACT OF POLLUTION

Health hazards

The hazardous ecological state of the natural environment in some regions of East–Central Europe has given rise to justifiable anxiety concerning links between environmental degradation and public health. Issues arising include the extent and causes of mortality, the pathology of reproductiveness and other medical circumstances. For example, several recent studies have linked air pollution with increased disease and mortality rates. The main ingredients are fine particles and ash from fossil fuel combustion, which are found in their more concentrated form usually in urban areas (Arden Pope 1993, Anon [i] 1995). Transport exhaust emissions also add to the air pollution cocktail. Also, globally, transport accounts for about a quarter of total CO_2 emissions, an important ingredient in the cocktail causing the greenhouse effect (Arp 1994). Moreover, noise mainly from road transport and around airports seriously affects human health and impedes the quality of life of many citizens. The transport infrastructure is devouring increasing amounts of land and helps destroy the beauty of historic monuments and natural scenery.

In Slovenia, inhabitants of urban areas are particularly vulnerable to the predicament of their environment. Most of Slovenia's urban centres and the country's manufacturing industries are located in narrow, often poorly aerated, headwater valleys, or sub-alpine basins. During winter domestic heating appliances and increased gas emissions (especially SO_2) are trapped in these cramped landscape conditions, where pollution intensity rises above permitted emission levels. Urban areas also face surface water pollution problems. In Slovenia, only about one-tenth of waste waters are subjected to adequate filtration methods, whilst the capital, Ljubljana, still does not possess any central filtering installation system. Urban areas also lack sufficient waste disposal sites. In the Zasavje region, municipal health standards are particularly affected by environmental pollution; here the Sava river flows through the industrial mining towns of Hrastnik, Trbovlje, and Zagorje, together with settlements in the Šaleška and Mežiška Valleys, and the basins of Celje and Ljubljana. In future, people's health standards will only improve if fossil fuel consumption declines and demands on Ljubljana's potable water supply are restricted (Zupančič 1994).

Since the fall of Communism in Bulgaria, there have been few changes made to the health system. Poorly equipped hospitals, chaotic management, dependency on state funding and an absence of a health insurance system (and hence lack of private medical care) are characteristic of the situation. The health care system has not been helped by the worsening economic crisis coupled with energy and water shortages (Christophorov 1994). Such a scenario is hardly reassuring for the nations's health standards faced with the threats of environmental degradation. For example, Burgas, on the Black Sea coast, is typical of a region under such threat (Pavlínek *et al.* 1994). Here the worst offender is the Burgas petrochemical works at Kameno, which releases hydrogen sulphide, lead, aerosols, phenol and hydrocyanogen into the environment. The refinery's purification equipment is woefully inadequate to deal with the 6,000 m^3/hour capacity, at 0.05 mg – 2 mg/litre petroleum concentration.

As a result, the health of inhabitants in the Burgas region mainly suffers from local petrochemical pollutants. Furthermore, the Black Sea coastal area is particularly susceptible to local wind breezes. When land breezes are prevalent (averaging ninety to one hundred days annually) they help circulate atmospheric pollution; some of the highest occurrences are recorded in Burgas and Varna bays with an average of ninety-two and one hundred and seven days per year respectively (i.e., the two most polluted areas along the coast). Recent research has shown that two-fifths of the recorded illnesses in the Burgas region are linked to the respiratory system (such as bronchitis, emphysema and asthma); the highest is concentrated at Kameno, site of the petrochemical works. Over a fifth (22 per cent) of infant mortality deaths in the region are linked to breathing problems, the Kameno district, with a third (33 per cent), having the highest. More generally, nearly two-thirds (62 per cent) of children up to fifteen years old are registered as having respiratory problems, whilst among adults the most prominent age group is forty-five to fifty-four years old, of which nearly three-fifths (57 per cent) are male (Nikolova and Rukova 1994).

As already noted, the environmental situation in the Czech Republic is very serious. It involves all the main environmental media, which are collectively linked and affect all regions, in some more seriously than in others (Dzúrová 1992). Two major regions can be identified: first, the industrial region of north-west Bohemia with its predominance of mining, raw material processing and heavy industry; here declining quality of life leads to the occurrence of seasonal 'emigration' of children to summer camps elsewhere in the republic. High mortality levels result from a mixture of environmental and social risks. Second, large urban areas such as Prague, Ostrava and Plzeň, are characterized by high population densities, with significant numbers employed in tertiary activities and there is considerable localized pollution.

Recent comparative research on the health situation in these two regions (Teplice project) has produced some surprising results. As expected, life expectancy in the Teplice Basin is much below the Czech average and pollution might reasonably appear to be the sole cause. Therefore, how can Prague's much higher life expectancy be explained, where pollution levels are just as high as parts of north Bohemia? Findings from the Teplice project suggest that people's unhealthy habits as well as the environment have led to this state of affairs, with smoking an important contributory factor. Further research will assess the relative values of 'environment versus bad habits' in the overall equation (Plamínková 1995: 5). Meanwhile in Prague, French energy companies (Gaz de France; Charbonnage de France; Compagnie Générale de Chauffe) have undertaken a study to reduce air pollution, which proposes replacing coal with gas and modernizing the city's domestic heating system. If accepted, the capital's winter smogs, caused by ash and soot, may eventually be eradicated. Finance for the project costing some $1 billion would be shared between the French government, the EU, the EBRD and the World Bank (Anon [j] 1994).

Laudable as the Teplice and Prague projects may be, high mortality may continue to be influenced by other factors. These include a poorly organized health service (e.g., inadequacy to prevent circulatory diseases and inefficient organization of emergency care), poor dietary habits (high fat and salt consumption leading to cardiovascular and cerebrovascular diseases) and stress factors related to social, economic and environmental conditions (Rychtaříková 1994).

In Slovakia, a recent government report has revealed that two-fifths (41 per cent) of the country's population live in a heavily polluted environment (Anon [k] 1994). For example, the controversial Slovak aluminium plant at Žiar nad Hronom has been polluting the environment for years, which has led to continued protests by the local populace. It is not surprising, therefore, that it has become one of the projects to be financially supported by the EBRD and Norsk Hydro. Inhabitants of the larger industrial towns are subjected to the worst living conditions from air pollution; people living in polluted areas suffered a greater incidence of disease and lower life expectancy, reaching as much as seven years below the European norm. In such areas there is a high morbidity and mortality from malignant neoplasms, especially amongst women (Dzúrová 1992: 103).

In Poland, environmental pollution, new stresses of life under capitalism and the notoriously unhealthy Polish diet have led to an increase in fatal diseases on a scale much greater than in west European countries. Environmentally, twenty-seven areas have been identified as hazardous to human health (Mazurski 1995). It is estimated that nearly 13 million people (i.e., every third person) in Poland inhabit these ecologically contaminated areas (Świecka *et al.* 1995). They are mainly concentrated in regions with the highest recorded pollution caused by industrial waste, dust, gases and sewage. The country's mortality pattern is territorially very clear; on a regional scale the division between western and eastern parts of the country is very significant, with heavily polluted areas having the highest recorded death rates; these are located particularly in the south west (especially Katowice voivodship), where pollution provides the greatest threat to human life (Mazurski 1993). In contrast, low mortality levels are found along a belt stretching north–south in Poland's eastern region (Mazurski 1994). Here the voivodships (especially Białystok) coincide with rural areas that have the least economic and hence pollution development. Nevertheless, deaths from cancer, while partly linked to increasingly environmentally related carcinogenic factors, have also been connected to lifestyles associated with diet, tobacco and alcohol consumption. For example, mortality rates from lung and laryngeal cancer, with the exception of environmental pollution, mostly occur in smokers, of whom Poles have been ranked among the world's heaviest participants for many years.

Witkowski (1993) refers to the poorer demographic performance of Poland's 'endangered towns', which include many of the older industrial centres. During the 1980s they registered higher mortality: 8.1 per thousand (compared with 7.1 for other towns); higher infant mortality (i.e., deaths within three weeks of birth = 14.6 per thousand live births compared with 12.9 in other towns) and lower net immigration (2.3 per thousand per annum compared with 9.6 in other towns). Polluted cities in the west of Poland do not offer higher living standards (such as better health care) in compensation, as is generally the case in the West (Okólski 1992).

Buildings and infrastructure

The historic towns of Eastern Europe have been much neglected over the last half century and at the same time they have suffered badly from intense pollution, which has literally eaten away at the fabric of such cities as Cracow and Prague. In the former GDR especially there was not only a low priority given to the restoration of public buildings but the perceived cost advantages of housing in apartment blocks prevented the renovation of derelict town houses often 'donated' to the local authorities by private owners whose income from the (state-regulated) rents of tenants was quite inadequate to allow repairs to be carried out. However, properties in a poor state of repair are still available for refurbishment unless properties have been demolished and the sites redeveloped. And it should not be overlooked that some major works of reconstruction have been carried out, such as the old town in Warsaw, which was left totally devastated at the end of World War II. Furthermore, there has been much attention to individual historic buildings and conservation areas across the region. However, living conditions in all residential areas in polluted cities were unpleasant. Apartment buildings need annual sand blasting to stop them turning black (this is becoming too expensive and buildings are now being painted). To some extent problems were increased by poor planning: thus the Tysiąclecie Estate built in Katowice in the Communist period used unsuitable sites close to industries like the long-established Kosciuszko steel works and the 'Jan' coalmine.

Infrastructure is adversely affected in areas of mining subsidence. Railway speed restrictions apply in the Katowice region because of the danger of subsidence relating to old mine shafts. Water shortages are exacerbated by pollution. In Bulgaria there is strict rationing of supplies in Sofia due to the inadequacy of supplies from the Iskar reservoir. There are also problems on parts of the Danubian plain and in the dry southwest during the summer months. There has been increasing use of water since the 1980s when drought has been occurring more persistently (D.T. Gergov 1991). There is great variability in run-off and while Bulgaria's river should carry a run-off of 23. 9 km^3 in one year out of four they will also carry less than 10. 5 km^3 with the same frequency (a level close to total water use during the mid-1980s; Mikhova and Pickles 1994). The cost of water is now being passed on to the consumer while some relief has arisen through the reduced level of consumption by industry. Further progress could be made by pipeline replacement and by a larger scale of water diversion, to bring water from the Rila Mountains to meet Sofia's needs. This option was discussed in 1968 but abandoned in 1971 because of opposition from ecologists; and political changes in 1989 were brought about to some extent through Ecoglasnost demonstrations against the revival of the Rila proposal (Knight 1995).

THE RURAL ENVIRONMENT

Rural areas are not so badly affected by pollution although they generate some problems for themselves through the degradation of farm land (through monoculture); pollution of water supplies through excessive fertiliser applications and the accumulation of slurry. Tipping is a problem in mining areas. However, air and water pollution generated in the major industrial regions has a serious impact on some adjacent agricultural areas because since the 1960s and 1970s toxic heavy metal dust emissions and gases have been changing the soil character in the Katowice area of Poland. As the dust penetrates the soil to a depth of 24 cm gardens became unsuitable for growing vegetables. The outlook for the forests remains rather bleak in many parts of the Continent (Table 10.3). Accurate information is difficult to obtain and the international comparisons which relate to units of woodland with more than 25 per cent defoliation should be treated with some reserve. However, while it is clear that serious problems are

Table 10.3 Results of forest damage surveys: percentage of trees with more than 25 per cent defoliation (classes 2–4). Averages for 1990–3

Percentage Band	Countries in the band with percentages in ascending order
0–10%	*Russia 4.6; France 7.7; Austria 7.9; Spain 9.4; =Yugoslavia 9.8
10–20%	*Ukraine 14.7; Ireland 16.4; *Liechtenstein 14.0; Sweden 14.9; *Romania 15.6; +Croatia 17.4; Finland 15.7; Belgium 16.4; +Italy 17.3; *Slovenia 17.7; Switzerland 18.0; Greece 18.4
20–30%	Lithuania 22.3; *Luxembourg 21.7; Hungary 20.9; Norway 22.2; Portugal 22.5; Netherlands 23.3; Bulgaria 24.3; Estonia 24.2; *Germany 25.1; Denmark 27.6
30–40%	Slovakia 35.9; *Latvia 36.0; *Belarus 34.2
40–50%	United Kingdom 42.7; Poland 45.5
50–60%	=Moldova 50.8; +Czech Republic 54.7

* Only three figures available;
+ Only two figures available
= Only one figure available
Source: *Acid News*, no. 5 (1993), p. 9; no. 5 (1994), p. 9

to be found in Western Europe as well as Eastern Europe, there are major problems in the Czech Republic, Poland and Slovakia where there are major areas of damaged trees, covering several thousand hectares (Elvingson 1994). In recent years, forest ecosystems have undergone rapid changes, many for the worse as evidenced in countries like Hungary and the Czech Republic (Szépeszi 1994; Černy 1994).

A third of the Czech Republic is forested (i.e., 2.4 million ha: 33.4 per cent). Unfortunately, forests in northern Bohemia suffer from the direct impact of SO_2 originating in thermal power plants. Attempts are under way to create a 'North Bohemian Green Zone' around Liberec and include the former Soviet military base at Ralsko (Anon [q] 1993). Like north Bohemia, forest in north Moravia and parts of western and southern Bohemia are also in decline from acid deposition. This process creates negative soil changes and disturbs forest nutrition, which has been further hampered in recent years by drought. It is hoped that results from forest monitoring programmes will provide a better and more detailed knowledge of those processes and factors which have accounted for Czech forest decay (Černy 1994: 88).

Intensive environmental pollution also affects forests in northern Slovakia; this is especially prevalent under certain prevailing metreological conditions, where forests between 700–1000 m (a.s.l.) have been particularly badly damaged. Recent research on leaf defoliation and needle loss at sample sites exposed to prevailing winds in the Beskydy Mountains of northwest Slovakia have proved conclusively that pollution emission is responsible. Such studies of forest ecosystems are invaluable for comparative purposes (Oszlányi 1994). The results may also prove useful for a projected scheme on 'Ecological Management of Forests in the Slovak Republic' funded by a proposed $70 million World Bank loan (Anon [r] 1994). Meanwhile in Hungary, where 18 per cent of the land is forested (i.e., 1.7 million ha) and woodlands continue to expand, the percentage of primary forest is receding. This is because of the preponderance of quick-profit commercial activity and heavy cutting rates, amongst other things. Estimates suggest that one million m³ of oak forest are decimated each year, mainly in the northeast. In woodland areas, broadleaved varieties are prevalent; already in 1991, every fifth tree was suffering from more than a quarter defoliation (Andersson 1993).

Polish forests continue to deteriorate, with fir trees (*Abiesita*) suffering the most, having already become extinct in the Sudety/western Carpathian Mountains as well as in Silesia; they join stone-pine (*Pinus cembra*) and yew (*Taxus baccata*) in a growing list of vanishing tree species there, the latter ironically first placed under royal protection in 1420. Meanwhile, the number of firs continues to decline in the Holy Cross and Beskidy Mountains. Spruce (*Picea*) and pine (*Pinus*) growing there may also be affected (Dmuchowski and Wawrzoniak 1994). More generally, trees of all ages are dying in southwest Poland; for example, over nine-tenths of the area under coniferous forest has been damaged in the Jelenia Góra and Wałbrzych, nearly nine-tenths (88.6 per cent) in the Katowice and over fourth-fifths (81.3 per cent) in the Cracow voivodships. Only 1.8 per cent of forests are healthy in the whole of the Sudety region, with about 13,500 ha of forest having already disappeared. In spite of attempts on several occasions at re-afforestation, about 3,600 ha of land above 800 m (a.s.l.) still remain deforested (Rykowski 1993). Further, recent research on Scots Pine (*Pinus sylvestris*) needles and common moss (*Pleurozium schreberi*) has confirmed the presence of SO_2, NO_X and heavy metal emission sources over various regions of Poland (Grodzińska *et al.* 1994). Meanwhile, logging continues in many Polish forests to supply the insatiable demand of the country's growing furniture-making industry (Stachowiak 1994).

ACTION TO SOLVE ENVIRONMENTAL PROBLEMS

Industrial change

Between 1990 and 1993 there have been improvements in the Czech Republic with an 11 per cent reduction in carbon dioxide and a one-fifth reduction in sulphur dioxide; also a reduction of two-thirds in the production and consumption of ozone depleting gases. Lower demand for electricity has helped to reduce pollution problems in lignite mining areas. In north Bohemia mining decreased by a third because of falling demand during the first three years of the transition (to reach 53.4 million tonnes in 1992). Mining should end totally in the Ustí nad Labem region by 1996 on account of the adverse effect on settlement. But in the region as a whole production is estimated at 32–45 million tonnes by the end of the century. The range of possibilities results from alternative scenarios with respect to power generation: either a minimal reduction in capacity but with a major programme of desulphurization; or a major reduction in capacity with desulphurization at the principal power stations only. The latter would involve heavy job losses, although it would pave the way for radical economic reconstruction including more efficient resource exploitation. Meanwhile the closure of lignite quarries opens up the prospect of reclamation through either non-toxic waste storage or conversion into wetlands for tourism.

Further improvements in the Black Triangle are occurring as a result of renovations at power stations. In the Turoszów pocket of southwest Poland the Turów plant will be subjected to a modernization programme begun in November 1994, which will take seven years to complete. Costs of more than $1 billion will be covered by Poland's National Environmental Protection and Water Management Fund and its Eco-Fund. Western transfer technology is involved (Borys and Mazurski 1993). A joint Swedish–Swiss–American consortium will carry out work on the thirty-year-old plant, which produces one-tenth of the country's electricity needs and employs some 10,000 people. The programme's priority will be to control air emission levels; SO_2 emissions will be halved, NO_2 compounds reduced by a third and dust will be completely eliminated. Waste products from the desulphurization process will be buried in old mine shafts (Ratajczyk 1994; Templin 1995). However, the local power station was found to be responsible for only a quarter of the pollution found in the area: the rest came from the Czech Republic and Germany, where the coal has a high sulphur content of some 3 per cent compared with 0.4 to 0.8 per cent at Turów. Some power stations in adjacent areas of Germany have been closed (such as Hirschfelde). Greater efficiency in the area could arise through the piping of Turów's surplus steam to Gorlitz, Liberec and Zittau. Wider use of gas is proposed.

In the Czech Republic, the government has a long-term strategy for reducing brown coal dependency, substituting it with nuclear power, gas and oil. It also hopes to close down by late 1997 twenty-three coal-fired power generators (combined capacity: 2,000 megawatts) located at Tisová, Prunéřov, Tušimice, Ledvice, Mělník, Hodonín and Poříčí. At stations where this is considered impracticable (i.e., forty-two generating units: 9,000 megawatts), it is intended to install air pollution control equipment to meet emission limits imposed by the Czech Clean Air Act, which will operate from 1998 onwards. Alternative fuels, especially natural gas, more stringent methods of cleaning flue gases and reducing electricity production are amongst attempts being made to improve air quality. Meanwhile the worst polluter in Eastern Europe, the brown coal-fired Maritsa power plant in south eastern Bulgaria, will receive aid from the EBRD in the form of a 40 million ECU loan. This sum will contribute towards total costs of 114 million ECUs for completing the construction of a 210 megawatt lignite-fired generation unit, installation of a flue gas desulphurization unit and expansion of an ash disposal dump (EBRD 1994).

Environmental policies are usually a ministerial responsibility, though not always given high priority. In Croatia, the Ministry of Environment, Physical

Planning and Housing is responsible for the promotion and enforcement of environmental legislation, collecting and storage of environmental information, education and environmental assessment and more generally to act in an overall advisory capacity. In Poland, the authorities have declared their intention of investing substantial funds for improving the environment. There are incentives for Polish industry to invest in new environmental protection equipment. Enterprises increased their investment in both systems and equipment, but it was not done on entirely altruistic grounds. The severity of potential financial consequences for firms failing to comply to Poland's environmental protection laws was also a decisive factor (Anon [m] 1993). This new approach is now impacting on various industrial branches. For example, the Polish technical gases industry now has comparable facilities with most Western European countries; privatized Polish companies have benefited from notable investment in new equipment and significant Western know-how particularly regarding distribution methods and development of new appliances. Urgent plans are also being made to modernize the Polish petroleum industry, partly through the threat of environmental pollution and partly from the fear that cheap west European refinery imports could lead to domestic bankruptcies (Morka 1995).

Enterprises producing detergents, cosmetics and toiletries, once plagued with very old equipment and low productivity (30–50 per cent below Western levels) have, since becoming privatized, cleaned up their environmental standards and modernized plants (Anon [n] 1993). More efficient recycling schemes are under way and the provision of sets of bins in contrasting colours is now common in the former GDR, although there is only limited scope in Balkan rural areas. Reference may also be made to the use of plastic returnable bottles by Pepsi Cola and Coca Cola. These can be cleaned and refilled up to twenty-five times before being recycled. Progress is being made over motor vehicle emissions since all cars sold after October 1993 in former Czechoslovakia must be equipped with catalytic converters while former Czechoslovakia, along with Hungary and Poland, has introduced policies which favour the importation of such vehicles.

Local authorities and town planning

Local authorities have also been active in environmental protection, as for example in Koszalin province, Poland. Here considerable success has been achieved in returning local rivers (e.g., the Błotnica) to acceptable standards of water purity. Local councils have removed smaller coal-fired boiler heating plants in preference to larger ones fuelled by gas or oil. Most municipal funds have been allocated to the construction of new sewage treatment plants (e.g., at Bobolice, Drawsko Pomorskie, Połczyn-Zdrój, Polanowo and Mielno), the largest at Jamno (54,000 m^3 of sewage per day) serving Koszalin city (Jurkiewicz 1994). Modernization of the Wałbrzych heating plant is contributing to a better environment in the Black Triangle. There is a major sewage programme for Warsaw: the Czajka station opened in 1989 (for the right side of the Vistula) with Pancerz and Siekierki to follow. Initiatives have been taken over traffic management and Cracow has banned vehicular traffic within the walls.

In urban areas historic buildings and monuments have often been neglected but investment can now go ahead in a conservation-conscious era; rescuing buildings and entire architectural groups which, in a Western situation, would probably have been lost beyond recall through the comprehensive redevelopment ethos of the 1960s and 1970s. There is no doubt about the desire to conserve in the Czech Republic, where the drive comes from the President downwards. The problem is to operate a system at a time of ideological change, financial stringency and introduction of a new taxation system. There is the clean-up in Prague among contaminated derelict buildings, expedited by the growth of small businesses, although it is not clear if this will be an effective answer to maintenance problems in general. There are also pressures from developers looking for quick profits and consequently building styles need to be harmonized: new car showrooms do not blend easily with the older buildings. New buildings are being proposed for key sites, including one massive tower to symbolize Prague's westward leanings. But this would alter the skyline. Meanwhile, air quality in Prague is now adversely affected by increasing numbers of cars. This leads to pressure for more roads although some projects are affected locally by the NIMBY (Not In My Backyard) syndrome. Tourist pressure in Prague is serious in this respect.

There are some major schemes of urban reconstruction (e.g. Głogów) under way, including the plan for the rebuilding of the old town at Elbląg (northeast Poland), which was devastated after World War II. The 15 ha site was initially grassed over and subsequently opened up for archaeological work, but since 1975 a plan has evolved for the reconstruction of some three hundred historic tenement houses and one hundred and thirty have so far been completed. The most important buildings will be faithful reconstructions while the others will simply blend in. Money will be raised by the sale of the houses and the attraction of funds from sources other than the town's budget. However, finance is a serious problem in Albania where there are many inappropriate uses for historic buildings. A seventeenth-century 'hammam' (bath house) in Elbasan, restored under Communism (along with its fountain), is now a disco and bar fitted out with plywood and plastic. This is because the restitution law of 1992 has not made allowance for historic monuments: the state cannot buy back historic buildings because there is no money. Car repairs are being done in a valuable nineteenth-century town house in Shkoder while the homeless are being accommodated in the amphitheatre in Durrës, and this is unfortunate in terms of future tourism. In Prague there is a strategy of selling off historic buildings by local authorities who lack the means to look after them. But this is being shelved until an adequate system of conservation control is in place. Of course war damage in Dubrovnik and Sarajevo has greatly increased the scale of the problem in these cases.

Pozes (1991) refers to the neglect of traditional buildings in the village cores in parts of Slovenia where between a third to a half of the housing is derelict: the result of upheavals at the end of World War II and also the perceived inconvenience of the older properties. It is suggested that remaining agricultural activities should be moved out and that property should be refurbished; partly for permanent residents (where properties can be combined) and partly for second homes. There needs to be some financial assistance. Museum villages in Bulgaria constitute a valuable resource in the context of conservation and tourism. Arbanassi near Veliko Turnovo developed as a prosperous community of craftsmen and merchants enjoying privileges because of their high status amongst the feudal authorities. There are eighty preserved houses and five churches designated as monuments of culture. Etura on the outskirts of Gabrovo also gives insights into a flourishing centre of craftsmanship from the eighteenth and nineteenth centuries. Wooden houses of the National Revival period predominate in Kotel, a village in the Balkan Mountains, which enjoyed the status of 'privileged soldiers' village'; while the houses of Koprivshtitsa provide the setting for the folk festival held every four years. Nessebur on a small offshore island, linked with the mainland by a 400 m long isthmus was a port in Thracian times and provides additional legacies of its Greek and Byzantine past.

Environmental education

Much importance is being attached to environmental education and the principle of sustainability (Jordan and Tomasi 1994). There is an environmental profile in Polish primary and secondary schools and there are also environmental science departments in some Polish universities. Such provision is an important complement to legislation over pollution limits and countryside management. There is particular emphasis in specialist agricultural schools on ecological farming (Staniewska-Zatek 1993). This encourages a growing interest in organic crops on smaller Polish farms. Rejection of artificial chemical fertilizers and increased demand by the general public for organic products may offer a future for Poland's small farmers (Przybylak 1994). Some observers believe that this lead in Poland's ecological farming could set pioneering environmental standards that West European agriculturists would do well to consider (Pudlis 1994). Adjustments are being made in agriculture in sympathy with the concept of environmentally sensitive areas. Slovakia is modifying its large open fields through the restoration of biolines such as belts of trees and bushes along streams. This will reduce erosion and restore some of the variety in flora and fauna which has been lost. A biofuel venture is being launched in the Czech Republic between a US company Seven Rock Mining Jersey and Agro-Bio Moravia Přerov to produce biofuel from rapeseed oil.

Moreover, the latest tendency in Polish forest management is to utilize rigorous accounting techniques, long-term planning methods and management logistics. This 'total quality management' (TQM) strategy is already being applied. State forest employees attend TQM training sessions and during the next few years there are hopes to expand the programme so that each regional State Forest Directorate is subjected to a five-level training session (Fronczak 1995). Poland has a Programme for the Protection of Forest Biodiversity which has attracted help from the Global Environmental Facility in setting up a Forest Gene Bank at Kostrzyca near Jelenia Góra to store seed that can be used to rebuild the devastated southern forests. The Programme is also concerned with the protection of the primeval Puszcza Białowieska forest in the northeast.

The World Conservation Union (IUCN) is keen to encourage better management of protected areas and a wider strategy of sustainable use for the mountains as a whole in order to maintain biodiversity. These objectives are threatened by instability in the Balkans arising from the political situation, including changing patterns of land ownership. Hence, the Guidelines for Mountain Protected Areas Programme included the setting up of an Association for Conservation of the Balkan Mountains, in September 1992. Its main responsibilities were the maintenance of a list of regional experts and receipt of project ideas and proposals for incorporation into future activities (Anon [p] 1992). The regional assignment would focus on training schemes in management and language, while the national element would deal with site-based surveys and management control. Lists of protected areas (less than 100 ha) would be catalogued, a unified biogeographical classification created and species/habitats identified.

Protected areas

Whether classified as national parks, nature parks, regional parks, nature reserves or landscape protection areas, East–Central Europe's protected areas are today facing a period of change with many potential problems. Protected areas already in existence must be shielded from the possible perils of land redistribution (Mundy 1993) and likely threat of rapid capitalist development with all their inherent pitfalls. The present period provides a 'unique but brief window of opportunity to protect valuable natural areas and environmental resources before they are opened to private economic claims' (Andrews 1993: 9). It is therefore opportune that a plan entitled 'Parks for Life: Action for Protected Areas in Europe' initiated by the IUCN was launched in Bulgaria, the Czech Republic, Hungary, Poland, Romania, Slovakia and Slovenia during September 1994 (Synge 1994). Now is a good time to schedule additional protected areas. As interests supportive of a market economy gain ground large areas have come under increasing development pressures.

Bulgaria contains thirteen major national parks, seventeen biosphere reservations and three protected zones of international significance. Some of them, like the Pirin Mountains, are subject to conservation by the above association (Borisova 1992; Danchev 1993). In fact, the whole Rhodope Mountains are covered by this scheme. The Rhodopes cover an area of 18,000 km², of which 82 per cent (14,737 km²) lies in Bulgaria, and contain nearly a quarter (23 per cent) of the country's forested area. Work is underway on a complete inventory of Rhodopean flora, which contains nearly half (46.7 per cent) of Bulgaria's floral gene pool and numbers about 2,250 vascular plant types. About fifty verte-brates living in the Rhodopes deserve special conservation; these include mammals (e.g., bear, marbled polecat, Balkan chamois and several bat species), birds (black stork, black vulture, griffon vulture, Levant sparrow-hawk), reptiles and amphibians (pond turtle, various lizards, sand-boa and Alpine newt, a relict from the Ice Age). As a result of the variety and richness of flora and fauna, contemporary protected areas in the Rhodopes (eighteen strictly controlled reserves and twenty-three protected sites) are considered inadequate for ensuring their long-term preservation (Nikolova *et al.* 1994).

The position of Slovenia offers a wide variety of biogeographical regions ranging from Alpine and Dinaric to Mediterranean and Pannonian landscapes. The first area to be protected (Kočevsko) dates back to 1892, whilst certain plants like the Edelwiess (Leontopodium alpinum) have been shielded since 1896. The Republic has twenty, mainly glaciated, lakes, whilst woodland covers half the country's total area. New legislation on nature conservation was included in the 1991 Constitution and more recently acts have been passed for protecting threatened animal species (1993) and fungi (1994). A long-term Nature Conservation Strategy was developed in 1994 based on such inter-national documents as Global Biodiversity Strategy, Caring for the Earth, and so forth (Skoberne 1995).

One fifteenth (6.5 per cent) of Hungary's territory is now under protection, and covers 415 plant and 619 animal species. Hungary contains five national parks, 423 landscape protection areas and 118 nature conservation areas, as well as some 2,500 protected caves (Szalay-Marszó 1993). In spite of this care, these areas are occasionally subjected to major environmental pollution (Retvari 1994). Most of them are located either on state farm property, in agricultural cooperatives or under forest management, with only a few controlled by the nature conservation authority. In future, the nature conservation agency under the Ministry for Environment and Regional Policy will take over the ownership/ management of these areas (Lakos 1993). In the meantime, organizations like the Göncöl Foundation, which amongst its many environmental commitments, is involved in environmental impact assessment and management.

Nature protection in the Czech Republic has a long tradition dating back to medieval times; the first Forestry Act was passed in 1754. All forests are now under legal supervision, and protection has been extended to include large game (e.g., bears) and ancient tree stands. In February 1995, the Czech Institute of Nature Conservation was replaced by two new organizations, the Agency for Nature and Landscape Protection, and Administration of Protected Landscape Areas (Kučera, 1995). Earlier designated nature reserves remain, while their successors include areas of natural forest. Today, protected areas cover 58,597 ha (i.e., 0.7 per cent of former Czechoslovakian state territory) consisting of 1,225 separate localities. The first national park (Krkonoše [Giant] Mountains) was established in 1963 in north Bohemia, an area badly affected by air pollution (Žák 1994). More recently, two new national parks (Šumava and Podyji) were created in 1991. The country now has twenty-four protected landscape areas (e.g., Broumovsko on the Czech–Polish border) covering some 13 per cent of the Czech Republic (Urban 1993). The Slovak Republic is also continuing a long history of nature conservation (Vološčuk 1995).

Poland also has a long history of nature conservation dating back to the eleventh century; the first act to protect forest resources was passed in 1347. Today, forests cover over a quarter (28 per cent) of state territory, with meadows and pastures a further tenth (12 per cent) and marshland areas 4.5 per cent (Denisiuk 1994). Poland has over 2,000 natural monuments (old/historic trees, glacial remnants like erratics etc.), which are under the protection of individual voivodships. By early 1995 Poland's protected areas included twenty national parks (totalling 267,650 ha) spread across the country, together with seventy-seven landscape parks (1,466,683 ha) and 1,063 nature reserves (126,673 ha). They covered 0.78 per cent, 4.75 per cent and 0.40 per cent respectively of state territory. National parks vary in size from the largest, Biebrza Valley (59,223 ha) to Ojców (1,592 ha) the smallest. They contain a great variety of flora and fauna, including 325 plant communities and 200 bird species, together with numerous collections of insects, reptiles, amphibians and fish. Forests dominate many landscapes and cover nearly half (141 355 ha: 52.8 per cent)) of the total area designated to national parks. Landscape parks are more recent, again varying in size from 54,393 ha (Popradzki) to 700 ha (Ksiażański) and aimed mainly at landscape management. Nature reserves, though most numerous, are small in size, averaging only 119 ha with forested areas covering half the total. A smaller number are devoted to specialist flora and fauna, the rarest linked with steppe and salt marsh vegetation (Denisiuk 1994: 3; Okołów 1993a). Poland also has a number of biosphere reserves formed from a mixture of all three types of protected area (Breymeyer 1994). Four of them date back to 1977 including the Białowieża national park, famous for its European bison (*Bison bonasus*).

Many protected natural areas in Poland are facing a growing influx of tourists, and several of the national parks and reserves are unprepared for such a growth in numbers. However, providing the natural resources of these areas are wisely managed, all involve benefit, not least from an economic viewpoint. Carrying capacity has proved a useful tool in planning tourism, as well as models of ultimate environmental thresholds (UET). The latter has been applied to the Tatra National Park, where a maximum of one hundred and twenty stop-points have been demarcated, together with a comprehensive analysis of the various tourist activities undertaken (Kozłowski and Baranowska-Janota 1993). In the glaciated Tatra Mountains National Park (1954), where tourism followed the railways into rural areas previously concerned with pastoralism and mining, problems of erosion have occurred. Hence the need to close some routes to allow vegetation to recover, with fences and barriers to reduce further erosion. New forest plantings of mixed woodland are replacing softwood monoculture. Closing paths is not effective in the long run because it puts pressure on other routes. Suggestions include closure of the mountains at night (eliminating the tourist chalets) and the imposition of a tourist tax to generate income for footpath maintenance. There is a need for more effective control in the development of Zakopane where few traditional buildings remain and improved infrastructure is needed to cover sewage disposal and diversified tourist facilities for all weathers (Wiska and Hindson 1991). In the Tatra and elsewhere there is also scope for conservation in rural settlements where the buildings and cultural manifestations offer resources for community-based ecotourism.

In Romania international efforts are being mounted in the Danube Delta, which is a particularly sensitive area. The need to protect the fauna and the delta landscapes was appreciated before World War II and the initial list of nature reserves drawn up in 1940 was increased in 1956 by the Romanian Academy's Natural Monuments Commission. However, conservation work was compromised by the economic development programmes previously described. Since the revolution a 'Danube Delta Biosphere Reserve' of 5,900 km^2 has been created for the delta itself and for the extensive Razim-Sinoie lacustrine complex to the south; also for the sea coast, the Danube channel and the floodplain upstream of Tulcea to Isaccea and Galaţi. International support has come from UNESCO, the World Conservation Union (IUCN) and the World Wildlife Fund (Gastescu 1993).

CONTEXTUAL ISSUES

Environmental legislation

Further progress will be conditional on changes in environmental legislation. Many East–Central European countries suffer from a pre-transition legacy of poor quality environmental legislation. The post-transition period has failed to eradicate some of this inheritance. In some instances this has been due simply to inadequacy, in others due to over elaboration and frequent contradictions. Post-transition environmental matters in these countries are often assimilated into other legislation are a serious problem. This often leads to difficulties in the execution and implementation of environmental legislation. The present situation is partially due to a lack of efficient personnel, but also can be linked to public attitude. People are reluctant to take personal responsibility, an attitude inherited from former Communist times. Moreover, in the post-transition period, not only is there often a deficiency of local finance for environmental projects, but also a lack of administrative and human resources. This can lead to a breakdown in coordination and collaboration on environmental matters between various public bodies.

In spite of this, environmental legislation is now moving forward in all East European countries, although there is a continuing discrepancy between eventual aims and short-term achievements (DeBardeleben and Hannigan 1994). 'Although society was assured that a leading priority for the new legislators was saving the environment, in fact very little work was devoted to integrating environmental concerns in the transitional policy framework' (Georgieva 1992: 73). However, certain principles are now widely accepted: that polluters should be legally responsible for the damage they cause (i.e., Polluter Pays Principle), that standards should be raised, that environmental impact should be fundamental in the consideration of new development and that the public should have access to information. Evidently, greater progress is being made in the northern half of the region where there is a relatively developed technical and scientific infrastructure. There is also the stimulus of membership of the EU to accelerate progress in environmental legislation. Adoption of current EU standards is being widely seen as a precondition for progress in the negotiations. However, within this northern group there are contrasts between the gradualism in Hungary and Poland where effective environmental movements were active in the 1980s and FCS where much of the activity necessarily followed on from the Velvet Revolution. In the Balkans there are varying degrees of success in developing a legislative programme and attention has been drawn to 'an impressively simple, yet strong, environmental protection act' in Bulgaria (Stec 1993: 116), inspired to no small degree by the public protest already noted. Despite the long-term nature of the problem there should be enforceable standards, backed by both penalties and economic incentives, wherever possible.

The former Czechoslovakia had a plethora of environment laws. After 1989, inherited pre-1989 federal laws on waste management, air protection, land-use planning and general environmental protection seem to have been accepted by the Czech Republic. Hopefully, new Czech environmental laws will be more effective. A significant landmark in this respect concerns new decrees for the North Bohemian and Sokolov regions (Govt. of Cz. Rep. 1992a). This 1992 act distilled earlier laws on these two, of the so-called seventeen 'afflicted regions', all of which were considered to have suffered considerable environmental damage (Anděl 1994). In 1993, these two regions underwent further legal refinement with a decree giving environmental targets for the 1991–5 period and proposed totals for the year 2000 (Godalovič 1993) Other environmental laws passed included internationally imposed limits on various chemical uses, including freons (chlofluoromethanes), a series of fluorocarbon products used amongst other things in refrigeration and air-conditioning equipment (Govt. of Cz. Rep. 1992b). Nature conservation also experienced new environmental legislation. The Nature Conservation and Landscape Protection Act of 1992 was prepared and passed through parliament (Govt. of Cz. Rep. 1992c). It catalogued all the republic's protected areas, listed those plant and animal species needing special preservation and created a system for securing ecological stability at local, regional and national levels.

The legislative process continues to gain momentum, but care has to be taken on the choice of background material. For example, in Slovakia, two reports on the Mochovce nuclear power plant have appeared. The first suggests that use of this plant is not the most effective formula for meeting Slovakia's future energy needs; it also stresses the urgent closure of Jaslovské Bohunice's dangerous reactors at its existing plant (Pavlovec 1994). The second report provided similar advice regarding Slovakia's least-cost option for heat and power production (Bedi 1994; Anon [y] 1995). However, both were commissioned by environmentalists, campaigning for the closure of both Mochovce and Buhunice. Thus, they can hardly be used for formulating future environmental legislation without discretion, as they fail to give an unbiased/neutral viewpoint.

Meanwhile in Hungary there has been a significant increase in the number of voluntary conservation organizations and citizens movements in recent years, which suggests that nature conservation legislation may be in need of better implementation (Szalay-Marzsö 1993: 6). The 1990 nature conservation law recognized only five national park directorates, together with just two technical and four sub-units accountable for the safety of wildlife, forest and area management control, custody of geological features/caves and landscape protection (Govt. of Hungary 1990). In Romania the management of protected areas remains inadequate. While some shortcomings arise from poor organization and lack of finance, which places heavy reliance on volunteers, the fact remains that the Danube Delta is the only protected area with its own administration. Legislation in respect of a series of national parks approved in principle in 1990 is long overdue. Meanwhile, the National Conservation Strategy is contemplating financial alternatives (e.g., user charges, nature tax on commercial operations). Monetary benefits from such activity would be allocated to protected areas via a special 'Nature Protection Fund' (Dinu et al 1994).

East–Central European countries often look to their counterparts in capitalist democracies for advice on policy vis-à-vis environmental legislation (Carter 1992). While nuclear safety is a legislative matter, committment to nuclear power production as such is not. With regard to nuclear safety there are basically two main choices; many West European societies use the US model, which entails establishing pollution and planning standards and then incorporating them into decrees. The UK model differs from this system, because norms are negotiated between the administrative authorities and economic enterprises; this places great faith in the administrators. Overall, however, in reality disparities between the two models are not so huge because the US system also relies on selective enforcement and is open to many exclusions (Yeager 1991). In both models, therefore, those economic enterprises being monitored can sway the monitoring operation. However, they both reflect the 'Polluter Pays Principle' and prices are now being set to reflect the cost of environmental damage. Poland imposed fines for illegal emissions before 1989 and the levels are increasing faster than inflation; although in some cases they are still too low to provide powerful incentives to modernize. There is growing interest in the use of permits to authorize pollution under specific circumstances (the extent, nature and timing of pollution – also water temperature in the case of water pollution). The system pays for monitoring and makes good damage if the terms are not met (i.e., 'command and control'). But, in order to encourage investment it was agreed in 1992 that the government of the Czech Republic would assume responsibility for the environmental damage handed down to new owners.

However the 'command and control' principle could be extended through a more sophisticated system of 'incentive-based control'. Maximum levels are set and the total volume is distributed among the polluters who can then trade their emission rights. There could be a market in permits: if there was too much pollution the market value of permits would increase and those enterprises most able to reduce pollution would have an incentive to do so. 'The system thus encourages polluters with the smallest pollution abatement costs to make the greatest reductions' (Toman 1993: 18). Of course the system assumes that polluters will respond to economic incentives and that there is an adequate system of monitoring and enforcement of standards; not to mention a legal system that would allow emission trading. Over the long term however, the system provides the possibility of controlling emissions of sulphur dioxide across the continent as a whole. It also 'helps to internalise, in the prices of goods and services, the costs to society of pollution emitted after firms have complied with environmental laws' (Burtraw 1993: 25): higher costs are passed on to consumers with a consequent reduction in demand and an overall increase in the efficiency of resource allocation. But there are legal problems over the ownership of permits at the moment when pollution occurs. There could be economic instruments through lower prices for unleaded petrol, farmers could be paid to reduce intensification but it needs an enforcement culture. These issues are being discussed in Eastern Europe. Up to 1989, Poland's policy-makers preferred emission charges, in spite of their known inadequacy, in preference to alternative non-fiscal mechanisms. Although this strategy is still in force under the transition, in the belief that new market forces will give the best results, enough evidence exists to suggest that the quantity-based and administrative methods, including application of transferable discharge units, could be beneficially applied (Sleszyński 1992).

Finally, it is evident that legislation is being framed with international standards and norms in mind. Poland's Nature Conservation Act controls legal and other problems related to the country's natural heritage (Govt. of Poland 1991). It fully supports and applies any legal resolutions decreed by the World Conservation Union and promotes important international policies like the World Conservation Strategy, UNESCO's 'Man and the Biosphere' programme, and the International Council for Bird Preservation (Denisiuk 1994: 5). Moreover, in November 1994, environment ministries in both Poland and Hungary discussed how to ensure that their environmental legislation would be in line with EU laws in time for proposed formal admission in 2000. During 1995, a PHARE financial project will work out a detailed costing programme for adapting Polish environmental laws to EU rulings. Hungary and the Czech Republic will also receive PHARE aid for policy development and compatibility of environmental laws (Anon [z] 1995).

Environmental green parties and NGOs

In a recent book, Alain Lipietz has argued a strong case for political ecology as a 'green' alternative to traditional doctrines and political movements. He claims that the ideals of Socialism and Communism are now unable to realize the public's desire for change. He believes this function must now be performed by political ecology, which will create a coherent system of values based on social, economic and political change. Lipietz sees political ecology as a means for promoting a different way of existing together (Lipietz 1995). In East–Central Europe such a philosophy may well have some appeal. In several countries environmental movements played an important role in expressing public concern over environmental issues. Popular environmental activism led to the demise of former Communist governments after 1989 and continued to be a potent force during the early phase of national elections. Since 1989 green parties have contested elections with some success, but are not yet strong enough to exert any influence on mainstream political thinking. Green movements have been successful as protest groups, while some have become political parties prepared to share in government (Redclift 1989). Damaging splits have occurred, as in the case of Ekoglasnost in Bulgaria. The latter course may indicate greater responsibility, but there are risks of declining credibility if little is achieved in the short term (Jancar 1992).

Although environmental movements have multiplied since 1989, they were effective in some cases during the last years of Communism, particularly in Hungary and Bulgaria. In Hungary, where the regime was receptive to pressure from environmentalists, several major issues emerged during the 1980s such as the scare over water pollution at Vac linked with the storage of hazardous wastes by the Chinoin enterprise; the problems of dust emissions from the local power station at Ajka and controversy over proposals to install an incinerator for hazardous wastes at Dorog (Szirmai 1993). Campaigns against nuclear power were successful in getting improved safety provision at the Paks power station. And a sustained opposition was maintained by the relatively affluent and articulate German community in Ofalu (Baranya county) against the dumping of nuclear waste (Juhasz 1993). Concerted opposition arose over the Gabčikovo-Nagymáros hydropower project mobilized by Janos Vargha's 'Danube Movement', which became a formidable organization extending way beyond the Budapest intelligentsia. The organization helped to undermine the Communist government in 1989 since the energy and water lobbies were seen as representing Stalinist structures.

Meanwhile in Bulgaria it was possible for environmental organizations to emerge because of the special concern over the problems on the Danube at Ruse which was badly hit by the chlorine factory on the Romanian side of the river at Giurgiu. The incidence of lung disease rose from 9.7 per thousand of the population in 1975 to 173.9 in 1985. A 'Committee to Save Ruse' was created and since the Bulgarian authorities were not directly attacked 'to a certain extent it was possible for party members and organizations to declare their solidarity with the ecological protest' (Baumgartl 1993a: 162). The Ekoglasnost movement was founded in 1989 and concerned itself with environmental damage throughout the country, including the question of nuclear energy. Later in the same year a green party was formed and subsequently played a part in post-Communist politics in Bulgaria as part of the UDF. In April 1994, Ekoglasnost organized a new wave of protest meetings in Svishtov when representatives of a parliamentary commission visited the town. This was against renewed construction of the nearby 'Belene' nuclear power station, the country's second nuclear power plant, where building activity had to be suspended in 1990 in the wake of local opposition and representations from Ekoglasnost as well as lack of money. But, four years later, the residents of Svishtov had revised their opinions, favouring the reactor for the job opportunities it would bring (Anon [u] 1994; *Duma* 1994).

The Brontosaurus Movement, though still in existence, was not really an independent NGO as none legally existed in the former Czechoslovakia before 1989. Founded in 1974, it was part of the Socialist youth movment (SSM) and thus directly controlled by the government. It was mainly directed towards instilling a responsible outlook towards nature and the environment through education, appreciation of conservation, and help in the restoration of monuments. Young members enjoyed holiday summer camps, organized for work in protected landscape areas, helping to restore cultural monuments and learning about environmental education. Weekend work and study trips were arranged during the school year, whilst technical and educational activities involved setting up exhibitions, conferences etc. (Anon [w] 1995).

The Czech Union for Nature Conservation, (ČSOP) was founded in 1979. At present it is the republic's largest environmental NGO with 7,000 adult members dispersed amongst 420 local groups. Emphasis is placed on two major activities; first, nature and landscape conservation and second, greater awareness of environmental education. The latter is seen as particularly important for young people and is accomplished through specific schemes, projects and campaigns, organized at local, regional and national levels. For example, recent activities have been involved with management of modestly-sized protected areas, organization of summer nature conservation camps for children, and the upkeep of forests. There is close cooperation between ČSOP and governmental nature and landscape protection agencies, especially in the administration of district offices, national parks and protected landscape areas. ČSOP makes a significant contribution to nature conservation in the Czech Republic, even though it does not attract the outside attention of some Czech NGO counterparts (e.g., anti-nuclear energy groups). In many places where commissioned to do so by government bodies, ČSOP practically runs the management of nature reserves and nature monuments. Moreover it delineates boundaries *in situ*, suggests potential areas for protection and operates emergency preservation and breeding stations (Damohorský 1994).

Such laudable organizations are often overshadowed by more sensational environmental events. According to the Czech Greenpeace representative recent confrontation occurred over nuclear power plans. With reference to the Temelín nuclear plant in southern Bohemia, it has been intimated that the Czech government and Westinghouse Power Company have totally disregarded local opposition. The government was asked by representatives from fifty-four out of the sixty towns and villages in the area not to go ahead with the plant. Their pleas were ignored and there has been no public participation in any part of the decision-making process, no vote in parliament, no public hearings and no environmental impact assessment. Meanwhile, there has been friction in parliament between the Environment Minister (František Benda) and some of his fellow ministers, who consider he is 'too green'. This reputation arises from Benda's opposition to plans for a new cement factory in Tmaň, central Bohemia, and his hard line approach against foreign waste imports. He is considered a moderate by many environmental groups such as Hnutí Duha (Rainbow Movement) and Děti Země (Children of the Earth), who see his policies as beneficial to the environment, in contrast to his political opponents, who see them as bad for business and market economics (Knox 1994; Tuček 1994b).

In Slovakia, as in the Czech Republic, it is the nuclear energy problem which attracts most commotion. This was amply manifested in a 200-strong demonstration in February 1994 at the planned completion of the two reactor units at Mochovce by the Slovak Energy Company (Slovenský Energetický Podník) and its western partners. Delegates from environmental groups in Austria, Germany and Slovakia, as well as local citizens from Levice, a town 15 km from the nuclear plant site, were in attendance. Representatives from Slovakia's Greenpeace, Children of the Earth, Ipel Union, SZOPK, Tree of Life and Association Slatinka were all present. Greenpeace Slovakia believes that an environmental impact assessment has to be carried out; alternative energy production means should be explored, together with greater energy savings and the implementation of progressive legislation. Environmental NGOs intend to watch progress of the assessment closely, add their own interpretations and encourage public opposition to the plant (Anon [w] 1994).

Postponement of Poland's nuclear power programme until well into the next century has provided little scope for NGO activity on this topic. However, the tradition of nature preservation organizations has existed in the country since the mid-nineteenth century (Gliński 1992). Poland's environmental movements, like other political sectors, have undergone radical change since 1989 (Kabala 1993). Their number has risen from less than one hundred before 1989 to more than six hundred by 1995. Environmental groups – collectively known as the Polish 'Greens' – are also having to modify their activities in the post-1989 era. Previous techniques like street demonstrations, protests and condemning government environmental reports are being superseded by useful ideas for practical projects which will hopefully improve the quality of Polish life (Glińksi 1994). NGOs (e.g. PTTK and LOP) there now realize they must adapt to the needs of contemporary society and no longer act like an anti-Communist opposition (Pudlis 1995d).

Poland's 'Greens' do not support existing ecological parties at present, in spite of overtures to try and elicit their backing. Maybe some time in the future the Polish 'Greens' will develop into an appropriate political Green Party, but at present they lack a forceful charismatic leader. If supportive Western funds continue after 1997, a genuinely political movement of ecological parties may well emerge. This would inevitably involve the Polish Ecological Club (PKE), founded in 1980, and one of Poland's oldest environmental groups. Today the PKE, like other ecological organizations, is looking for a new identity in contemporary Poland. Its popularity in the country after 1989 has waned, although it continues to maintain its political neutrality. Only the future will tell if the PKE remains the major force it once was amongst the Polish 'Greens' (Pudlis 1995e; Janecki 1995).

The public is becoming less supportive of environmental objectives, given the economic problems, and there is a widespread appreciation that the problems will only be solved over the long term, given the scale of the inheritance from the Socialist past. Genov (1993) points out that economic and social problems (including crime and unemployment) have increased to put environmental problems in a different perspective, while the fall in production has in itself brought some noticeable diminution of pollution hazards. There may also be some cynicism sparked off by a traditional distrust of politicians and their ability to take decisive action even after a revolution. But although economic issues seem to have overtaken those of their environmental counterparts, environmental degradation is still omnipresent in many parts of Eastern Europe. Unless the public remains activeand alert, decrees passed by post-1989 governments on environmental matters, will have little chance of success (Caddy 1995). Hence the crucial role to be played by green parties and environmental non-governmental organizations (NGOs) of all kinds. In environmental matters their function is primarily to supply reliable data on local circumstances, arouse popular feeling, motivate public participation in decision-making and ensure that legislation is correctly administered (Anon [s] 1994; Anon [t] 1993).

The NGOs have problems raising finance and it is a pity that because, in principle, the EU does not dispense 'small' financial amounts, it is not well adjusted to provide support. Between December 1992 and June 1993, only one of the eighty-five projects approved within the PHARE programme was related to environmental issues. In any case the EU usually does not award full grants, so the successful applicant even then has to contribute a proportion of the costs. Cooperation with international networks/organizations appears the best way for NGOs to overcome these problems, particularly as such bodies would have some experience in dealing with EC funds (Rubin and Kaspar 1993: 18). Nevertheless, the PHARE programme has provided some support for national and regional programmes in East–Central Europe. In Hungary the 'Somogy' Nature Conservation Organization succeeded in receiving equipment for a project in the

Boronka wetland. Some Civic Sector Programmes also include support for NGOs; for example, the 'Support for Civic Dialogue' project in Poland was able to allocate money for information and legal services, educational and training programmes and grant-aided projects to NGOs. Faced with such financial stringencies many NGOs can only resort to condemnation tactics.

Foreign aid and technology transfer

Concern over environmental reconstruction in East–Central Europe was well-recognized in April 1993 at a European ministerial conference held in Lucerne, Switzerland. Although the urgent need for financial investment to fight environmental degradation was acknowledged, the full fiscal backing necessary has been less forthcoming (Baumgartl 1993b). The tendency has been to favour those environmental proposals that provided most economic advantage from investments made, or were fiscally less demanding. Furthermore, it should be remembered that although introducing western environmental methods to East–Central Europe can initiate many improvements over the short term, they may also bring problems. In spite of some policy achievements and its own advanced technology, the West is still a long way off solving some of its own environmental issues. Further, West European concern is partly motivated by self-interest, associated with problems of transboundary pollution. However, the idea of 'debt-for-nature/ecology-swaps' has proved an enterprising move, whereby East–Central European countries can be temporarily relieved of debt repayments, allowing such money to be invested in solving environmental problems.

Western countries have established a number of bi- and multilateral aid programmes for East–Central Europe, mainly in an attempt to establish new democratic institutions there. Most of the funding was provided through the PHARE programme, originally targeted at helping to reconstruct the economies of Poland and Hungary, but later extended to Albania, Bulgaria, the Czech Republic, Romania, Slovakia and some states in the former Yugoslavia (e.g., Slovenia) (Fontaine 1994). Environmental protection, although one of the five priority areas, was often undertaken on a limited and *ad hoc* basis. In general it was felt that there had been too many feasibility studies/reports and insufficient action on investment decisions. More specifically from an environmental viewpoint, too little effort has been made to combat problems such as air pollution, especially from transport and smokestack industries, whilst local expertise had been under used.

In May 1993, thanks to an international pollution conference in Lucerne, proposals were made for better use of western assistance; in future a more integrated approach (including an institutional infrastructure) would be made; sectorial working groups were to be confined to nuclear energy, macro economic assistance and environmental projects, while other issues would be undertaken on an individual country basis (Rubin and Kaspar 1993). EU countries with borders adjoining the former Communist states had a greater sense of urgency and practical application. For example, Austria offered aid for projects that would prevent external air and water pollution sources entering its territory; Italy signed a mutual cooperative agreement with Slovenia for solving transboundary water-borne pollution problems; Germany wished to advise on environmental policy, law, administration/management and nature protection projects, through training and institutional programmes (Pape 1993b). In the long run, however, the real ecological challenge for PHARE is to actively integrate the environment into all aspects of its overall programme and not merely pursue damage prevention policies.

Nevertheless, part of the PHARE environmental programme strategy is dedicated to establishing monitoring systems and providing equipment through technology transfer. Also, the EU has encouraged cooperation in science and technology with East–Central European countries. However, its major function is only that of coordination and providing administration and organizational costs. Prospective countries must apply to take part and, if successful, can utilize EU research and technological development programmes to help execute and complete their own projects, which they themselves choose. Recent experience of environmental mismanagement has shown the need to think globally and act locally. Several research institutes have been established worldwide to concentrate on global thinking, but few consider the relevant research requirements for acting locally. Such an EU approach towards East–Central European countries helps redress this imbalance.

However, it is not only a case of giving money, for there is an element of technology transfer. Evidence of technology transfer varies for individual East–Central European countries. In 1990, immediately after the collapse of Communism, Western responses were forthcoming to problems of environmental degradation. However, they were mainly of a short-term character, reacting to most pressing requirements. In Bulgaria's case, this resulted in the provision of monitoring equipment for air pollution. After the 1991 elections in Bulgaria, a period of financial austerity resulted in closure of the Centre for Environmental Monitoring, one of the country's few expert units. This suggested that countries like Bulgaria are unable to pursue their own environmental programmes and are totally dependent on foreign aid. If West European states do not give precedence to environmental problems in such countries, there is scant opportunity for reconstruction of the environment to take place in the immediate future. However, motivation can be stimulated when West European concern becomes apparent.

For example, modern technology and advice has been forthcoming for Bulgaria's ailing Kozloduy nuclear power plant. Western aid was essential in overcoming safety problems and with the help of a EU PHARE funding programme of 20 million ECUs, Electricité de France established a scheme to improve its safety features and invited Bulgarian nuclear operators to experience western projects at first hand through the twinning of Kozloduy with Bugey, a French nuclear power plant. Also, advice has been given on Kozloduy's two oldest units and the safety level considerably improved. In 1993 the EBRD allocated the plant a further 24 million ECUs to purchase necessary safety equipment, whilst the EU continues to finance a European safety authority consortium, under French–German control, to visit Bulgaria (Baret 1995). Furthermore, Bulgaria is receiving a loan of 40 million ECUs from the EBRD under the PHARE programme in order to reduce sulphur dioxide emissions at Maritsa-Iztok II thermal power station. The money will also contribute towards a total cost of 114 million ECUs for installation of a flue gas desulphurization unit and expansion of an ash disposal dump (EBRD 1994).

The Czech Republic was also an early recipient of western short-term environmental help. In 1990, it received basic engineering services to establish a hazardous waste disposal network. The EU has funded a sludge burning incineration system to treat waste previously dumped in the Vltava and this will help to reduce the amount of polluted material reaching the Labe at Mělník. Money is also being provided for a water monitoring system (Kabala 1992). But, unlike the Bulgarians, the Czechs appear to have some finance for joint environmental projects. For example, Czech scientists, in cooperation with a Norwegian firm, have developed a new type of clean coal (Gravimelt) for burning in power stations. It does not release any SO_2, has less than one tenth of the sulphur content and produces up to five times the energy found in ordinary brown coal. The first Gravimelt plant will be completed in 1996, and the manufacturers claim it will cut energy production costs in power stations by up to two-thirds (Bond 1993). Soviet troops left large amounts of fuel and oil behind in places like Ralsko base near Milovice which have subsequently drained into the ground. A firm using a sophisticated western system of pumps and filters has been employed to render the area more attractive for sale to private investors (Rocks 1994). In northwest Bohemia authorities are consoled by the fact that it is possible to improve the local ecology with improved western technology, but unless foreign aid materializes, costs may remain prohibitive (Tuček 1994a). In Slovakia, potential funding for improving the Mochovce nuclear plant is promised by the EBRD.

In Hungary, transfer technology through Geographical Information Systems (GIS), is being used to identify critical areas of environmental degradation and compare results with ecological stability/sensitivity within particular areas (Bohn 1992). Use of LANDSAT TM images of Hungary provided a detailed base to measure ecostability. Research revealed the existence of critical environmental areas (CEA) in both the central part of the Great Hungarian Plain and Transdanubian Mountains. These areas suffered from the negative effects of pollution, together with two further CEAs in the national parks of the Bükk Mountains and Hortobágy (Mezósi and Mucsi 1993). Furthermore, the Danube River Basin has been the recipient of an intensive environmental programme. The EBRD funded a plan to target major areas requiring crucial action, which included pollution.

In the context of acid rain damage, the EIB is providing a twenty-year loan for afforestation in Poland over 17,000 ha; this is part of a five year programme by the World Bank to safeguard Poland's forests. There will eventually be more timber for wood processing industries and better employment prospects in depressed rural areas. In 1992 the World Bank donated money for filters on the chimneys of Nowa Huta steelworks and Skawina power station in Cracow. More generally, a conference on industrial development in Eastern Europe was held in Zaborów, Poland, in November 1992, supported by the United Nations Development Programme (UNDP). It encouraged speedy economic recon-struction, based on an itemized and phased strategy of development, and sought to harmonize plans based on the lowest possible use of energy, protection of natural resources, disposal of hazardous wastes and high recycling efficiency standards. If adhered to this would help maintain the principle of economic sustainability (Baumgartl and Stadler 1993). For Poland, this conference provided some guidance for future environmental policy (Niehörster 1993). Almost immediately after the conference, Polish environmental authorities publicly declared their intention of investing substantial funds into solving the country's environmental problems. Proposed aid from foreign agencies, including restructuring and debt cancellation, were directly tied into investment plans by the Polish government and industry to improve environmental structures.

Scandinavian firms (e.g., Kemiry Kemi AB, from Sweden) are also bringing greater awareness to ecological problems through joint ventures, particularly new technology for improving drinking water quality, and providing sewage pipe networks for communal/industrial waste (Wątróbski 1994). The long-term Action Programme for Norwegian Assistance to East–Central European Countries, is the main vehicle for Norwegian–Polish environmental cooperation. Poland is one of the states given priority status under the programme to help finance projects on the environment, as well as for agriculture, manufacturing, energy, health care, and education (Anon [o] 1993; Pudlis 1995a).

Not all western links are so positive to the Polish environment. For example, West European lorries conveyed thousands of tonnes of waste to Poland annu-ally, because it was either too expensive, or inconvenient, to dispose of in the West. Their contents included hazardous smelting dust, sludge, rubble, used hospital syringes, amputated limbs, expired paints and chemicals, shredded clothing, broken furniture, rusty refrigerators and scrap tyres for burning, which were all part of the growing environmental waste trade. In spite of strict border laws against such trade and vigilant frontier guards, such waste continued to be dumped in countries like Poland (Coll 1994; Bachmann 1994), but since 1994 it has been forbidden in Poland. However, a happier scenario for Poland has been the increasing success of the Debt-for-Nature Swap (DNS) campaign (Schreiber 1989). Three of Poland's creditors, the USA, Switzerland and France, have agreed to the eco-conversion of part of their Polish debt; other nations, including Germany, Italy and Scandinavia, have so far remained non-committal. Poland's Eco-Fund Foundation has emerged from the DNS initiative and with surprising rapidity has become one of the country's major sources for ecological investment projects. Since its inception two years ago, the Eco-Fund has contributed more than US $3,400 to sixty projects. In 1995 it will donate about the same amount annually up to AD 2010 for projects which have supra-regional significance. The Eco-Fund is seen by the Poles as a pragmatic and positive solution for solving their various debts with other wealthier western countries (Pudlis 1995b; Pudlis 1995c).

International cooperation

There is already a high level of cooperation between East European countries, especially in former Czechoslovakia and Hungary. In January 1993, notes were sent to the Hungarian Ministry of Foreign Affairs by the newly formed Czech and Slovak Republics stating that they would sanction any international agreements concluded by the former Czechoslovakia (Anon [x] 1993). These totalled some one hundred agreements, among them those of environmental significance. But further cooperation is needed to achieve common goals. The waste disposal issue should be addressed in the light of legislative shortcomings that encourage the export of waste from Germany (with its tough disposal laws) to Poland and the Balkan states. It will need a regional plan and a measure of international assistance (Merritt 1991). Another opportunity for international cooperation lies in the repair of buildings at historic sites like Dubrovnik, Cracow and Prague, where the problem is beyond the resources of both local and national governments. Likewise, there must be agreed goals in terms of emissions. The original sulphur protocol of the UN–ECE Convention on Long Range Transboundary Air Pollution, signed in 1985, aimed at reducing sulphur emissions by 30 per cent between 1980 and 1993. The proposed new protocol diverges from its predecessor by allowing different targets for each country as opposed to a flat-rate percentage reduction previously set for all European countries. While Western European states were given until 2000 to achieve their goals, the East European republics were initially allowed a little longer (i.e., until 2005), which was later extended for a further five years. The protocol was expected to be signed in 1993, but difficulties arose when some countries bettered their bids, while others states did the reverse. It was becoming obvious that even by 2010 some East–Central European countries would not reach their stipulated totals. Amongst this category were Bulgaria, Croatia and Hungary, while Romania has failed to return any estimates. Nevertheless, it will presumably be included in this same group (Table 10.4). Ironically Albania, with a figure lower than its proposed limit, will be allowed to increase its sulphur emission total (Ågren 1994).

The Baltic Sea Environment Protection Committee dates back to 1960, with the aim of reducing pollution and creating a task force for programme implementation. A new convention was agreed in 1994, which provides for more stringent monitoring and control; also provisions for one hundred and thirty-two hot spots (most seriously threatened by pollution) and a wider action programme to run for twenty years. In 1991 a Union of Baltic Cities was formed after the

Table 10.4 East–Central Europe: commitment to sulphur reduction (% over 1980 total)

Country	Protocol %	2000 %	2010 %
Bulgaria	−74	−33	−45
Croatia	−40	−11	–
Czech Republic	−72	−50	−72
Hungary	−68	−45	−60
Poland	−66	−45	−66
Romania	−41	NC	–
Slovakia	−72	−60	−72
Slovenia	−45	−45	−70

NC No commitment given to reduce emissions
Source: C. Ågren, 'Sulphur: The making of a protocol', *Acid News*, no. 1 (1994), p. 11.

idea had been floated by the authorities in Gdańsk. Improvements will depend on reducing the dumping of sewage, but there is also oil leaking from ships. Hence the designation of a tanker route further from the coast to prevent oil slicks from reaching the beaches. A Vessel Traffic Service is being introduced as a safety measure and rescue procedures to cope with tanker breakdowns are also being improved. There is a role for NGOs such as the Coalition Clean Baltic in the programme which is supported by EBRD and other financial institutions; and debt conversion schemes are operating to enable Poland to carry out projects such as the Hel Peninsula waste treatment plant.

Poland is involved in cooperation with neighbouring countries as part of the 'Green Lungs of Europe' project to coordinate activities in networks of national parks and other protected areas (Degórska 1993). The project was developed in 1993 from the earlier 'Green Lungs of Poland' project to include Belarus, Estonia, Latvia, Lithuania, Russia and the Ukraine. It is hoped such international contacts will bring stronger political support for the development of Poland's own national parks and landscape protected areas, in a country where no real nature conservation lobby exists (Okołów 1993b). There is close cooperation between Poland and Belarus over the Białowieża National Park, but more significant is the East Carpathian Biosphere Nature Reserve established in 1993 to cover 164,000 ha in Poland, Slovakia and Ukraine. The Polish section includes the Bieszczady National Park and the Ciśniańsko-Wetliński and Dolny San Landscape Parks, while the Slovak part includes the East Carpathian Region of Protected Landscape and the Ukrainian section covers the Stuzhitsa Nature Reserve. The concept was first discussed in the 1960s in the context of a small area of some 2,760 ha but only now have the signatories agreed to comply with international standards for biosphere reserves. There are core, buffer and transition areas as well as corridors across the frontiers that migrating animals can use. The reserve includes the largest stand of beech forest in Europe as well as distinctive mountain meadow country called 'poloniny'. There are many rare species including the brown bear, Carpathian deer, lynx, wildcat and wolf; also rare birds such as the Ural tawny owl and golden eagle and plants like Carpathian spurge, Dacia violet and Taurida aconite. The population is light (less than five people per km²) but the cultural landscape of wooden Orthodox churches and wayside shrines perpetuates the traditions of the Boyko and Lemko peoples, sadly dispersed by the Polish Communist regime after World War II for pro-Nazi collaboration. A nature conservation plan for the Bieszczady Park will probably serve as a model for use in the other countries. The region could well become a mecca for nature lovers in the years ahead; PHARE programme money is being made available for a tourist complex at Wetlina in the buffer zone with a capacity of 1,000 people.

Poland's most recent park (Magurski; 19,962 ha) was established in January 1995 near Gorlice, its southern edge bordering on the Polish–Slovak frontier. It was founded to protect extensive forest consisting mainly of Carpathian beech (*Dentario glandulosae-Fagetum*) and is rich in large fauna (e.g., lynx, wolf, bear, red deer, and wild boar) (Dąbrowski 1995). Poland's Karkonosze and Tatra National Parks involve cooperation with the Czech Republic and Slovakia respectively. There is also cooperation between Hungary and Slovakia for a nature heritage and conservation area in the Danube–Ipoly area (Kiszel 1993). It covers the Naszaly mountain near Vac, with a wide range of habitats including dry grassland (on limestone) and *Tilio-Fraxinetum* forests. The objective is sustainable development with close monitoring of some areas. Meanwhile, the Czech Republic has transboundary cooperation programmes operating in national parks and protected landscape areas on the frontiers with Austria and Germany.

The case of Gabčikovo-Nagymáros

This is a particularly protracted and controversial case of international cooperation which is now well-documented (Fitzmaurice 1995; Galambos 1993; Persanyi 1993). The plan involved two hydropower and navigation complexes on the Danube where the river formed the frontier between FCS and Hungary. The two countries agreed to cooperate during the Communist period but from the very outset controversy raged over potential environmental damage resulting from the dam becoming operational. For the energy-deficient Slovakia, the issue is primarily economic; for the environmentally-conscious Hungarians, however, the project spelt disaster for the fauna and flora, especially in northwestern Hungary (Szigetkoz inland delta area) and the decay of several reservoirs. The Hungarians also believed there would be devastating flooding should the dam burst (S. Fisher 1993).

After the revolution public pressure forced the Hungarian government to abandon the project and this effectively meant that work at Nagymáros had to stop. But FCS (Slovakia after the break-up of the federation) was determined to proceed and the situation at Gabčikovo was such that unilateral action could be taken. The Slovaks moved the diversion point from Dunakiliti to Cunovo so that it would run entirely through Slovak territory. The river was diverted and the hydropower station opened in 1992. The whole dispute now awaits a ruling by the International Court of Justice in the Hague and it will be some time before a decision is made. There has been much discussion over the percentage allocation of water between the old river bed and new artificial channel (Verseck and Herre 1994). The Hungarians want to see the old channel adequately supplied while the Slovaks naturally want to maximize the discharge through the power station. A provisional water management system was based on the London Accord, under which former Czechoslovakia had undertaken to re-channel back into the main Danube River at least 95 per cent of the original used water volume. However, the quantity to be left to negotiate the old channel remains unsettled.

In sum, the whole project has been dogged by mishaps created by a mixture of bad luck, bad blood and bad planning. Even as recently as March–April 1994 the project's canals were closed to shipping for a month due to an accident caused by faulty locks (Drew 1994; Marsh 1994). But there are hopeful signs for a more harmonious future as regards international relations in general and conservation in particular. The concern over the hydro plant focused on the danger that the inland delta, consisting of forests threaded by river branches and pools, would be desiccated as less water became available. But the trees are healthy and the wells are full. Indeed Slovak engineers claim that the canal is an ecological boon, returning the forest to a state of wetness not seen for three decades. The old Danube was eroding its bed so deeply (2.5 m over thirty years) that there was rarely enough water to flood the wetland under natural conditions. But the problem has been solved by diversion of a tenth of the water (30–40 m^3/second) from the artificial (higher) channel on to the wetland at Dobrohost. At the same time the wetland at Cunovo will replenish underground water reserves; so that the water table has risen by some three metres at Bratislava. Although Hungarians think that water quality will deteriorate, Slovak engineers argue that the river is now less polluted and that pollution is not penetrating underground to the gravel layers from which drinking water is taken because of layers of mud which intervene. They also claim that water from the reservoir will flush out nitrates and other chemicals. Hungarian demands for more water in the old channel are therefore seen as an irrelevance.

It now seems that Hungary would like to settle the dispute and secure a greater flow of water along the old channel in order to save the Szigetkiz region. It has been indicated that the Skovaks have increased the flow from 200 to 400 m^3/second and although environmentalists see 800 m^3 second as the minimum, many Hungarians no longer give the issue a high priority. But the longer-term implications are interesting. The deeper incision of the river in the delta is attributed to dam construction in Austria in the 1950s, holding back gravel that was formerly carried down to the delta. Carrying less material, the river eroded more effectively. The impact was exacerbated by dredging carried out by the

Slovaks at Bratislava (to avoid flooding and improve the port) in the 1970s and 1980s. Thus the river was progressively cut off from the wetland. But river regulation will continue to have a knock-on effect and if Gabčikovo has become necessary as a consequence of Austrian projects then in the future the Nagymáros scheme which the Hungarians have cancelled may in turn become unavoidable because erosion will make navigation increasingly difficult (Pearce 1995).

CONCLUSION

Environmental problems have by no means been fully solved. Even though sulphur dioxide emissions in the Czech Republic have been reduced by one fifth since 1990, it was still possible for monitoring in November 1993 to record 3,438 micrograms of sulphur dioxide for each cubic metre of air at Chomutov (twenty-three times the allowable maximum of one hundred and fifty micrograms). Three per cent of deaths in the Czech Republic are still considered to be due to air pollution. There are still 300,000 flats in Budapest which are not connected to a sewerage system. In Poland, pessimistic assessments see emissions from Upper Silesia continuing to rise at least until the year 2000, with variations in the target date arising from different levels of commitment. Tighter regulations are inevitable but there will also be high costs and hence control must be adequately linked with means to sustain the cost and appropriate systems of control. The enormous cost of the environmental clean-up (for example $300 billion in Poland to meet EU standards) threatens businesses with claims for damage arising in the past under different management. Uncertainty over liability is making for a cautious approach over the acquisition of assets that could attract claims for compensation in respect of environmental damage.

The uncertainty is deepened by civil law rather than common law enforcement which 'is defined neither by precedent nor by a consistent application of judicial principles' (Boyd 1993: 2). An effective liability scheme may defer fresh cases of pollution but if such a scheme is applied retroactively (as is the case with legislation in the USA) it could stifle investment and result in negative asset values if properties were assessed to take account of the cost of retroactive liability. And if enforcement is delayed by an overburdened legal system, there is no incentive to reduce pollution, especially where closure is the likely outcome. Hence the alternative strategy of setting up public funds to finance the clean-up where the greatest ecological and health threats arise. Fractions of up to the entire purchase price of properties are now being placed in funds for the treatment of pollution damage. Regarding motor vehicles, the introduction of additional regulations could affect forecasts quite radically. While regulations for the fitting of catalytic convertors will not affect carbon dioxide emissions (which may even rise as traffic increases) there will be a major reduction in the case of carbon monoxide. Reductions in air pollution (sulphur dioxide and lead) to meet EU standards could achieve considerable benefits, estimated to exceed more than a tenth of 1988 GNP in Bulgaria and Hungary (Krupnik, Harrison, Nickell and

Toman 1993: 11). Poland's National Fund for Environmental Protection and Water Management has called for bids to build the country's own zeolite plant (providing a substitute for sodium triphosphate) so that the import of detergents with a high phosphate content can be stopped.

An environmental policy has to be based on a fair balance of resources between conservation and development. In the past this technology was purchased in order to increase the quality of industrial production, but there was relatively limited interest in using hard currency to acquire equipment for the control of pollution and production of such equipment, which within Eastern Europe was not a high priority. However, in periods of uncertainty like the present there is a temptation to place development issues above the need for environmental protection. If there is growing export pressure to help service loans then environmental problems could get worse before they get better. However, there are areas where progress can be made at reasonable cost. The cost of improving river water quality to EU standards has been estimated (in 1992 values) at $49 thousand per capita in former Czechoslovakia, followed by 3.8 in Bulgaria, 2.1 in Hungary, 1.4 in Romania and 1.2 in Poland (Paulsen 1993: 29). While the use of the best available technology to minimize discharges would be extremely expensive, experiments have shown that substantially higher standards (measured in terms of concentrations of dissolved oxygen) could be achieved at relatively modest cost. Only the first stages of improvement can be detected in the Balaton area (Retvari 1994). Following experience with biotechnology in the US Lemna Corporation it is hoped to clean up Bulgaria's waste water by using an ecofilter of duckweed-type plants to absorb pollutants in the Burgas area.

Finally, it is clear that the landscape will only change slowly and the legacy of central planning will remain in the buildings for the indefinite future (though some of the poorly-built apartment blocks may disappear relatively quickly). The same situation applies to the environment which was a major issue in the rising tide of opposition to Communism. But each state will have to determine its own priorities because perceptions vary and there are different attitudes to nature. In Slovenia's national park of 'Notranjski' around Postojna there is a need to reconcile the strong pro-conservation mentality of the Slovenes with the more exploitative attitude evident among Italian visitors who behave in a more predatory manner towards the environment. This is part of a wider issue over national parks and other protected areas: should access be strictly controlled or should they be developed to cope with heavier visitor pressure? But the final point should insist on the quality and charm of the East European landscape. Discussion of environmental problems should not lead to the supposition that the whole region is devastated. There is great variety in the fauna and flora of Eastern Europe. Many of the region's plants are much sought after in the West and some are now appearing in British seed catalogues after a long absence. The scenic attractions of many rural areas remain substantial. Despite severe damage, the highlands of North Bohemia are still most attractive, especially the section adjacent to Germany's 'Saxon Switzerland'; likewise the mountains of South Bohemia and the Tatra in Slovakia. In the Balkans the scenic resources (along with the archaeological monuments and popular culture) are even more arresting and a future for the region in ecotourism is an obvious possibility.

BIBLIOGRAPHY

2 ALBANIA

Adami, J. (1983) *Rrugë dhe objekte arkeologjike në Shqipëri*, Tirana, 8 Nëntori.

Adhami, S. (1988) 'Historical monuments, their restoration and use', in Anon, *Legacy of Centuries*, Tirana, 8 Nëntori.

—— and Zheku, K. (1981) *Kruja dhe monumentet e saj*, Tirana, 8 Nëntori.

Ahmataj, H. and Ceku, K. (1984) *Bimët aromatike*, Tirana, Botimi Shtëpisë së Propagandës Bujqësore.

Alia, Z. (1990) *The Family and its Structure in the PSR of Albania*, Tirana, 8 Nëntori.

Albturist (n.d.) *The Tourist Map of Albania*, Tirana, 8 Nëntori.

—— (1989) *Hotels and Tourist Centres in Albania*, Tirana, 8 Nëntori.

Almond, M. (1988a) 'A Balkan journey: part 1', *The Rock Garden* 20 (4), 459–66.

—— (1988b) 'A Balkan journey: part 2', *The Rock Garden* 21 (1), 76–81.

AMC (1991) 'L' Albania alla vigilia delle elezioni tra tensioni sociali e crisi economica', *Est-Ouest* 22 (1), 16–21.

Anagnosti, V. (1985) *The Terraces of Lukova*, Tirana, 8 Nëntori.

Anon (1971) *Atlas gjeografjk i Shqipërisë*, Tirana, Hamid Shijaku.

—— (1973a) 'Rregullore "mbi mbrojtjen, restaurimin dhe administrimin e qytetitmuze të Gjirokastrës"', *Monumentet*, nos 5–6, 211–17.

—— (1973b) 'Rregullore "mbi mbrojtjen dhe restaurimin e ansambleve dhe ndërtimeve të tjeva me vlerë historike të qytetit të Korçës"', *Monumentet*, nos 5–6, 219–22.

—— (1977) 'Rregullore "mbi mbrojtjen dhe restaurimin e qendrës historike të qytetit Elbasanit"', *Monumentet*, no. 13, 160–2.

—— (1982) *Bujqesia në Republikën Popullore Socialiste të Shqipërisë*, Tirana, 8 Nëntori.

—— (1985a) 'Albania comes in out of scientific cold', *New Scientist*, 30 May, p. 8.

—— (1985b) *The Enver Hoxha University of Tirana*, Tirana, 8 Nëntori.

—— (1990) 'Fruitful collaboration', *New Albania*, no. 6, 6–7.

—— (1991a) 'Albania; open for business', *East European Markets* 11 (16) 14–15.

—— (1991b) 'Albania's push for investment', *East European Markets* 11 (15), 11–13.

Atkinson, R.I., Bouvier, M., Hall, D. and Prigioni, C. (1991) 'Albania', in IUCN-EEP, Environmental status reports 1990, volume 2, Albania, Bulgaria, Romania, Yugoslavia, Cambridge, IUCN.

Baçe, A. and Çondi, D. (1987) *Buthrot*, Tirana, 8 Nëntori.

Baçi, I. (1981) *Agriculture in the PSR of Albania*, Tirana, 8 Nëntori.

Basha, H. (1986) *Tirana et ses environs*, Tirana, 8 Nëntori.

BBC (1990) 'Albania: environmental protection campaign', *Summaries of World Broadcasts*, 21 June, EE/W0133 A/6.

—— (1991a), 'Albania: France: Alb-Ducros joint agribusiness enterprise in operation', *Summaries of World Broadcasts*, 14 November, EE/W0205 A/1.

—— (1991b) 'Albania: government decree: lek exchange rate to be related to ECU', *Summaries of World Broadcasts*, 3 October, EE/W0199 A/1.

—— (1991c) 'Albania: Italy: trial cooperation with Fiatagri in crop farming successful', *Summaries of World Broadcasts*, 7 November, EE/W0204 A/1.

—— (1991d) 'Albania: need for caution in closing unprofitable enterprises', *Summaries of World Broadcasts*, 18 April, EE/W0175 A/4.

—— (1991e) 'Albania: review of offshore oil prospecting', *Summaries of World Broadcasts*, 14 November, EE/W0205 A/5.

—— (1991f) 'Albania: Yugoslavia: agreement between Pogradec and Skopje banks on construction projects', *Summaries of World Broadcasts*, 21 March, EE/W0171 A/1.

—— (1991g) 'Albanian People's Assembly Session', *Summaries of World Broadcasts*, 15 May, EE/1072 C1/1–11.

Business Eastern Europe (BEE) *Business International*, Vienna and London (fortnightly).

Bërxholi, A. (1985) *Vlora and its Environs*, Tirana, 8 Nëntori.

—— (1986) *Durrës and its environs*, Tirana, 8 Nëntori.

—— (ed.) (1990a) *Atlas gjeografik i RPS të Shqipërisë*, Tirana, Shtëpia Botuese e Librit Shkollor.

—— (1990b) 'On the surface – flowers and leaves, underground – silver and gold', *New Albania*, no. 5, 6–7.

——, Memisha, O. and Konomi, F. (1988) *Gjeografia e Shqipërisë*, Tirana, Shtëpia Botuese e Librit Shkollor.

—— and Qiriazi, P. (1986) *Albania: a Geographic Outline*, Tirana, 8 Nëntori.

——, —— (1990) *Republika Popullore Socialiste e Shqipërisë: harte fizike*, Tirana, Shtëpia Botuese e Librit Shkollor.

Bogliani, G. (1987) 'C'e'del verde in Albania – riscopriamo una nazione che cambia', *Airone* 74, June.

Borisov, V.A., Belousova, L.S. and Vinokusov, A.A. (1985) *Okhranyaemye prirodnye territorii mira*, Moscow, Agropromizdat.

Bouvier, M. and Kempf, C. (1987) 'La nature en Albanie', *Le Courier de la Nature*.

Carp, E. (1980) *A Directory of Western Paleartic Wetlands*, Cambridge, IUCN-UNEP.

Carrier, C. (1991) 'Pace of Albanian reforms picking up, slowly', *Business Eastern Europe* 20 (44), 389.

Cikuli, Z. (1984) *Health Service in the PSR of Albania*, Tirana, 8 Nëntori.

Çobani, J. (1988) 'Problems of state in protection of the atmospheric environment', in Balkan Scientific Conference on Environmental Protection in the Balkans, *Abstracts*, Varna.

—— (1989) 'Conservation of the environment in Albania', *Albania Today*, no. 109, 47–9.

Dede, S. (1983) *The Earthquake of 15 April 1979 and the Elimination of its Consequences*, Tirana, 8 Nëntori.

—— (1986) *Rruga e gjeologjisë Shqiptare*, Tirana, 8 Nëntori.

Demiri, M. (1983) *Flora ekskursioniste e Shqipërisë*, Tirana, Shtëpia Botuese e Librit Shkollor.

Dingu, L. (1988) 'The protection and enrichment of the vegetation genetic fund', in Balkan Scientific Conference on Environmental Protection in the Balkans, *Abstracts*, Varna.

Dollani, K. (1988) 'Control of radioactivity', in Balkan Scientific Conference on Environmental Protection in the Balkans, *Abstracts*, Varna.

Donovan, P. (1991) 'A drip-feed for the East', *The Guardian*, 15 April, p. 23.

Dyvik, E. (1991) Development Banking, European Bank for Reconstruction and

Development, Infrastructure, Energy and Environment Department, London, Personal communications, October–November.

EBRD (1991) *How to Work with the European Bank for Reconstruction and Development*, London, EBRD.

East European Markets (EEM) *Financial Times*, London (fortnightly).

East European Newsletter (EEN) (1990) 'Albania: reform?', *East European Newsletter* 4 (10), 1–3.

—— (1991a) 'Albania: companies and carpet-baggers', *East European Newsletter* 5 (8), 4–5.

—— (1991b) 'Albania: the morning after', *East European Newsletter* 5 (7), 3.

EIU Economist Intelligence Unit (EIU) (1990) *Romania, Bulgaria, Albania: Country Report No. 3, 1990*, London, EIU.

—— (1991a) *Bulgaria and Albania: EIU Country Profile 1991–92*, London, EIU.

—— (1991b) *Romania, Bulgaria, Albania: Country Report No. 1, 1991*, London, EIU.

Ellenburg, H. and Damm, K. (1989) *Albanien 1989*, Berlin, Institut für Geographie, Technische Universität.

Frashëri, M. (1988) 'Hydric resources in the context of Albania's environmental protection problems', in Balkan Scientific Conference on Environmental Protection in the Balkans, *Abstracts*, Varna.

Gardiner, L. (1976) *Curtain Calls: Travels in Albania, Romania and Bulgaria*, London, Duckworth.

Gjiknuri, L. (1988) 'Fauna and some protection problems', in Balkan Scientific Conference on Environmental Protection in the Balkans, *Abstracts*, Varna.

Greenberg, S. (1991) '"Green" board to oversee East European revival bank', *The Guardian*, 19 January.

Grimmett, R.F.A. and Jones, T.A. (1989) *Important Bird Areas in Europe*, Cambridge, International Council for Bird Preservation.

Gurashi, A. and Ziri, F. (1982) *Albania Constructs Socialism Relying on its Own Forces*, Tirana, 8 Nëntori.

Haigh, W.E. (1925) *Malaria in Albania*, Geneva, League of Nations Health Organisation.

Hall, D.R. (1984a) 'Albania's growing railway network', *Geography* 69 (4), 263–5.

—— (1984b) 'Foreign tourism under socialism: the Albanian "Stalinist" model', *Annals of Tourism Research* 11 (4), 539–55.

—— (1984c) 'Tourism and social change: reply to Romsa', *Annals of Tourism Research* 11 (4), 608–10.

—— (1985) 'Problems and possibilities of an Albanian–Yugoslav rail link', in J. Ambler, D.J. Shaw and L. Symons (eds), *Soviet and East European Transport Problems*, London, Croom Helm.

—— (1987a) 'Albania', in A.H. Dawson (ed.), *Planning in Eastern Europe*, London, Croom Helm.

—— (1987b) 'Albania's transport cooperation with her neighbours', in J.F. Tismer, J. Ambler and L. Symons (eds), *Transport and Economic Development – Soviet Union and Eastern Europe*, Berlin, Duncker & Humblot.

—— (1990a) 'Albania', in J.A.A. Sillince (ed.), *Housing Policies in Eastern Europe and the Soviet Union*, London, Routledge.

—— (1990b) 'Albania: the last bastion?' *Geography* 75 (3), 268–71.

—— (1990c) 'Change closes in on Europe's last communist bastion', *Town and Country Planning* 59 (9), 251–4.

—— (1990d) 'Eastern Europe opens its doors', *Geographical Magazine* 62 (4), 10–15.

—— (1990e) 'Introduction: geographic dimensions of change', *Geography* 75 (3), 239–44.

—— (1990f) 'Stalinism and tourism: a comparative study of Albania and North Korea',

Annals of Tourism Research 17 (1), 36–54.

—— (1991) 'Albania', in D.R. Hall (ed.), *Tourism and Economic Development in Eastern Europe and the Soviet Union*, London, Belhaven.

Hoda, P. (1991) Lecturer in Botany, Faculty of Natural Science, University of Tirana, Personal interviews, April/August.

Hoxha, D. (1988) 'Assessment and protection of land', in Balkan Scientific Conference on Environmental Protection in the Balkans, *Abstracts*, Varna.

IUCN (1967) *Liste des Nations Unies des parcs nationaux et reserves analogues*, Brussels, IUCN.

—— (1990) *United Nations List of National Parks and Protected Areas*, Cambridge, IUCN.

IUCN-EEP (1990) *Protected Areas in Eastern and Central Europe and the USSR*, Cambridge, IUCN.

Kabo, M., Kristo, V., Qiriazi, P., Krutaj, F., Gonda, G., Meçaj, N. and Bërxholi, A. (eds) (1985) *Studime gjeografike*, Tirana, Akademia e Shkencave e RPSSH.

——, Bërxholi, A., Krutaj, F., Gonda, G., Meçaj, N., Qiriazi, P. and Kristo, V. (eds) (1987) *Studime gjeografike 2*, Tirana, Akademia e Shkencave e RPS të Shqipërisë.

——, ——, ——, ——, ——, ——, —— (eds) (1988) *Studime gjeografike 3*, Tirana, Akademia e Shkencave e RPS të Shqipërisë.

Karadimov, I. (1989) *Balkan Mayors on the Environment*, Sofia, Sofia Press.

Karaiskaj, G. and Baçe, A. (1975) Kalaja e Durrësit dhe sistemi i fortifikimit përreth në kohën e vonë antike, *Monumentet*, no. 9, 5–33.

KPS (Komisioni i Planit të Shtetit, Drejtoria e Statistikës) (1988) *Vjetari statistikor i R.P.S. të Shqipërisë*, Tirana, Komisioni i Planit të Shtetit.

—— (1989) *Vjetari statistikor i R.P.S. të Shqipërisë*, Tirana, Komisioni i Planit të Shtetit.

—— (1990) *Vjetari statistikor i R.P.S. të Shqipërisë*, Tirana, Komisioni i Planit të Shtetit.

Kraja, E. (ed.) (1976) *Shkodra Almanak*, Tirana, 8 Nëntori.

Kristo, D. (1991) Senior Lecturer in English, Faculty of History and Philology, University of Tirana, Personal communication, November.

Kusse, P.J. and Winkels, H.J. (1990) *Remarks on Desalination and Land Reclamation in the Coastal Area of the People's Socialist Republic of Albania*, The Hague, Dutch Ministry of Agriculture, Nature Management and Fisheries.

Lloshi, X. (1990) *Academy of Sciences of the PSR of Albania*, Tirana, Publishing Council of the Academy of Sciences of the PSR of Albania.

McDowall, L. (1991a) 'Albania learns the art of wrecking', *New Statesman & Society*, 13 December, 18–19.

—— (1991b) 'Albania's desperate poor die in the fight for bread', *The Guardian*, 16 December.

—— (1991c) 'Democrats pull out of Albania coalition', *The Guardian*, 5 December.

—— (1991d) 'New prime minister installed as 32 die in Albanian food riot', *The Guardian*, 11 December.

McKinley, E.B. (1935) *A Geography of Disease*, Washington, DC, The George Washington University Press.

Mara, M. (1988) 'Problems of the state and protection of the natural environment in the PSR of Albania', in Balkan Scientific Conference on Environmental Protection in the Balkans, *Abstracts*, Varna.

Mason, K., Myres, J., Winterbotham, H.S.L., Longland, F., Davidson, C.F., Turrill, W.B., White, N. and Mann, S.E. (1945) *Albania*, London, Naval Intelligence Division.

Meksi, A. (1983) *Arkitektura mesjetare në Shqipëri*, Tirana, 8 Nëntori.

Miho, K. (1987) *Trajta të profilit urbanistik të qytetit të Tiranës*, Tirana, 8 Nëntori.

Muka, A. (1978) 'Banesa fshatare në malësinë e Tiranës gjatë shek. XIX dhe fillimit të shek. XX', *Etnografia Shqiptare* 9, 209–312.

Murphy, T. (1991) Senior Environment Consultant, European Bank for Reconstruction and Development, London, Personal communication, November.

Nano, T. (1987) *Bimet mjaltese*, Tirana, 8 Nëntori.

North, C. (1990) 'Into Albania', *The Rock Garden* 22 (1), 63–71.

Omari, L. and Pollo, S. (1988) *The History of the Socialist Construction of Albania*, Tirana, 8 Nëntori.

Ostreni, A. (1972) *Gjeografia e Shqipërisë*, Tirana, Shtëpia Botuese e Librit Shkollor.

Protected Areas Database Unit (PADU) (1990) *Information from Protected Areas Database*, Cambridge, World Conservation Monitoring Centre.

Pano, N. (1990) 'Ecological balance in the Mediterranean Sea', *New Albania*, no. 5, 28.

Papajorgji, H. (1989) 'Triumph of the correctness of the Party on the countryside', *Albania Today*, no. 109, 42–6.

Parkin, S. (ed.) (1991) *Green Light on Europe*, London, Heretic Books.

Petrela, S. (1990) 'A pearl of nature', *New Albania*, no. 5, 32–3.

Piperi, R. and Kajno, K. (1990) *Flora mjekësore e Korçës*, Tirana, Drejtoria e Arsimit Shëndetësor.

Pollo, S. and Puto, A. (1981) *The History of Albania*, London, Routledge & Kegan Paul.

Polunin, O. (1980) *Flowers of Greece and the Balkans*, Oxford, Oxford University Press.

Prifti, P.R. (1978) *Socialist Albania since 1944*, Cambridge, Mass., MIT Press.

PSR of Albania (1977) *Constitution of the People's Socialist Republic*, Tirana, 8 Nëntori.

Puka, V. (1988) 'Physical-chemical qualities of river water', in Balkan Scientific Conference on Environmental Protection in the Balkans, *Abstracts*, Varna.

Rilindja Demokratike ('Democratic Revival'), Tirana (daily).

Riza, E. (1971) 'Banesa e fortifikuar Gjirokastrite', *Monumentet*, no. 1, 127–48.

—— (1975) 'Studim për restaurimin e një banese me cardak në qytetin e Kruja', *Monumentet*, no. 9, 107–25.

—— (1978) *Gjirokastra: Museum City*, Tirana, 8 Nëntori.

—— (1981) 'Banesa popullore në qytetin-muze të Beratit', *Monumentet*, no. 21, 5–35.

Sandström, P. and Sjöberg, O. (1991) 'Albanian economic performance: stagnation in the 1980s', *Soviet Studies*, 43 (5), 931–47.

Shilegu, H. (1991) Director, Tirana Travel Service, Personal interview, April.

Shkodra, G. and Ganiu, S. (1984) *The Well-being of the Albanian People and Some Factors and Ways for its Continuous Improvement*, Tirana, 8 Nëntori.

Sjöberg, O. (1989) 'A note on the regional dimension of post-war demographic development in Albania', *Nordic Journal of Soviet & East European Studies* 6 (1), 91–121.

—— (1991) *Rural Change and Development in Albania*, Oxford, Westview.

—— and Sandström, P. (1989) *The Albanian Statistical Abstract of 1988: Heralding a New Era?*. Working Paper 2, Department of Soviet and East European Studies, Uppsala University.

Skarço, K. (1984) *Agriculture in the PSR of Albania*, Tirana, 8 Nëntori.

Strazimiri, G. (1987) *Berati: qytet muze*, Tirana, 8 Nëntori.

Tartari, T. (1988) 'Principal aspects of the protection of the genetic fund of flora and fauna', in Balkan Scientific Conference on Environmental Protection in the Balkans, *Abstracts*, Varna.

Temo, S. (1985) *Education in the PSR of Albania*, Tirana, 8 Nëntori.

Thomo, P. (1988) *Korça: urbanistika dhe arkitektura*, Tirana, Akademia e Shkencave e RPS të Shqipërisë.

Toci, V. (1971) 'Amfiteatri i Dyrrahit', *Monumentet*, no. 2, 37–42.

Toçka, J. (ed.) (1980) *Korça Almanak* 2, Tirana, 8 Nëntori.

—— (ed.) (1981a) *Elbasani Almanak* 1, Tirana, 8 Nëntori.

—— (ed.) (1981b) *Fieri Almanak 3*, Tirana, 8 Nëntori.

—— (ed.) (1982) *Saranda Almanak 2*, Tirana, 8 Nëntori.

—— (ed.) (1985) *Librazhdi Almanak 1*, Tirana, 8 Nëntori.

Trojani, V. (1991) Professor of Geography, Faculty of History and Philology, University of Tirana, Personal interview, April.

Turrill, W.B. (1929) *The Plant Life of the Balkan Peninsula*, Oxford, Clarendon Press.

Vejsiu, Y. (1990) 'The vitality of a population', *New Albania*, no. 6, 14–15.

Xherahu, Q. and Baruti, V. (1975) *Gjeografia e Shqipërise*, Tirana, Shtëpia Botuese e Librit Shkollor.

Zanga, L. (1991a) 'Albania: the state of the press', *Report on Eastern Europe* 2 (45), 13.

—— (1991b) 'Albania: the woeful state of schools', *Report on Eastern Europe* 2 (41), 1–3.

Zarshati, F. (1982) *Monuments of Culture in Albania*, Tirana, 8 Nëntori.

Zeri i Popullit ('Voice of the People'), Tirana (daily).

3 BULGARIA

Anon (1987) 'Fall out prompts Black Sea Study', *World Water*, November, p. 38.

—— (1990) 'Ecoglasnost about nuclear power stations', *Panos Feedback 90 (Budapest)*, September p. 7.

Ashley, S. (1987) 'Bulgaria', in *The Environment in Eastern Europe*, Radio Free Europe Research Background Report 42 (EE), 20 March, p. 5.

—— (1988) 'Bulgarian Politburo announces "New" ecological policy', Radio Free Europe Research, RAD, 11 May, p. 3.

Atanassov, V. (1989) 'Kremikovtsi: a geologist's view', *Sofia News*, 23–9 November, p. 7.

Bachvarov, M. (1991) 'Yuni 1990 – Geografija na izbornite rezultati', *Geografija* 45 (1), 22–7.

Bl'skova, D. (1983) *Klimat i Mikroklimat na Sofiya* (BAN), p. 127.

——, Zahariev, V., Lanzov, I., Modeva, Z. and Teneva, M. (1979) 'Nyakoi rezultati ot eksperimenta izsledvaniya na zam'rsyavaneto na atmosferata v gradovete K'rdzhali, Burgas i Sofiya', *Problemia na Geografijata*, 2, Sofia, 22–35.

Bondev, I. (1991) *Rastitelnostta na B'lgarija*, Sofia.

Bozhanov, S. (1991) 'Ecological policy', *Bulgarian Quarterly (Sofia)* 1 (2), 102–110.

Carter, F.W. (1973) 'Changements fonctionnels de l'apres-guerre dans la conurbation de Sofia', *Geographie et recherche*, no. 8, 25–39.

—— (1976) 'Four countries develop their own energy', *Geographical Magazine*, no. 49, 16.

—— (1978) 'Nature reserves and national parks in Bulgaria', *L'Espace Geographique*, no. 1, 69–72.

—— (1979) 'Prague–Sofia: an analysis of their changing internal city structure', Ch. 15 in R.A. French and F.E.I. Hamilton (eds), *The Socialist City: Spatial Structure and Urban Policy*, Chichester, J. Wiley & Sons.

—— (1984) 'Balkan historic cities: pollution versus conservation', *Proceedings of the Anglo-Bulgarian Modern Humanities Symposium, London 1982*, l, ii, 20.

—— (1987) *The City and the Environment in Socialist Countries*, Wissenschaftszentrum Berlin fur Sozialforschung (IIUG -pre 87–5), 13–16.

—— (1989) 'Bulgaria's dirty smoke stacks', *Earthwatch*, no. 36, 9.

Crampton, R.J. (1988) ' "Stumbling and dusting off", or an attempt to pick a path through the thicket of Bulgaria's new economic mechanism', *Eastern European Politics and Societies* 2 (2), 333.

Dakov, M. (1976) *Environmental Management in the People's Republic of Bulgaria*, Sofia Press.
—— (1984) *Chervena kniga na N.R.B'lgariya*, I (Rasteniya) (BAN), Sofia.
Dampier, W. (1983) 'Smoke gets in your eyes', *Sweden Now*, no. 3, 29.
Daneva, M. (1984) 'Nyakoi landschaftno-ekologichni problemi na ratsionalnoto izpolzuvane na pozemlenite resursi v N.R.B'lgariya', *Problemi na Geografiyata*, no. 1, 34–44.
—— and Vaptsarov, I. (1978) 'Vseobkhvatna strategicheska programa za opazvane na prirodnata sreda v N.R.B'lgariya', *Problemi na Geografiyata*, no. 1, 3–9.
Dempsey, J. (1990) 'Bulgaria's ecologists hope to clean up in Poll', *Financial Times*, 6 June, p. 6.
Dimov, N. (1989) *Geografiya na chimicheskata promischlenost v B'lgariya* (Izd. 'Kliment Ohridski'), Sofia.
Dimova, V. (1964) 'Zam'rsen Vodi' *Geografiya*, 14 (7), 1–4.
Donchev, D. (1979) 'The industrial development of the Bulgarian Black Sea coast and the environmental problem', Paper presented at the First British–Bulgarian Geographical Seminar, Norwich.
—— (1983) 'Trends in the regional development of Bulgarian industry', Ch. 5 in J.H. Johnson (ed.), *Geography and Regional Planning*, Norwich, Geo Books.
D'rzhaven S'vet (1983) *Osnovih Ekologichni Iziskvaniya pri Razvitieto na Selskoto Stopanstvo v N.R.B'lgariya*, Sofia.
D'rzhaven Vestnik (1967), no. 47, 16 June.
Economist Intelligence Unit, (1988) *Country Report*, no. 3, p. 20.
—— (1989) *Country Report*, no. 2, p. 21.
Ekimova-Melnishka, M. (1990) 'Bulgaria: Ecoglasnost: a breath of fresh air?', *Panascope*, no. 18 (May), p. 13.
Gavrilov, V. (1990a) 'Energy crisis looming', *Report on Eastern Europe* (RFE) 1 (34), 3–4.
—— (1990b) 'Bulgaria: environmental damage creates serious problem for government', *Report on Eastern Europe* (RFE) 1 (21), 4–12.
Gencheva, I. (1990) 'The Kouklen children still have a chance', *Bulgaria* (July–Aug.), 22–3.
Gerzilov, P. (1989) 'Kremikovtsi: bleak future or gold mine?' *Sofia News*, 23–9 November, p. 7.
Gorunova, D. and Mourgova, M. (1988) 'S'stojanie i dinamika na gorskite ekosistemi i promishlenya kompleks Varna-Devniya', *Problemi na Geografiyata*, no. 1, 22–32.
Hall, D.R. (1989) 'Planning Bulgaria's uncertain future', *Town & Country Planning*, December, p. 348.
Hanley, M.C. (1990) 'Can the Black Sea be saved?', *World Development* 3 (2), 6.
Hordijk, L. (1988) 'A model approach to acid rain', *Environment* 30 (2), 19.
Hristov, H. (1991) 'S'stojanie na pov'rhnostnite i podzemnite vodi', *Godishnik za S'stojanieto na Pridonata Sreda na Republika B'lgarija–1989*, Sofia, 127–68.
Hristov, P. (1966) 'Temperaturni inversiya v Sofiisko Pole', *Hidrologiya i Metrologiya* 15 (3), 19–28.
Ikonomicheska Misl (1985) pp. 55–7.
Izvestiya (1960) no. 74, Sofia, 13 September.
Kal'chev, P. (1989) 'Rekultivatsiya na zemite v Gorno Trakiiskata Nizina', *Geografiya* 43 (5), 4–6.
Khristov, T. (1983) 'Geografski problemi na yadrenata energetika', *Problemi na Geografiyata*, no. 2, 3–17.
—— and Popov, P. (1976) 'Sesiya po problema "Teritorialno razpredelenie na promishlenostta i v'prosite po opazvane na prirodata"', *Problemi na Geografiyata*, no. 2, 91–4.
—— and Stankov, G. (1982) 'Territorial'noe razmeshtenie khimicheskoi promishlenosti v Bolgarii i okhrana prirodni Sredi', *Acta Facultatis Rerum Naturalium Universitatis Comeniane (Geographica)*, 20, 15–22.

—— and Dancheva, N. (1983) 'Toploenergetikata, metalurgiyata i khimicheskata promishlenost i deystvieto na ekologichniya faktor', *Problemi na Geografiyata*, no. 4, 29–38.

Kovachev, S. (1991) 'Zakonodatelstvo i normativi v opazvaneto na prirodnata sreda', *Godishnik za S'stojanieto na Prirodnata sreda na Republika B'lgarija – 1989*, Sofia, 354–62.

Kr'stev, T. (1988) 'Aktualni problemi na izuchavaneto i usvojaneto na B'lgarskiya sektor na Cherno More', *Geografiya* 42 (10), 14.

Kŭndev, T. (1981) 'V'zmozhnosti za razshirjavane na mrezhata ot zashtiteni prirodni teritorii v B'lgarija', *Geografija* 33 (4), 17.

Literature Front (1988), no. 34, Sofia, 18 November.

Manolov, A. (1985) 'Kiselnnite d'zdove unishozhavat gorite', *Geografiya* 40 (6), 20.

Markov, G. (1978) 'D'lgosrochno prognozirane na vodopotreblenieto pri polivnoto zemedelie v N.R.B'lgariya', *Problemi na Geografiyata*, no. 3, 14–23.

Mihailov, T. (1991) 'Shte se realizirat li proektite "Rila" i "Mesta"', *Geografija* 45 (3), 19–22.

Mitchev, N. (1983) 'The demographic transition and regional variations in birth and death rates in Bulgaria', Ch. 13 in J.H. Johnson (ed.) *Geography and Regional Planning*, Norwich, Geo Books.

Montias, J.M. (1988) 'Industrial policy and foreign trade in Bulgaria', *Eastern European Politics and Societies* 2 (3), 522.

Onchev, N. (1991) 'S'stojanie na obrabotvaemite zemi i zemnite nedra', *Godishnik za S'stojanieto na Prirodnata Sreda na Republika B'lgarija-1989*, Sofia, 169–217.

Oschlies, W. (1986) 'Balance Akt zwischen Plan und Pest', *Rheinischer Merkur*, 31 May, p. 3.

—— (1987) *Schwefelstaub auf Rosenbluten Umweltsorgen in Bulgarien*, Köln-Wien, Bohlan Verlag.

Painton, F. (1990) 'Darkness at noon', *Times International*, no. 15, 9 April, p. 26.

Panev, I. (1972) 'Perspektivnoe razvitie Sofii do 2000 – go goda i problemi, svjazann'e s sistemoi ozeleneniya', *Urbanistica Contemporalis* (Aedes Academiae Scientiarum Hungaricae), Budapest, pp. 503–14.

Petrov, P. (1980) 'Landschaftno rayonirane na B'lgariya', *Godishnik na Sofiyskiya Universitet*, no. 71, 121–36.

—— (1981) 'Opit za landschaftno rayonirane na B'lgariya', *Geografiya* 31 (4), 2.

Pehe, J. (1990) 'The Green Movements in Eastern Europe', *Report on Eastern Europe* (RFE) 1 (11), 35.

Pogled (1987), no. 34, Sofia, 24 July.

Popilieva, P. (1989) 'National Assembly talks environment', *Sofia News*, 23–9 May, p. 6.

Popov, V. (1976) 'Geografsko razprestranie na rezervatite i narodnite parkove v N.R.B'lgariya' *Geografiya* 25 (2), 1–4.

—— (1989) 'Zashiteni prirodni teritorii i obekti v oblastite na Severna B'lgariya', *Problemi na Geografiyata*, no. 3, 65–70.

Popov, R., Naidenova, R. and Donchev, D. (1983) 'Development and location of industry in the Stara Planina mountain region and the solution to some social and economic problems', Ch. 10 in J.H. Johnson (ed.); *Geography and Regional Planning*, Norwich, Geo Books.

Radio Free Europe Research (1987a) *Bulgaria: Situation Report/10*, Item 2, 4 November.

—— (1987b) *Bulgaria: Situation Report/5*, Item 5, 8 July.

—— (1989a), *Bulgaria: Situation Report/1*, Item 5, 5 December.

—— (1989b) *Bulgaria: Situation Report/7*, Item 3, 7 August.

Radoeva, T. and Ivancheva, I. (1987) 'On "acid rain" and "white roofs": how to organize them', *Bulgaria*, 39, Sofia, July/August pp. 18–20.

Rosencranz, A. (1986) 'The acid rain controversy in Europe and North America: a political analysis', *Ambio* 15 (1), 48.

Slama, J. (1986) 'Internationaler Vergleich der Emission von Stickoxiden', *Working Papers* (Osteuropa-Institut), Munchen, no. 117, Table A, p. 18.

Smit, H. (1988) 'Europe: how much needs to be reduced', *Acid News*, no. 3, 6–7.

Sobell, V. (1988) 'The CMEA's post-Chernobyl nuclear energy program', Radio Free Europe Research, Background Report/19 (EE), 15 February, pp. 4–5.

Spasovski, M. and Hristev, V. (1983) 'General principles for the prevention of occupational hazards from chemicals in Bulgaria', *Industry & Environment* 6 (4), 8–9.

Stankov, G., Vodenska, M., Tishkov, H., Bachvarov, M., Stanev, P. and Ebrev, P. (1985) *Geografija na otdiha i turizma v Bulgarija: Problemi na teritorialnata organizatsija* (Nanko i Izkustvo), Sofia.

Starshel (1987), no. 2/176, Sofia, 23 October.

Starzewska, A. (1988) 'The legislative framework for EIA in centrally planned economies', Ch. 12 in P. Wathern (ed.), *Environmental Impact Assessment: Theory and Practice*, London, Unwin Hyman.

State Council of People's Republic of Bulgaria (1977) *Guidelines for the Protection and Reproduction of the Environment in the People's Republic of Bulgaria* (Approved, 29 June 1977) Sofia, p. 19.

Stoilov, D. (1980) 'Bulgarskite rezervati-etaloni na nepromenena priroda', *Geografiya* 30 (2), 20–5.

Stojanov, S. (1989) 'Problem'it za kiselinnite d'zhdove', *Geografiya* 43 (7), 2.

Sullivan, S. (1985) 'Eastern Europe: a dying landscape *Newsweek*, 14 January, pp. 14–15.

Tishkov, K. (1974) 'Thermal inversions and industrial pollution of the near-surface air of some depressions in Bulgaria', in *Man and Environment* (Studies in geography in Hungary, no. 11), Budapest, pp. 197–202.

—— (1980) 'Rekreatsionite resursi v Staroplaninskata fizilkogeograska oblast', *Geografiya* 30 (2), 1–4.

Vecherni Novini (1990), Sofia, 9 February, p. 2.

Velev, S. (1986) 'Klimat na grad Sofiya', *Problemi na Geografiyata*, no. 5, 10–16.

—— (1992) 'Transgranichen prenos na zam'rsiteli na v'zduha i vodnata i geopoliticheskoto polozhenie v B'lgorija', in S. Karastoyanov et al. *Geografija–Geopolitika (P'rvakniga)*, Sofia, pp. 153–61.

Vitkov, T. (1991) 'S'stojanie na atmosfernija v'zduh i problemni raioni', *Godishnik za S'stojanieto na Prirodnata Sreda na Republika B'lgarija – 1989*, Sofia, pp. 39–126.

Yaneva, M. (ed.) *Generalna Shema za kompleksno izpolzvanei opozvone na vodnite resursi*, Sofia 446 pp.

Zabojnikov, M. (1984) 'Priroda uz nepocka', *Pravda*, Bratislava, 25 May, p. 4.

Zyapkov, L. (1974) 'Economic activities influencing the fluvial regime of the Pazardzhik and Plovdiv Plains', in *Man and Environment* (Studies in geography in Hungary, no 11), Budapest, pp. 157–61.

4 CZECHOSLOVAKIA

Albrecht, C. (1987) 'Environmental policies and politics in contemporary Czechoslovakia', *Studies in Comparative Communism* 2 (3–4), 291–302.

Alm, H. (1989) 'Emissions are falling ... but is it enough?', *Ambio* 18, (8), 6–7.

Anon (1986) *Czechoslovak Bibliography 1985: Vol. 27 General and Environmental Hygiene, Hygiene of Children, Adolescents and Food Hygiene*, Prague/Bratislava.

—— (1987) 'Irrigation success boosts Czech agriculture', *World Water*, November, p. 32.

—— (1988) 'Air pollution – a very serious problem in Czechoslovakia', *Environmental Policy Review* 2 (2), 40–6.

—— (1989) 'Informační leták skupiny aktivistů "Greenpeace"' *Ekologicky Bulletin*, 14, 42–3.

—— (1990) 'Dramatic statistics about the northern Czech brown-coal basin', *Panos Feedback 90*, September, pp. 15–16.

Ashley, S. (1989) 'More independent groups emerge', *Radio Free Europe* RAD/SR 2, 9 March, p. 10.

Atlas Životního Prostředí a Zdraví obyvatelstva CSFR, Brno/Prague, 1992.

Barták, S. (1988) 'Potrebujeme skutky pre Bratislavu', *Pravda*, 11 November, p. 5.

Brchanova, M. (1988) 'Trendy znečišťování ovzduší v ČSSR a vazby na evropsky region', *Statistika*, 10, 459–60.

Brozyniak, B. (1988) 'Mniej siarki nad glowa', *Slowo Polskie*, no. 227, 29 September, p. 4.

Budaj, J. (ed.) (1987) *Nahlas Bratislava* (UPI), Bratislava, 26 November.

Burger, H. (1985) 'Prag setzt auf den Atomstrom', *Süddeutsche Zeitung* 79, 3 April, p. 11.

Carter, F.W. (1984) 'Pollution in Prague: environmental control in a centrally planned socialist country', *Cities* 1 (3), 258–73.

—— (1985) 'Pollution problems in post-war Czechoslovakia', *Transactions of the Institute of British Geographers* (New Series) 10 (1), 17–44.

—— (1986) 'Nuclear power production in Czechoslovakia', *Geography*, vol. 71, pt 2, no. 311, pp. 136–9.

—— (1988) 'Czechoslovakia: nuclear power in a socialist society', *Environment and Planning C: Government and Policy* 6 (3), 269–87.

—— (1989) 'Czechoslovakia's ecological crisis', *Earthwatch*, no. 86, 8–11.

Cerný, J. (1989) 'Baseflow buffering against acid atmospheric output', *Ambio* 18 (5), 280–3.

Ceřovský, J. and Podhajská, Z. (1989) 'Czechoslovakia', in *Environmental Status Reports* (IUCN East European Programme – 1: Czechoslovakia, Hungary, Poland), Cambridge, pp. 3–58.

Cervenová-Bakošová, M. (1991) 'Vymedzenie ekologických nákladov a problémy ich úhrady', *Ekonomický časopis* 39 (4), 279–84.

Cestrová, D. (1989) 'Situace téměř normální?' *Průboj*, 28 January, p. 19.

Chobotský, P. (1990) 'Vznáším žalobu za lidi i jablünky', *Květy* 12 (4), 7.

Davies, G. (1991) 'Danube conflict sharpens', *The Times*, 15 July, p. 3.

Dočekal, B. Sevčík, L. and Stěpánek, J. (1989) 'Co není ekologické, to není ekonomické', *Mladá fronta*, 4 January, p. 3.

Doležal, Z. (1990) 'Czechoslovakia: country feeling its way', *Acid News*, vol. 4, pp. 12–13, 152.

Dykast, J. (1989) 'Inverze znamená totéž co "zákaz vycházení pro školky"? – o inverzích a znečištění ovzduší v kraji', *Průboj*, 28 January, p. 21.

Echikson, W. (1987) 'Environmental problems are explosive issue for East bloc', *Christian Science Monitor*, 16 November, pp. 5–7.

Frank, M. (1989a) 'Nordböhmen: Wo sich ein Graufilm über das Leben legt', *Süddeutsche Zeitung*, 10 January, p. 5.

—— (1989b) 'Erfolgreicher Aufstand gegen Schwindelkulisse', *Süddeutsche Zeitung*, 18 November, p. 3.

Fronkova, A. (1988a) 'Záležitost nás všetkých', *Lud*, 5 April, p. 1.

—— (1988b) 'Pôde treba vrátiť život', *Lud*, 1 December, p. 4.

Genillard, A. (1991) 'Road littered with hurdles', *Financial Times*, 3 July, p. 3.

Glenny, M. (1987) 'Living in a socialist smog', *New Scientist*, 24 September, 41–4.

Hesek, F. Rak, J. and Závodský, D. (1979) 'Znečištenie ovzdušia Bratislavy', Ch. 15 in

M. Koníček (ed.), *Klima a bioklima Bratislavy*, Bratislava, Veda.

Hospodářské noviny (1988a), no. 30, 29 July, p. 3.

Hospodářské noviny (1988b), no. 17, 29 April, p. 15.

Klapáč, J. (1989a) 'Právo životného prostredia v troch nových zákonoch (Všeobecna časť)', *Právny obzor* 72 (3), 234–46.

—— (1989b) 'Právo životného prostredia v troch nových zákonoch (Osobitna časť')', *Právny obzor* 72 (5), 417–28.

—— (1990) 'Právo životného prostredia v štyroch nových zákonoch (Systematika zákonov, právne pojmoslovie a nazvoslovie)', *Právny obzor* 73 (7), 601–28.

Klášterská, I. (1991) 'Bohemian problem bared', *Acid News*, no. 1, March, 1–3.

Klub (1988) (émigré monthly), Vienna, April, p. 6.

Konopka, J. (1988) 'Osud lesov varuje', *Rolnícke noviny*, 19 April, p. 16.

Kopečný, Z. (1988) 'Zachráníme vodu, lesy a vzduch?', *Hospodářské noviny*, no. 45, 11 November, p. 1.

Kulich, J. (1989) 'Does brontosaurus sleep in winter?', *Czechoslovak Life*, 2, 12–13.

Lokvenc, T. (1988) 'Sněžka není jen vrchol', *Pochodeň*, 20 November, p. 1.

Loudová, I. (1989) 'Anketa k životnímu prostředí Prahy', *Nika*, nos. 5–6, p. 18.

Medliak, J. (1990) 'K súčasnému stavu trestnoprávnej ochrany životného prostredia', *Právny obzor* 73 (1), 47–56.

Medvedev, Z. (1990) *The Legacy of Chernobyl*, Oxford, Blackwell.

Merta, O. (1988) 'Jak zachovat Křivoklátsko: Elektřina z dolů', *Hospodářské noviny*, no. 2, p. 6.

Mikloš, L., Lisický, M. and Kozová, M. (1989) 'Ecological evaluation of the territory of the Gabčikovo–Nagymáros dam project', *Ekológia (CSSR)* 8 (2), 167–77.

Moldan, B. (ed.) (1990) *Životní prostředí České republiky: Vývoj a stav do konce roku 1989*, Praha, Academia.

—— (1991a) *Environment of the Czech Republic*, Pts 1, 2, Prague Brázda.

—— (ed.) (1991b) *Environmental Recovery Programme for the Czech Republic* (Rainbow Programme, Ministry of Environment of the Czech Republic), Prague, Academia.

Nesrovnal, I. (1990) 'Spoločenskě vzťahy starostlivosť a životné prostredie ako objekt trestnoprávnej ochrany', *Právny obzor* 73 (3), 226–37.

Novák, I. (1972) *Ochrana čistoty ovzduší v ČSR*, Prague, MLVH.

Ondráčková, J. (1991) 'Czechoslovakia's nuclear power industry', *East European Reporter* 4 (4), 59–60.

Ostry, D. (1988) 'The Gabčikovo-Nagymáros dam system as a case study in conflict of interest in Czechoslovakia and Hungary', *Slovo* 1 (1), 11–24.

Patakyová, M. (1988) 'Ekologická funkcia hospodárskeho práva', *Právny obzor* 72 (7), 570–80.

Pehe, J. (1990) 'The ecological damage caused by Soviet troops', *Report on Eastern Europe* (RFE) 1 (31), 28–31.

Pirč, J. (1988) 'K právnym aspektom jadrovej bezpečnosti a starostlivosť o životné prostredie', *Právny obzor* 72 (7), 581–95.

Pohl, F. (1985) 'Environmental deterioration in Czechoslovakia', in V. Mastny (ed.), *Soviet and East European Survey 1983–1984*, Durham, Duke University Press.

Pomíchal, R. (1990) 'On the Slovak side of the dam', *Panos Feedback 90*, September, pp. 10–12.

Práce (1988) Weekend Supplement /B, 28 August, p. 1.

Prochazka, K. (1988) 'Správa o stave lesa', *Technické noviny*, no. 18, 3 May, p. 17.

Rak, J. (1989) 'Problematika kvality ovzdušia hl. m. SSR Bratislavy', *Ochrana ovzduší* 21 (4), 87–93.

Schonherr, O. (1988) 'Böhmens und Mährens Umwelt aus der Nahe gesenhen', *Europäische Rundschau* 16 (1), 145–51.

Seger, J. (1986) 'Příspěvek k budování soustavy ukazatelů statistiky životního prostředí v ČSSR', *Statistika*, no. 4, 155–63.

Senjuk, V. (1988) 'Když umírá les', *Svět práce*, no. 12, 14–17.

Skoda, L. (1981) 'Zlepšovat životní prostředí v severočeském kraji', *Plánované hospodářství* 7, 38–42.

Sláma, J. (1987) *Umweltschutz im RGW*, (Insbesondere Kooperationmoeglichkeiten in wirtschaftlicher Hinsicht zwischen der Bundesrepublik Deutschland und der UdSSR, der DDR, Polen und der ČSSR,) München, Osteuropa-Institut, Appendix Table 1.

Sniegon, T. (1990) 'Czechoslovakia: privatising forests and the nuclear dilemma', *Panascope*, no. 18, p. 11.

Staszova, M. (1988) 'Komu má pomôcť predstieranie? Ekologická kontrola v Rudňanoch', *Nove slovo*, no. 46, p. 5.

Straškraba, M., Žižková, E., Guth, J., & Dlouhá, J., (199 *National Report of the Czech and Slovak Federal Rep* (U.N.C.E.D.), Prague, 141 pp.

Stursa, J. (1988) 'Pražská Stromovka ohrožena', *Tvorba*, no. 13, p. 5.

Vavroušek, J. (1989a) 'Soustava vodních děl Gabčíkovo–Nagymáros', *Nika*, no. 1–2, 37–40.

—— (1989b) 'K'otázkam sociálno-ekonomického prognozovania', *Ekonomický časopis* 37 (2), 141–56.

—— (ed.) 1990 *The Environment in Czechoslovakia*, Prague, Federal Committee for the Environment.

—— and Moldan, B. (1989) (eds) *Stav a vývoj prostředí v Československu*, Prague.

Vondráčková, E. (1988) 'Místo života Praha', *Večerní Praha*, 24 November, p. 5.

Waller, M. (1989) 'The ecology issue in Eastern Europe: protest and movements', *Journal of Communist Studies* 5 (3), 303–28.

Wolchik, S.L. (1991) *Czechoslovakia in Transition: Politics, Economics & Society*, London/New York, Pinter.

Zeman, J. (1989) 'Ekologické škody v soustavě plánovitého řízeni', *Ekonomický časopis* 37 (12), 1095–1107.

5 HUNGARY

Balló, M. (1975) MTA Matematikai és Természettudományi Közlemények, Budapest, MTA-Hungarian Academy of Sciences.

French, H. (1990) *Green Revolutions: Environmental Reconstruction in Eastern Europe and the Soviet Union*, Washington, DC, Worldwatch Institute.

Hinrichsen, D. (ed.) (1986) *World Resources 1986*, New York, Basic Books, pp. 203–25.

—— (1989) 'Blue Danube', *Amicus Journal*, 11, 6–8.

—— and Enyedi, G. (eds) (1990) *State of the Hungarian Environment*, Budapest, Statistical Publishing House.

Hock, B. (1987) 'The quality of surface water and changes thereof' (in Hungarian), Budapest, Hungarian Academy of Sciences.

—— and Somlyódy, L. (1990) 'Freshwater resources and water quality', in D. Hinrichsen and G. Enyedi (eds) *State of the Hungarian Environment*, Budapest, Statistical Publishing House, pp. 67–88.

Láng, I. (1978) 'Hungary's Lake Balaton: a program to solve its problems', *Ambio*, vol. 7, 164–8.

—— (1991) 'Environmental challenges for science in Eastern and Central Europe' in B. Nath (ed.), *Proceedings of the International Conference on Environmental Pollution*, Lisbon, Interscience Enterprises Ltd, pp. 16–24.

Lean, G., Hinrichsen, D. and Markham, A. (eds) (1990) *Atlas of the Environment*, New York, Prentice-Hall, pp. 81–4.

Lesenyei, J. (1975) 'Water quality in Hungarian streams', *Water Resources of Hungary* Budapest, VITUKI.

Nemeth, G. (ed.) (1981) *Hungary: A Comprehensive Guide*, Budapest Athenaeum Printing House.

Soloman, A. and Kauppi, L. (eds) (1990) *Toward ecological sustainability in Europe*, Luxemburg, International Institute for Applied Systems Analysis, pp. 127–31.

Stefanovits, P. (1984) 'Agricultural production and the environment', *Ambio*, vol. 13, 99–100.

United Nations Population Fund (1990) *State of the World Population 1990*, New York, United Nations Population Fund.

Várallyay G. (1990) 'Soil quality and land use', D. Hinrichsen, and G. Enyedi (eds), *State of the Hungarian Environment*, Budapest, Statistical Publishing House, pp. 101–16.

Várkonyi, T. and Kiss, G. (1990) 'Air quality and pollution control', in D. Hinrichsen and G. Enyedi (eds), *State of the Hungarian Environment*, Budapest, Statistical Publishing House, pp. 44–69.

Vukovich, G. (1990) 'Trends in economic and urban development and their environmental implications', in D. Hinrichsen and G. Enyedi (eds), *State of the Hungarian Environment*, Budapest, Statistical Publishing House, pp. 13–34.

6 POLAND

Agren, C. (1991) 'Poland: projecting car emissions', *Acid News*, no. 2, June, p. 4.

Albinowski, S. (1990) 'The debate on nuclear energy in Poland', *Panos Feedback 90*, September, pp. 4–5.

Andersson, C. (1987) 'A sad forester speaks', *Acid News*, no. 3, October, p. 8.

Anon (1981a) 'Ochrona środowiska i gospodarka wodna 1981', *Statystyka Polski-Materiały statystyczne*, no. 3, Warszawa, GUS, pp. 146–51.

—— (1981b) 'Ograniczenie produkcji Huty', *Krajowa Agencja Informacyjna I*, no. 3/1095, 19–25 January, p. 6.

—— (1983) 'Report from Poland – health and health protection of the Polish population – experience and the future', *International Journal of Health Services* 13 (3), 487–513.

—— (1987) 'Nuclear power stations', *Contemporary Poland*, no. 9, p. 11.

—— (1989a) 'Uber das Fundament noch nicht hinausgekommmen: Polens Kernkraftwerkbau hinkt de Planung weit hinterher', *Der Tagesspeigel* (W. Berlin), 10 January, p. 2.

—— (1989b) 'Czechoslovakia, GDR and Poland signed ecological agreement', *Information from Czechoslovakia*, no. 2, 3–5.

—— (1990a) *Poland 1990*, Warszawa, PAP.

—— (1990b) 'De Bucarest a Varsovie: Le Face Sinistrée de l'Europe', *Greenpeace* 15 (1), 9–10.

—— (1991) 'Co trzeci Polak w zatrutym środowisku', *Gazeta Wyborcza*, no. 27, (February), p. 2.

Atlas Miej. Woj. Krakowskiego (1979), Kraków.

Atlas Miasta Krakowa (1988) Warszawa–Wrocław, PPWK (47 maps).

Białobek, S. and Rachwał, L. (1981) 'Studies on tolerance variability of trees and shrubs to air pollution and utilization of the results in landscape establishment', *Archiwum Ochrony Środowiska*, no. 2–4, Wrocław–Warszawa–Kraków–Gdańsk–Łódź, pp. 101–6.

Boyes, R. (1986) 'Nuclear fear hits Poland's tourism', *The Times*, 27 August, p. 3.

Brzeziński, W. (1974) *Legal Protection of Natural Environment in Poland*, Wrocław–Warszawa–Kraków–Gdańsk, PAN.

Budnikowski, A. (1991) 'Upper Silesia: in need of aid', *Acid News*, no. 1 (March), 4–5.

Carter, F.W. (1982) 'Historic cities in Eastern Europe: problems of industrialisation, pollution and conservation', *Mazingira: The International Journal for Environment and Development* 6 (3), 62–76.

—— (1987) *The City and the Environment in Socialist Cities*, International Institute for Environment and Society (IIUG pre-1987–5), Berlin.

—— (1989) 'Air pollution problems in Poland', *Geographica Polonica*, 56, Warszawa, pp. 155–77.

Chorązy, M. (1990) 'Zagrożenia biologiczne skażeniami powietrza z terenu województwa Katowickiego', *Jutro* 2 (3), 6–7.

Clough, P. (1990) 'One in four Poles will have cancer by 2000', *The Independent*, 23 February, p. 3.

Cofala, J. and Bojarski, W. (1988) 'Sulphur and nitrogen oxides emission resulting from energetic utilization of fuels – the case of Poland', *Archiwum Ochrony Środowiska*, no. 3–4, 7–42.

Cooper, R., Schatzkin, A. and Sempos, C. (1984) 'Rising death rates among Polish men', *International Journal of Health Services* 14 (2), 289–302.

Council of Ministers (1990) *Minutes*, no. 44/90, 19 November.

Curzydło, J. (1990) Wpływ Kombinatu Metalurgicznego Huty im. Lenina na zanieczyszczenie gleby i roślin metalami ciężkimi oraz plan rolniczego zagospodarowania terenów skazonych', in M. Guminska and A. Delorme, (eds), *Klęska Ekologiczna Krakowa* (Polski Klub Ekologiczny), Kraków.

Dawson, A. (1984) 'City profile: Kraków', *Cities* 2 (4), 12.

Delorme, A. (1989) *Green Parties in Poland*, Kraków.

Dienes, L. (1974) 'Environmental disruption and its mechanism in east–central Europe', *The Professional Geographer* 26 (4), 375–81.

Domański, B. (1989) 'Public attitudes towards industry in Cracow and its region', in H. Forster and B. Kortus (eds), *Social-Geographical Problems of the Cracow and Upper Silesian Agglomerations*, Paderborn, F. Schoningh.

Fura, Z. (1985) 'Institutions: the Polish Ecological Club', *Environment* 27 (9), 4–5, 43.

Gesing, A. (1986) 'Zatrucie Odry olejem', *Trybuna Ludu*, 13 December, p. 5.

Gieruczkiewicz-Bajtlik, M., (1987), *Degradacja Jezior Ewingi, Znińskie Duże i Tuczno*, Warszawa, PWN.

Gilewska, S. (1964) 'Changes in the geographical environment brought about by industrialization and urbanization', *Geographica Polonica*, 3, Warszawa, PWN, pp. 201–10.

Górny, G. (1991) 'Modlitwa o wiatr: Lament ekologiczny ze Śląska', *Gazeta Wyborcza*, no. 27, 1 February, p. 3.

Greszta, J. (1974) 'The recultivation of post-industrial territories', in W. Michajłow (ed.), *Protection of Man's Natural Environment*, Warszawa, PWN.

Grzesiak, M. (ed.) (1990) *Raport o stanie, zagrożeniu i ochronie środowiska 1990*, Warszawa, GUS.

Hagerhall, B. (1990) 'Saving the Baltic: a race against mankind', *Our Planet* 2 (2), 8–10.

Hasiński, W. (1987) 'Les forêts en Pologne', *Hommes et Terres du Nord*, no. 2, 121–4.

Hess, M.T., Niedźwiedź, T. and Obrębska-Starklowa, B. (1989) 'Bioklimat Krakowa', *Prace Geograficzne*, no. 73, Warszawa–Kraków, PWN.

Hinrichsen, D. (1988) 'Poland's chemical cauldron', *Amicus* 10 (2), 3–7.

Horwath, L. (1988) 'Przebudowa równowaga ewakuacja', *Kraków Magazyn Kulturalny*, 17 January, p. 4.

Jastrzębski, L. (1990) *Prawo ochrony środowiska w Polsce*, Warszawa, PWN.

Jędrychowski, W. Becher, H., Wahrendorf, J. and Basa-Cierpialek (1990) 'A case-control study of lung cancer with special reference to the effect of air pollution in Poland',

Journal of Epidemology & Community Health, 44, 114–20.

Kabala, S.J. (1989) 'The economic effects of sulfur dioxide pollution in Poland', *Ambio* 18 (4), 250–1.

—— (1991a) 'The environment and economics of Upper Silesia', *Report on Eastern Europe* 2 (33), 12–23.

—— (1991b) 'The environment in Cracow: great concern and modest action', *Report on Eastern Europe*, 2 (25), 20–3.

Kacprzyk, J. and Zurek, J. (1990) *Quality of the Environment in Poland*, Warszawa, Institute of Environmental Protection.

Kalicka, S. *et al.* (1991) *Eko Indeks '91*, Warszawa, 219 pp.

Kamiński, B. (1990) 'Poland's environmental problems and priorities', *Polish Foreign Trade*, no. 3 (293), 5.

Kasina, S. (1978) 'Precipitation acidity in the Kraków region', Paper presented at a workshop on Ecological Effects of Acid Precipitation, Galloway (CEGB Laboratories), mimeo.

Kassenberg, A. (1988) 'Poland: opening up to cooperation', *Acid News*, October, pp. 9–10.

—— and Rolewicz, C. (1985) *Przestrzenna diagnoza ochrony środowiska w Polsce*, Warszawa, PWEk.

Klimek, K. (1990) 'Poland', *Environmental Status Reports 1988/1989*, Cambridge, IUCN.

Kormondy, E.J. (1980) 'Environmental protection in Hungary and Poland', *Environment* 22 (10), 10–20.

Kondracki, J. and Leszczycki, S. (1989) 'Atlas zasobow walórów i zagrożeń środowiska przyrodniczego Polski', *Acta Universitatis Carolinae* (Geographica), no. 2 (Prague), 5–11.

Kowalczyk, A. (1987) 'Accessibility and availability of health care services in Poland: a case study of the Sierpc medical district', *Geographica Medica*, 17, 47–62.

Kowalczyk, T. (1984) 'Biosphere and anthroposphere in the Płock district', *Village and Agriculture*, Warszawa, PAN, pp. 15–25.

Lemiński, K. (1990) 'Poland: the environmental crisis', *Report on Eastern Europe*, 31 August, p. 25.

Leszczyńska, D. and Vetter, R. (1987) 'Informationsmangel und Proteste: Die Auswirkungen der Havarie von Tschernobyl auf Polen', *Osteuropa*, no. 1, 19–21.

Liszewski, S. (1985) *Zmiany w środowisku przyszłego Bełchatowskiego Okręgu Przemysłowego. Stan z okresu przedinwestycyjnego i wstępnej fazy pracy elektrowni*, Warszawa–Lódź, PWN.

Luchter, L. (1988) 'Skutki wykorzystania wysoko zasiarczonego węgla kamiennego w elektrowniach Górnoslaskiego Okręgu Przemysłowego', *Folia Geograficzne*, 72, (Kraków), pp. 93–107.

Lukaszuk, L. (1989) 'Wspólpraca państw nadbałtyckich na rzecz ochrony środowiska morskiego', *Sprawy Międzynarodowe*, 42 (Warszawa), pp. 77–88.

McCarthy, M. (1990) 'Pope angry at pollution of Poland', *The Times*, 23 April, p. 3.

Malisz, B. (1978) 'Gospodarowanie ziemia ze szczególnym uwzględnieniem optymalnego jej wykorzystania na cele rolnicze lesne, przemysłowe i turystyczne oraz problem rekultywacji gruntów', *Ekspertyza*, no. 7, pp. 80–91.

Mańczak, H. (1979) 'Ochrona wód Wisły w świetle kompleksowego zagospodarowania jej dorzecza', *Gospodarka Wodna* 39 (10–11) (Warszawa) pp. 305–12.

Marcinkiewicz, J. (1987) 'S.O.S. Poland', *Environment Now*, October/November, pp. 48–50.

Maziarka, S. (1987) 'Zarys prognostycznych zagadnień zdrowotnych związanych z zanieczyszczeniem i ochroną środowiska', *Perspektywy zdrowotne Polska. Polski 2000*, (Ossolinum), 2, 95–106.

Mazurski, K.R. (1986) 'The destruction of forests in the Polish Sudetes Mountains by industrial emissions', *Forest Ecology and Management*, 17, 303–15.

—— (1987) 'Die Rolle der Kleingarten in den polnischen Stadten', *Forum Stadt – Hygiene*, 38, 96–9.

—— (1988a) 'Dewastacja i rekultywacja gruntów w Polsce', *Ochrona Terenow Górniczych*, 22, 36–8.

—— (1988b) 'Ökologische Lage der polnischen Stadte', *Forum Stadt – Hygiene*, 39, 86–9.

—— (1989a) 'Znečištení ovzduší polských Sudet', *Opera Corcontica*, 26, 51–9.

—— (1989b) 'Przemysłowe zagrozenia lasów Dolnósląskich' (Wiadomości Geograficzne), *Czasopismo Geograficzne* 40 (2), 191–4.

—— (1989c) 'Counteracting soil erosion in Poland', *Applied Geography*, 9, 115–21.

—— (1990) 'Industrial pollution: the threat to Polish forests', *Ambio* 19 (2), 70–4.

—— (1991a) *Problemy ochrony środowiska w Jeleniógorskiem*, Jelenia Góra.

—— (1991b) 'Ökologische Probleme der polnischen Landwirtschaft', *Geographische Rundschau* 43 (12), 709–12.

Medvedev, Z. (1990) *The Legacy of Chernobyl*, Oxford, Blackwell.

Millard, L.F. (1982) 'Health care in Poland: from crisis to crisis', *International Journal of Health Services* 12 (3), 497–515.

Nauka Polski (1983) Warszawa, April, p. 20.

Niemtur, S. (1982) 'The content of Zn, Pb, Fe, Mg, Ca, and K in the needles of various families of Scots pine grown in the experimental area near the zinc smelter', *Archiwum Ochrony Srodowiska*, no. 2–4, Wrocław–Warszawa–Kraków–Gdańsk–Łódź, pp. 17–28.

Nowe drogi (1978) Warszawa, July, p. 20.

Nowicki, M. (ed.) (1990a) *Stan zanieczyszczenia atmosfery w Polsce w 1987 roku*, Warszawa, Instytut Ochrony Srodowiska.

—— (1990b) 'Poland: power clean-up envisaged', *Acid News*, no. 1, March, pp. 6–7.

OECD (1991) *Energy Policies: Poland; 1990 Survey* (International Energy Survey), Paris.

Orval, J. (1989) 'Les Polonais ont peur de l'energie nucleaire', *La Libre Belgique*, 30 March, p. 4.

Osterberg, K. and Teknik, N. (1989) 'Air put first in Polish Plan', *Acid News*, no. 3, July, p. 7.

Paczka, J. (1990) 'Letter and activism against Żarnowiec plant', *Panos Feedback 90*, September, p. 6.

Pape, R. (1990a) 'Environment Minister: Poland's coming needs', *Acid News*, no. 3, September, p. 15.

—— (1990b) 'Poland: indignation aroused', *Acid News*, no. 3, September, pp. 14–15.

—— (1991) 'Poland: environmental fund', *Acid News*, no. 4, December, p. 12.

Papuzinska, M. (1990)'Greenpeace nad Wisłą', *Gazeta Wyborcza* (Warszawa), no. 153 (320), p. 3.

Parkin, S. (1989) 'Poland's Green Party', *Index on Censorship*, no. 6–7, 65.

Partek, T. (1990) 'Na odsiecz Bałtykowi', *Rzeczpospolita* (Warszawa), no. 4, May, p. 3.

Paszyński, J. (1959) 'Investigation of local climate in the Upper Silesian industrial district', *Prace Geograficzne* 25, Warszawa, IGPAN, pp. 41–53.

Pawłowski, L. and Kozak, Z. (1988) *Chemiczne Zagrozenie środowiska w Polsce*, Lublin (Politechnika).

Pehe, J. (1990) 'The Green Movements in Eastern Europe', *Report on Eastern Europe*, 16 March, pp. 35–7.

Perłowski, A. (1988) 'Poland: the Catholic Church warns of an ecological catastrophe', *Environmental Policy Review* 2 (1), 26–31.

Polish Government Decree (1987) (Sejm) *Dziennik Ustaw*, no. 4, 20 February, pp. 29–30.

—— (1989) (Sejm) *Dziennik Ustaw,* no. 3, 11 February, p. 1.

Polish People's Republic (1983) *Environmental Impact Assessment System: Final Technical Report,* Warsaw, Ministry of Administration, Local Economy and Environment Protection.

Polski Klub Ekologiczny (1981) *Quo Vadis Cracovia? Jaki jest Kraków, a czym i jaki ma być w przyszłości?* (Problemy ekologiczne Krakowa, no. 5), Kraków.

—— (1990) *Klęska Ekologiczna Krakowa* (M. Gumińska and A. Delorme, eds), Kraków.

Ponard, P. (1987) 'Katastrofa ekologiczna w Zatoce Gdańskiej', *Tydzień Polski,* (1) 18 July, p. 3; (2) 25 July, p. 39 (London).

Poskrobki, B., (ed.) (1992) *Dziataność Gospodarcza a Ochrona środowiska Przyrodniczego,* Biatystok, 296 pp.

Praca zbiorowa (1981) *Ocena stanu środowiska w Polsce 1980,* Centralny Ośrodek Badań i Kontroli Srodowiska, Warszawa, PIOS.

Pudlis, E. (1989) 'Once there was a queen . . .', *Earthwatch,* no. 36, pp. 3–4.

—— (1991) 'Environmental protection: on the brink', *Voices* (Warsaw), 9 June, pp. 8–9.

Radio Free Europe Research (1986) *Situation Report/10, Poland,* 27 June, p. 17.

—— (1987) *Situation Report/5, Poland,* 15 May, p. 9.

—— (1988) *Situation Report/5, Poland,* 11 April, pp. 23–5.

Reeves, R. (1986) 'Industrialization is burying Poland in pollution', *International Herald Tribune,* 21 May, p. 3.

Rich, V., (1985) 'Poland – government denies pollution', *Nature,* no. 317, 19 September, p. 195.

—— (1987) 'Polish government getting the message on environment', *Nature,* no. 326, 30 April, p. 819.

Rosenbladt, S. (1988) 'Is Poland lost?: pollution and politics in Eastern Europe', *Greenpeace* 13 (6), 14–18.

Salter, M. (1988) 'Poland's environmental crisis and the green activists', *Journal of European Nuclear Disarmament,* October, p. 7.

Schreiber, H. (1985) 'Problems of air pollution in the People's Republic of Poland', *Acid Rain,* June, pp. 1–10.

—— (1989) 'Eastern Europe I: advantage in modernizing', *Acid News,* no. 3, July, 4–5.

Simmons, M. (1988) 'Pollution threat to 12 million Poles', *The Guardian,* 1 December, p. 3.

Strzuszcz, Z. (1982) *Oddziaływanie przemysłu na środowisko glebowe i możliwości jego rekultywacji,* Kraków–Warszawa–Gdańsk–Lódź, PAN.

Survey of World Broadcasts (1983) *Eastern Europe* (weekly economic reports), EE/W1232/A/9 (14 April), p. 9.

Szafer, W. (1973) 'History of nature conservation in the world and in Poland', in W. Michajłow (ed.), *Protection of Man's Natural Environment,* Warszawa, PWN-Polish Scientific Publishers, pp. 7–51.

—— (1974) 'History of nature conservation in the world and in Poland', in W. Michajłow (ed.), *Protection of Man's Natural Environment,* Warszawa, PAN/PWN.

Timberlake, L. (1981) 'Poland – the most polluted country in the world?' *New Scientist* 92 (1276), 248–50.

United Nations (1985) *Pollution Across Boundaries,* Brussels, Economic Commission for Europe.

von Urban, K. (1989) 'Atomgegner demonstriernen in Lublin', *Süddeutsche Zeitung,* 13 January, p. 4.

Wachacki, B. (1989) 'Gdy zginie ostatnie drzewo . . .', *Kraków Magazyn Kulturalny,* no. 2 (Kraków), p. 19.

Zięba, A. (1986) 'Conditions in Poland: commentary', *Environment* 28 (4), 3–5.

Żmuda, S. (1967) 'Vliv hospodářské činnosti človeka na geografické prostředí na příkladu Hornoslezské Průmyslové Oblastí', *Teorie a metody výzkum Ostravské průmyslová oblast ve 20 století* (Slezský ústav (CSAV) Opava.

—— (1980) 'Environmental development barriers of the territory of the Katowice province', *Folia Geographica*, 13 (Krakow), pp. 115–228.

Żyskowski, H. (1988) 'Druga obrona Helu: "ogon" Rzeczpospolitej', *Rzeczpospolita*, no. 88 (2522), 13–16 April, p. 8.

7 ROMANIA

Anon (1991) *Construcţii hidroenergetice în România 1950–1990*, Bucharest, Hidroconstrucţia SA.

Apostol, L. (1985) 'Observaţii microclimatice în perimetrul platformei întreprinderii de celuloza hîrţie și cartoane din Piatra Neamţ', *Lucrările Seminarului Geografic 'Dimitrie Cantemir'* 6, 119–25.

—— (1987) 'Consideraţii privind rolul precipitaţiilor în depoluarea atmosferi într-un areal urban', *Studii şi Cercetări: Geografie* 34, 47–55.

Baicu, T. (1990) 'Cîteva probleme actuale ale influentei pesticidelor asupra mediului înconjurâtor în România', *Mediul Inconjurător* 1 (1), 43–7.

Bally, R.J. and Stănescu, P. (1977) *Alunecările şi stabilitatea versanţilor agricoli*, Bucharest, Editura Cereş.

Bălteanu, D. *et al.* (eds) (1985) *Cercetări geomorfologice pentru lucrările de îmbunătăţiri funciare*, Bucharest, Institutul de Geografie.

Bălauţă, L. *et al.* (1973) 'Unele aspecte ale poluării aerului in RSR', *Buletin SRG* 3, 228–32.

Banarescu, P. *et al.* (1979–80) 'Viitorul parc national Semenic-Cheile Caraşului', *Ocrotirea Naturii şi Mediului Inconjurător* 23–4, pp. 99–104, 127–33.

Banu, A.C. (1964) 'Situaţia reservaţiilor naturale din Delta Dunării', *Ocrotirea Naturii şi a Mediului Inconjurător* 8 (1), 103–6.

—— *et al.* (1965) *Delta Dunării*, Bucharest, Editura Stiinţifică.

Barnea, M. and Ursu, P. (1969) *Protecţia atmosferei impotriva impurificari cru pulberi şi gaze*, Bucharest, Editura Technică.

—— and Papadopol, C. (1975) *Poluarea si protecţia mediului*, Bucharest, Editura, Stiinţifică şi Enciclopedică.

Berindan, C. (1975) 'Poluarea industrială a mediului şi implicaţiile acesteia în valorificarea turistică a teritoriului', in L. Badea *et al.* (eds), *Lucrările celui de-al II-lea colocviu naţional de geografia turismului*, Bucharest, Editura Sport-Turism.

Bleahu, M. (1987) 'Turismul şi protecţia peisajului', *Ocrotirea Naturii şi a mediului Inconjurător* 31, 14–24.

—— (1990) 'Ceva despre ocrotirea naturü', *Eco* 1 (12), 2.

—— *et al.* (1977) *Reservaţii naturale geologice din România*, Bucharest, Editura Tehnică.

Bobirnac, B. *et al.* (1984) *Rezervaţii şi monumente ale naturii din Oltenia*, Bucharest, Editura Sport-Turism.

Bogdan, O. (1987) 'Influenţa fenomenelor de secetă şi exces de umiditate asupra evoluţiei peisajelor de cîmpie din România', *Studii şi Cercetări: Geografie* 34, 5–11.

Bold, I (1973) 'Aspecte ale punerii în valoare a terenurilor degradate prin exploatari miniere', *Buletin SRG* 3, 212–18.

Boşcaiu, N. (1975) 'Problemele conservării vegetaţiei alpine şi subalpine', *Ocrotirea Naturii şi a Mediului Inconjurător* 19, 17–21.

Buza, M. *et al.* (1980) 'Alunecarea masivă de teren de la Daia Romana: combatere şi mod de folosinţa' (and other case studies), *Buletin SRG* 6, 189–209.

Buzea, V. (1989) 'Amenajarea pădurilor ocolului silvic Mediaş în condiţiile poluării industriale', *Revista Pădurilor* 104, 73–8.

Calin, D. (1988) 'Observaţii geomorfologice în unele perimetre afectate de degradări de teren din Munţii Paring-Căpăţînii', *Terra* 20 (2), 38–42.

Ceauşescu, N. (1973) *România pe drumul construirii societăţii socialiste multilateral dezvoltate* (vol. 7), Bucharest, Editura Politică. (Also referred to in Institutul Central de Cercetări Economice (1983), *Progresul economic al României socialiste în concepţia preşedinteliu Nicolae Ceauşescu*, Bucharest, Editura Politică, pp. 288–92.)

Chiţu, M. (1973) 'Funcţiile oraşului Ploieşti şi poluarea atmosferei', *Studia Universitatis Babes-Bolyai: Geologia-Geografia* 22, 41–8.

Chiţu, T. (1990) In delta dunării înca se aplică planurile ceauşiste', *Eco* 1 (11), 3.

Cîneţi, A. (1990) 'Apele subterane ale Aradului în pericol', *Eco* 1 (10), 2.

Ciobanu, M. *et al.* (1972) *Monumentele naturii din judeţul Neamţ*, Piatra Neamţ, Consiliul Popular al Jud. Neamţ.

Ciplea, I.I. and Ciplea, A. (1978) *Poluarea mediului ambiant*, Bucharest, Editura Technică.

Constantinescu, F. (1990) 'Disputa Giurgiu-Ruse', *Eco* 1 (32), 2.

Crowther, W.E. (1988) *The Political Economy of Romanian Socialism*, London, Greenwood Press.

Cuşa, E. (1990) 'Aprecieri asupra calităţii apelor din Romania în anul 1989', *Mediul Inconjurător* 1 (1), 27–30.

Doniţa, N. *et al.* (1978) *Ecologie forestieră*, Bucharest, Editura Cereş.

Drimer, D. (1990) 'Unele probleme actuale ale învăţămîntului superior pentru protecţia mediului', *Mediul Inconjurător* 1 (1), 61–2.

Dumitrescu, S. (1990) 'SOS Rastoliţa', *Eco* 1 (9), 6.

Erdeli, G. *et al.* (1987) 'The role of air dynamics in the atmospheric pollution of the Baia Mare depression', *Revue Roumaine: Géographie*, 31, 75–80.

Erhan, E. (1972) 'Impurificarea cu pulberi a atmosferei în zona oraşului Iaşi', *Analele Universităţii A.I. Cuza din Iaşi* IIc 18, 163–71.

Fischer, M.E. (1989) *Nicolae Ceauşescu: A Study in Political Leadership*, London, Lynne Rienner.

Fisher, D. (1990) *Developments Within the Environmental Movement*, London, Ecological Studies.

Florea, M. and Cătălin, B. (1987) 'Observaţii geomorfologice si biogeografice în parcul naţional Retezat', *Terra* 19 (1), 35–8.

Gâştescu, P. (1989) 'La station de recherches géographiques de Pătîrlagele', *Revue Roumaine: Géographie* 33, 85–6.

Gilberg, T. (1990) *Nationalism and Communism in Romania: The Rise and Fall of Ceauşescu's Personal Dictatorship*, Boulder, Colo., Westview.

Gîrlea, D. (1977) 'Necesitatea ocrotirii zonelor montane', *Ocrotirea Naturii şi a Mediului Inconjurător* 21 (2), 132–4.

Giurescu, D.C. (1989) *The Razing of Romania's Past: International Preservation Report*, Washington, DC, World Monuments Fund: United States Committee, International Council on Monuments & Sites.

Giurgiu, V. (1988) *Amenjarea pădurilor cu funcţii multiple*, Bucharest, Editura Cereş.

—— (1990) 'Salvaţi pădurile', *Eco* 1 (7), 4–5.

Grasu, C. and Turculeţ, I (1980) 'Rezervaţia Lacul Roşu-Cheile Bicazului: particularitaţi geologice şi geomorfologice', *Ocrotirea Naturii şi a Mediului Inconjurător* 24, 135–45.

Gruescu, S.I. and Vasile, I. (1985) 'Geografia hidroenergeticii româneşti: realizări şi perspective', *Terra* 17 (2), 29–34.

Gugiuman, I. (1976) 'Recherches de climatologie urbaine a l'appui des actions de protection de l'atmosphère des grands villes', *Revue roumaine: géographie* 20, 151–6.

Hâncu, S. (1990a) 'Dacă nu cîştigăm războiul ecologic vom pieri', *Eco* 1 (6), 1–4.

—— (1990b) 'Calitatea mediului înconjurător în România: perspective de îmbunătăţire', *Mediul Inconjurător* 1 (1), 5–9.

Hopkins, L. (1991) *Conservation Status of the Danube Delta* Cambridge, IUCN-World Conservation Union (East European Programme).

Horeanu, C. and Borcea, M. (1982) 'Ceahlăul: viitor parc naţional', *Ocrotirea Naturii şi a Mediului Inconjurător* 25 (2), 20–33.

Iacob, G. (1985) 'Zona carboniferă din estul Olteniei Subcarpatice', *Lucrările Seminarului Geografic 'Dimitrie Cantemir'* 6, 197–201.

Ianculescu, M. (1977) 'Efectele poluării atmosferei asupra ecosistemelor forestiere şi masuri pentru protejarea lor', *Ocrotirea Naturii şi a Mediului Inconjurător* 21 (2), 123–6.

Ianovici, C. (1978) 'Protejarea mediului înconjurător în strategia dezvoltării economico-sociale în RSR', *Viitorul social* 2, 252–7.

Ianovici, V. and Popescu, M. (1977) 'General conception of industralization urbanization and environmental protection in Romania', in V. Ianovici *et al.*, *Seventh International Fellows Conference*, Bucharest, Johns Hopkins University Center for Metropolitan Studies & Research.

Ionescu, A. (1982) *Agricultura ecologică*, Bucharest, Editura Cereş.

—— (1988) 'Realizari în protecţia şi amenajarea apelor', in A. Ionescu and M. Berca (eds), *Ecologie şi protecţia ecosistemelor*, Bucharest, Institutul Agronomic Nicolae Bălcescu.

—— *et al.* (1989) *Protecţia mediului înconjurător şi educaţia ecologică*, Bucharest, Editura Cereş.

Ionescu, D. (1988) 'The "Romanian Democratic Action" Group on the environment', *Radio Free Europe Situation Report* 8, 41–5.

Iordan, I. (1987) 'Modificari ale peisajului agrogeografic prin lucrări de îmbunătăţiri funciare', *Terra* 19 (2), 7–13.

Maiorescu, T.G. (1990) 'Constiinţa ecologica plantară', *Eco* 1 (7), 1–2.

Mărăşanu, N.G. (1990) 'Copii Sucevei au dreptul la viaţa', *Eco* 1 (5), 4.

Mărcuţă, P. and Serban, R. (1990) 'Aspecte privind calitatea aerului şi precipiţatiilor în anul 1989 pe teritoriul României', *Mediul Inconjurător* 1 (1), 15–19.

Mihăilescu, I.F. (1984) 'Poluarea cu pulberi a atmosferei în zona oraşului Bicaz', *Buletin SRG* 7, 212–17.

Mihu, D. (1990) 'Projectarea obiectivă a reţelor de supraveghere a calităţii aerului', *Mediul Inconjurător* 1 (1), 21–5.

Modorcea, G. (1990) 'Casa a dezvăluiri dintre-o construcţia faraonică, *Eco* 1 (4), 5.

Muică, C. (1987) 'Observaţii comparative asupra vegetaţiei de pe terenuri afectate de pornituri din împrejurimile Pătîrlagelor', *Studii şi Cercetări: Geografie* 34, 40–6.

Muică, N. (1986) 'Observaţii privind degradarea terenurilor în împrejurimile Pătîrlagelor', in O. Bogdan and D. Bălteanu (eds), *Cercetări geografice asupra mediul înconjurător în judeţul Buzău*, Bucharest, Institutul de Geografie.

Muja, S. (1984) *Spaţiile verzi în sistematizarea teritoriului şi localităţilor*, Bucharest, Editura Cereş.

Negulescu, M. (1990) 'Ploile acide', *Mediul Inconjurător* 1 (1), 11–14.

Niculescu, G. (1987) 'Acţiunea proceselor geomorfologice asupra drumurilor carosabile din Carpaţii Meridionali', *Studii şi Cercetări: Geografie* 34, 32–9.

Nimigean, D. *et al.* (1982) *Monumente geologice*, Bucharest, Universitatea din Bucureşti.

Pecsi, K. (1989) 'The extremist path of economic development in Eastern Europe', *Communist Economics* 1, 97–109.

BIBLIOGRAPHY

Pîrvulescu, I. (1985) 'Repartiția concentrațiilor noxelor din aer în funcție de amplasamentul surselor de poluare și dinamica atmosferei', *Lucrările Seminarului Geografic 'Dimitrie Cantemir'* 6, 127–32.

——— (1986) 'Relații între condițiile fizico-geografice si factori sociali-economici implicați în procesul poluării aerului', *Lucrările Seminarului Geografic 'Dimtrie Cantemir'* 7, 93–8.

Pop, E. and Sălăgeanu, N. (1965) *Monumente ale naturii din România*, Bucharest, Editura Meridiane.

Popovici, I. (1987) 'Rolul parcului național Retezat în promovarea cercetărilor de ecologie terestră în România', *Ocrotirea Naturii și a Mediului Inconjurător* 31, 39–46.

Posea, G. *et al.* (1974) 'Geografia și mediul înconjurător', *Buletin SRG* 4, 1–326.

Posea, G. *et al.* (1978) *Probleme geografice fundamentale ale terrei*, Bucharest, Editura Didactică și Pedagogică.

——— and Pisoța, I (1972) 'Poluarea apelor și a atmosferei', *Buletin SRG* 2, 51–63.

Preda, I. (1990) 'Uciderea lacului Greaca', *Eco* 1 (14), 5.

Pușcariu, V. (1972) 'Păduri parcuri naționale și a rezervații naturale din România', *Ocrotirea Naturii și a Mediului Inconjurător* 16 (2), 151–66.

Pușcariu, V. (1975) '25 de ani de activitate a Comisiei Monumentelor Naturii', *Ocrotirea Naturii și a Mediului Inconjurător* 19, 93–104.

Pușcariu, V. (1981–2) 'Viitorul parc național al Munților Apuseni', *Ocrotirea Naturii și a Mediului Inconjurător* 25 (2), 162–78; 26 (1), 5–13.

Răuță, C. and Cârștea, S. (1990) 'Starea calității solurilor agricole din România la finele anului 1989: principalele orientari pentru conservarea și sporirea capacitații lor în condițiile protecției și ameliorarii calității mediului înconjurător', *Mediul Inconjurător* 1 (1), 39–42.

Roman, D.A. (1990) 'Delta acum', *Eco* 1 (2), 4–5.

Romanian Communist Party (1988) *Nicolae Ceaușescu Epoch: Historical Chronology*, Bucharest, Editura Stiințifică și Enciclopedică.

Roșu, A. (1987) *Terra-Geosistemul-Vieții*, Bucharest, Editura Stiințifică și Enciclopedică.

——— and Ungureanu, I. (1977) *Geografia mediului înconjurător*, Bucharest, Editura Didactică și Enciclopedică.

Sălăgeanu, G. (1978) *Rezervații monumente și frumusești ale naturii din jud. Constanța*, Constanța, Comitetul de Cultura și Educație Socialiste.

Sandru, I. *et al.* (1960) 'Schimbari în repartiția geografica a centrelor de populație din Valea Bistriței prin formarea lacului de acumulare al hidrocentrelei "V.I. Lenin"', *Analele Stiințifice ale Universității A.I. Cuza din Iași II* 6 (suppl.), pp. 479–87.

Sărățeanu, L. (1990) 'Techirghiolul îsi recuperează virtuțile', *Eco* 1 (9), 5.

Scutăreanu, P. (1979) 'Combatarea integrate: o cale ecologică de refacere a echilibrului biologic în ecosistemele forestiere', *Ocrotirea Naturii și a Mediului Inconjurător* 23, 111–17.

Seghedin, T.G. (1977) 'Parcul național al Muntilor Rodnei'. *Ocrotirea Naturii și a Mediului Inconjurător* 21 (1), 13–22.

Seghedin, T.G. (1983) *Rezervațiile naturale din Bucovina*, Bucharest, Editura Sport-Turism.

Sencu, V. (1968) 'Muntele de sare la Slănic-Prahova', *Ocrotirea Naturii și a Mediului Inconjurător* 12 (2), 167–79.

Șerban, P. and Simionaș, S. (1990) 'Influența antropică asupra Lacului Techirghiol', *Mediul Inconjurător* 1 (1), 31–8.

Sturgen, B. (1965) *Ecologie generală*, Bucharest, Editura Didactică și Enciclopedică.

Surdeanu, V. (1984) 'Part of landslides in the filling process of the Izvoru Muntelui reservoir', *Analele Stiințifice A.I. Cuza din Iași IIb* 30, 63–6.

Teodosie, N. (1974) 'Protecția naturii și a mediului înconjurător în preocuparile RSR',

Ocrotirea Naturii şi a Mediului Inconjurător 18 (1), 37–48.

Toader, T. and Nitu, C. (1976) *Invitaţie la drumetie: ghidul pădurilor*, Bucharest, Editura Cereş.

Traci, C. and Costin, E. (1966) *Terenurile degradate şi valorificarea lor pe cale forestieră*, Bucharest, Editura Agro-Silvică.

Trufaş, V. and Trufaş, C. (1978) 'Cîteva aspecte privind poluarea termică a rîurilor', *Studia Universitatis Babes-Bolyai: Geologia-Geografia* 27, 43–53.

Tufescu, V. (1974) *România: natura om economie*, Bucharest, Editura Stiinţificǎ.

—— *et al.* (1978) *Geografia mediului înconjurător*, Bucharest, Editura Didactică şi Pedagogică.

—— and Tufescu, M. (1981) *Ecologia şi activitatea umană*, Bucharest, Editura Albatros.

Turnock, D. (1977) 'Industrial development in Romania from the unification of the Principalities to the Second World War, in F.W. Carter (ed.), *An Historical Geography of the Balkans*, London, Academic Press, pp. 319–73.

—— (1979) 'Water management resources in Romania', *GeoJournal*, no. 3, 609–22.

—— (1982) 'Romanian geography reunited: the integrative approach demonstrated by the conservation movement', *GeoJournal*, no. 6, 419–31.

—— (1988) 'Woodland conservation: the emergence of rational land use policies in Romania', *GeoJournal*, no. 17, 59–70.

—— (1991) 'Romanian villages: rural planning under communism', *Rural History: Economy Society Culture* 2, 81–112.

Ujvári, I. (1962) 'Rîurile mici de munte din ţară noastră ca surse hidroenergetice', *Natura* 14 (5), 14–20.

Ungureanu, I. (1982) 'Quelques aspects géographiques de la protection de l'environnement dans le plateau moldave', *Analele Stiinţifice ale Universităţii A.I. Cuza din Iaşi IIb* 28, 91–6.

Untaru, E. *et al.* (1986) 'Valorificarea prin impădurire a terenurilor excesiv degradate prin eroziune şi alunecari în zona Vrancei', *Terra* 18 (2), 39–42.

Vădineanu, A. (1990) 'Consideraţii asupra semnificaţiei abordării holiste a problemelor poluării cu metale grele şi radioactive', *Mediul Inconjurător* 1 (1), 55–8.

—— (1991) 'Environmental status report 1990: Romania', in Nature Conservation Bureau (ed.), *Environmental status reports 1990: Albania, Bulgaria, Romania, Yugoslavia*, Cambridge, IUCN East European Programme.

Vespremeanu, E.E. (1981) *Mediul înconjurător: ocrotirea şi conservarea lui*, Bucharest, Editura Stiinţifică şi Enciclopedică, pp. 83–131.

Vinicius, P. (1990) 'Dosarul "Sulina": o pista falsa', *Eco* 1 (3), 4–5.

Voicu-Vedea, V. *et al.* (1986) 'Alunecari recente în judeţul Sibiu: posibilitaţi de valorificarea', *Terra* 18 (2), 34–7.

Zamfir, G. (1973) Poluarea mediului ambiant, Iaşi, Junimea.

Zăvoianu, I. *et al.* (eds) (1981) *Studii geografici cu elevi asupra calităţii mediului inconjurător*, Bucharest, Editura Didactică şi Pedagogică.

8 FORMER YUGOSLAVIA

Brooks, H. (1986) 'The typology of surprises in technology institutions and development', in W.C. Clark and R.E. Mann (eds), *Sustainable Development of the Biosphere*, London, Cambridge University Press, pp. 325–50.

Cerović, D. (1991) 'Filter ne spasava Zenicu', *Oslobodjenie*, 19 January.

Davis, G.R. (1990) 'Energy for planet earth', *Scientific American* 263 (1), 55–6.

Federal Executive Committee (1980) *Statistical Pocket Book of Yugoslavia*, Belgrade, Federal Statistic Office.

Holling, C.S. (1986) 'The resilience of terrestrial ecosystems: local surprise and global change', in W.C. Clark and R.E. Munn (eds), *Sustainable Development of the Biosphere*, London, Cambridge University Press, pp. 310–13.

Jancar, B. (1985) 'Environmental protection: "the tragedy of the republics"', in P. Ramet (ed.), *Yugoslavia in the 1980s*, Boulder, Colo., Westview Press.

—— (1987) *Environmental Management in the Soviet Union and Yugoslavia: Structure and Regulation in Communist One-Party Federal States*, Durham, N.C., Duke University Press.

OECD (1989) *Environmental Data Compendium 1989*, Paris, OECD.

Orr, D.W. and Soroos, M.S. (1979) *The Global Predicament*, Chapel Hill, University of North Carolina Press.

Pirages, D.C. and Ehrlich, P.R. (1974) *Ark II: Social Response to Environmental Imperatives*, San Francisco, W.H. Freeman.

Rosenau, J.N. (1990) *Turbulence in World Politics: A Theory of Change and Continuity*, Princeton, N.J., Princeton University Press.

Stanković, I. (1991) 'Vazduha uprkos politicsi', Borba, 28 May.

World Resources Institute (1990) *World Resources 1990–1991*, London, Oxford University Press.

9 ENVIRONMENTAL ISSUES IN THE NITRA VALLEY OF SLOVAKIA

Andrews, R.N.L. (1993) 'Environmental policy in the Czech and Slovak Republics', in A. Vari and P. Tamas (eds), *Environment and Democratic Transition: Policy and Politics in Central and Eastern Europe*, Dordrecht, Kluwer Academic Publishers, 5–48.

Baráth, J. (1992) 'The identification of the characteristics of the depopulation process in the villages of the Levice District', *Nitra Geographical Studies*, no. 1, 98–112. (In Slovak with an English summary.)

Drdoš, J. and Szekely, V. (1994) 'Environmental quality and the possibilities of environmental promotion (Upper Nitra Region)', *GeoJournal*, no. 32, 225–9.

Drdoš, J. *et al.*, (1994) 'The Upper Nitra region: man and his environment', *Geograficky Casopis*, no. 46, 131–48.

Drgoňa, J. *et al.*, (1993) 'Regional analysis of the air contamination in the Slovak Republic', *Nitra Geographical Studies*, no. 2, 97–114.

Dubcová, A. and Kramáreková, H. (1992) 'The background to the development of travel and recreation in the Prievidza district', *Geograficke Studie Nitra*, no. 1, 69–96. (In Slovak with an English summary.)

Hornik, S. (1993) 'A contribution to the vegetation–geographical characterization of selected biocentres in the region of "Jihlavske Vrchy" in the Czech Republic', *Katedra Geografie Pedagogicke Fakulty Masarykovy Univerzity: Geografie*, no. 4, 15–25.

Ivanička, K. (1961) 'Geography of industry of the Upper Nitra', Bratislava, Acta Geologica et Geografica Universitatis Comenianae. (In Slovak with an English summary.)

Jakál, J. (1993) 'Effect of sludge beds on the landscape in the region of Chalmova in the Upper Nitra', *Geografický Časopis*, no. 45, 67–79. (In Slovak with an English summary.)

Jordan, P. (1992) 'Slovakia in the scope of Central European tourism: present state and outlook', *Geografický Časopis*, no. 44, 105–19.

Krajčír, A. (1993) 'The distribution of selected diseases in the Upper Nitra Valley region in 1986–1990', *Geografický Časopis*, no. 45, 53–65. (In Slovak with an English summary.)

Kramáreková, A. and Dubcova, A. (1993) 'Some environmental aspects of dumping sites in the Prievidza district', *Nitra Geographical Studies*, no. 2, 115–35.

Kris, J. *et al.*, (1995) 'Mineral and geothermal water in Slovakia', *GeoJournal*, no. 35, 431–42.

Lehotsky, M. *et al.*, (1993) 'A regional system model: the case of Spis in Slovakia', *L'Espace Géographique*, no. 22, 125–31. (In French with an English summary.)

Olas, G. (1991) 'Some remarks on the growth of Nitra as a centre of the Middle Nitra Region', *Acta Facultatis Rerum Naturalium Universitatis Comenianae: Geographica*, no. 31, 167–74. (In Slovak with an English summary.)

—— (1992) 'The development of the settlement hierarchy of the Nitra District', *Nitra Geographical Studies*, no. 1, 36–50. (In Slovak with an English summary.)

—— (1993) 'Conservation of nature aimed at the preserve areas in the district of Nitra', *Nitra Geographical Studies*, no. 2, 137–50.

Otahel, J., Zigrai, F. and Drgoňa, V. (1993) 'Landscape use as a basis for environmental planning: case studies of the Bratislava and Nitra hinterlands', *Nitra Geographical Studies*, no. 2, 7–83.

Paulsen, C.M. (1993) 'Cost-effective control of water pollution in Central and Eastern Europe', *Resources for the Future*, no. 113, 28–31.

Plesník, P. (1988) 'Biogeography in Slovakia: inventory and prospect', *Acta Facultatis Rerum Naturalium Universitatis Comenianae: Geographica*, no. 29, 27–44.

Seko, L. (1988) 'Towards the functional structures of high mountains in the West Carpathians from the point of view of nature and landscape protection', *Acta Facultatis Rerum Naturalium Universitatis Comenianae: Geographica*, no. 29, 107–21.

Smith, A. (1994) 'Uneven development and the restructuring of the armaments industry in Slovakia', *Transactions of the Institute of British Geographers*, no. 19, 404–24.

Szöllös, J. (1993) 'Analysis of functional and spatial structure of soft coal energy chain in the Upper Nitra Valley', *Geografický Časopis*, no. 45, 29–40. (In Slovak with an English summary.)

—— (1994) 'Energy industry of Horna Nitra and its position in Slovak energetics', *Geografický Časopis*, no. 46, 159–71.

10 A REVIEW OF ENVIRONMENTAL ISSUES IN THE LIGHT OF THE TRANSITION

Ågren, C. (1994) 'Sulphur: the making of a protocol', *Acid News*, no. 1, 11–12.

Amon, A. (1994) 'NGO hits out at Hungary's energy and industry policies', *Aegis: European Environmental Protection Newsletter*, no. 1, London, 5.

Anděl, J. (1994) 'Regions of environmental burden in the Czech Republic – Methods of definition', *Acta Universitatis Carolinae: Geographica*, vol. 29 (1), Prague, 111–25.

Andersson, M. (1993) 'Hungary: change is not easy', *Acid News*, no. 1, 1–4.

—— (1994) 'East–Central Europe: motorways, more motorways', *Acid News*, no. 3, 9.

Andrews, R.N.L. (1993) 'Environmental policy in the Czech and Slovak Republics', in A. Vari and P. Tamas (eds), *Environment and Democratic Transition: Policy and Politics in Central and Eastern Europe*, Dordrecht, Kluwer Academic Publishers, 5–48.

Anko, B, (1994) 'Kočevsko Natural Park (Slovenia) – a forester's initiative', *IUCN Newsletter: East Europe*, no. 5 (10), Prague, 4.

Anon (a) (1994) 'Katowice Province: coal, steel and much to be done', *Privatization Update*, no. 24, Warsaw, 8.

Anon (b) (1993) 'Danube River Basin Environmental Programme', *Environments in Transition*, Winter, (EBRD), London, 10–12.

Anon (c) (1994) 'Battle on for eastward expansion of western nuclear power industry',

Aegis: European Environment Protection Newsletter, no. 1, London 1.

Anon (d) (1991) 'Eastern Europe's nuclear power: buying peace', *The Economist*, 24 July, 21–6.

Anon (e) (1994) 'European Bank reserves judgement on Mochovce', *Aegis: European Environmental Protection Newsletter*, no. 1, 2.

Anon (f) (1994) 'NGO hits out at Hungary's energy and industry policies', *Aegis: European Environment Protection Newsletter*, no. 1, London, 5.

Anon (g) (1994) 'Lebe wohl, Krsko', *Aegis: European Environment Protection Newsletter*, no. 1, London, 3.

Anon (h) (1994) 'Cernavoda facing further delays', *Aegis: European Environmental Protection Newsletter*, no. 1, London, 2.

Anon (i) (1995) 'Particulates: A health risk', *Acid News*, no. 1, 5.

Anon (j) (1994) 'French bid to cut Prague's pollution', *Aegis: European Environment Protection Newsletter*, no. 1, London, 16.

Anon (k) (1994) 'Slovak government environmental report', *Aegis: European Environment Protection Newsletter*, no. 1, London, 6.

Anon (l) (1994) 'Ukraine oil threatens Slovakian drinking water', *Aegis: European Environment Protection Newsletter*, no. 1, London, 15.

Anon (m) (1993) 'Environmental Protection Sector: a safe bet for a healthy future', *Privatization Update: Monthly Digest of Investment Opportunities*, no. 10, Warsaw, 2–3.

Anon (n) (1993) 'Results via the sector approach: squeaky clean and smelling of roses', *Privatization Update: Monthly Digest of Investment Opportunities*, no. 8, Warsaw, 2–3.

Anon (o) (1993) 'Norwegian–Polish Cooperation', *Privatization Update: Monthly Digest of Investment Opportunities*, no. 10, Warsaw, 7.

Anon (p) (1992) 'Improving Mountain Conservation in East Europe', *Newsletter*, (World Conservation Union/East European Programme), no. 5, Prague, 6.

Anon (q) (1993) 'Teure Sanierung einer Militärbasis in Böhmen: Nutzungskonzept mit Schwerpunkt Ökologie', *Neue Zuercher Zeitung*, 29 May, 5.

Anon (r) (1994) 'The same old story ? The World Bank, the Environment and Central and Eastern Europe', *Euronature Background*, Rheinbach and Bonn, 8.

Anon (s) (1994) *Manual on Public Participation in Environmental Decision-Making: Current Practice and Future Possibilities in Central and Eastern Europe*, Regional Center for Central and Eastern Europe, Budapest, Aqua Press.

Anon (t) (1993) 'East–West NGOs: cooperating for reconstruction', *Acid News*, no. 1, 12–15.

Anon (u) (1994) 'Bulgaria says no to atomic moratorium', *Aegis: European Environment Protection Newsletter*, no. 1, London, 5.

Anon (v) (1995) 'Hnutí Brontosaurus – Brontosaurus Movement, Czech Republic', *IUCN Newsletter: Central and Eastern Europe*, no. 1 (14), Prague, 11.

Anon (w) (1994) 'Fierce NGO opposition calls for open Mochovce process' *Aegis: European Environment Protection Newsletter*, no. 1, Rheinbach and Bonn, 4.

Anon (x) (1993) 'Bohemia, Slovakia accept existing international agreements', *Newsletter*, no. 9 (Republic of Hungary, Ministry of Foreign Affairs, Press Dept) Budapest, 2.

Anon (y) 1995) 'Slovakia: 'No need for nuclear', *Acid News*, no. 2, 7.

Anon (z) (1995) 'East to harmonize', *Acid News*, no. 2, 7.

Arden Pope, C. (1993) 'Effects of air pollution on six American cities', *Scientific American*, November, 12–24.

Arp, H. (1994) 'European Community policies on transport and the environment: an overview', *Working Group of Environmental Studies Newsletter*, no. 12, Florence, European University Institute, 14–18.

Bachmann, K. (1994) 'Invasion der Müllöfen in Polen', *Die Tageszeitung*, 15 February, Berlin, 5.

Baret, J.P. (1995) 'Electricité de France et la centrale nucléaire de Kozloduy: histoire d'un partenariat réussi dans la durée', *Les Enjeux de l'Europe*, no. 15, Paris, 71–4.

Barrett, M. and Protheroe, R. (1994) *Sulphur emission from large point sources in Europe*, Swedish NGO Secretariat on Acid Rain, Gothenburg.

Bassa, L. (1994) 'What is next for the Hungarian environment?' in P. Jordan and E. Tomasi, (eds), *Zustand und Perspektiven der Umwelt im östlichen Europa*, Frankfurt am Main, P. Lang, 113–20.

Baumgartl, B. (1993a) 'Environmental protest as a vehicle for transition: the case of Ekoglasnost in Bulgaria', in A. Vari and P. Tamas (eds) *Environment and Democratic Transition: Policy and Politics in Central and Eastern Europes* Dordrecht, Kluwer Academic Publishers, 157–78.

––––– (1993b) 'West provides no new aid to clean up Eastern Europe', *RFE/RL Research Report*, vol. 2, no. 29, 41–7.

Baumgartl, B. and Stadler, A. (1993) 'East–West relations in change – some implications of the western aid regime for the political economy and ecology in Poland and Bulgaria 1989–1992', *Working Paper* (EPU no. 93/6), European Policy Unit, European University Institute, Florence, 65.

Bedi, E. (1994) *Possibilities for Combined Heat and Power Production in Slovakia*, Foundation for Alternative Energy, Bratislava, 30.

Bennetts, D.A. (1995) 'Modelling climate change 1860–2050', *The Globe*, no. 24, Swindon, 2–4.

Bialic, A. (1994) 'Waste management association: hot focus on a burning issue', *The Warsaw Voice*, December, G4.

Björklund, S. (1993) *The Black Triangle – A General Reader*, Air Pollution and Climate Series, no. 2, Swedish NGO Secretariat, Gothenburg, 25.

Bohn, P. (1992) 'Magyarország környezeti állapota', *Öko*, vol. 2–3, Budapest, 96–118.

Bond, M. (1993) 'Czech power cleans up its act' *The European*, 28–31 January, London, 3.

Borisova, P. (1992) 'Orbeluş, Perun, Pirin … Vechen', *EKO Sreda*, no. 2, Sofia, 18–19.

Borys, T. and Mazurski, K.R. (1993) *Air Pollution in the Border Areas of the Euroregion 'Neisse'*, Sudety, Wrocław, 24.

Boyd, J. (1993) 'The allocation of environmental liabilities in Central and Eastern Europe', *Resources for the Future* 112, 1–6.

Boyle, S. (1995) 'Eastern Europe: the need for energy efficiency', *Acid News*, no. 3, 9–11.

Breymeyer, A. (ed.) (1994) *Rezerwaty biosfery w Polsce*, (*PAN*), Warsaw, 156.

Burtraw, D. (1993) 'Tradable sulphur dioxide emission permits and European economic integration', *Resources for the Future* 113, 23–7.

Caddy, J. (1995) 'Public participation in environmental policy making: the case of Central and Eastern Europe', *Working Group of Environmental Studies Newsletter*, no. 13, European University Institute, Florence, 17–20.

Carter, F.W. (1992) 'Geographical aspects of East–West environmental policy', in M. Jachtenfuchs and M. Strubel (eds), *Environmental Policy in Europe*, Baden Baden, Nomos Verlagsgesellschaft, 177–96.

Černy, M. (1994) 'Monitoring of forests in the Czech Republic', in J. Solon, E. Roo-Zielińska and A. Bytnerowicz (eds), *Climate and Atmospheric Deposition Studies in Forests*, Conference Papers, no. 19, IGSO PAS, Warsaw, 87–100.

Christophorov, B. (1994) 'La santé en Bulgarie', *Les Enjeux de l'Europe*, no. 15, Paris, 104–5.

Coll, S. (1994) 'Free Market intensifies waste problem', *Washington Post*, 31 March, 7.

Crockford, R. (1993) 'Darkness at noon shrouds north', *Prague Post*, 10–16 November, 7–8.

Dąbrowski, P. (1995) 'New national park in Polish Carpathians', *IUCN Newsletter: Central and Eastern Europe*, no. 1 (14), Prague, 9.

Damohorský, M. (1994) 'Český svaz ochránců přírody- ČSOP (Czech Union of Nature Conservation)', *IUCN Newsletter: East Europe*, no. 6 (11), Prague, 9.

Danchev, J. 'Natural Resources of the Bulgarian Mountains – The Ecological Problems of the Rhodope Mountains' in *Umwelt in Osteuropo*, (Akademie für Natur- und Umweltschutz, Baden-Würtemberg) Stuttgart, 1993, 125–31.

DeBardeleben, J. and Hannigan, J. (eds) (1994) *Environmental Security and Quality after Communism: Eastern Europe and the Soviet Successor States*, Boulder–San Francisco–Oxford, Westview Press, 188.

Degórska, B. (1993) 'Problematyka ekologiczna wschodniego pogranicza Polski', in P. Eberhardt and T. Komornicki (eds), *Problematyka Wschodniego Obszaru Pogranicza*, PAN IGPZ, Warsaw, 51–73.

Denisiuk, Z. (1994) 'Poland: country profile', *IUCN Newsletter: East Europe*, no. 7 (12), Prague, 3–5.

Dinu, A., Romanca, G. and Vadineanu, A. (1994) 'Romania: country profile', *IUCN Newsletter: East Europe*, no. 8 (13), Prague, 6–8.

Dmuchowski, W. and Wawrzoniak, J. (1994) 'Spatial distribution of sulphur and nitrogen content in needles of Scots Pine (Pinus sylvestris L.) as related to air pollution and tree stands vitality in Poland', in J. Solon, E. Roo-Zielińska and A. Bytnerowicz (eds), *Climate and Atmospheric Deposition Studies in Forests*, Conference Papers, no. 19, IGSO PAS, Warsaw, 177–86.

Drda, A. (1994) 'Kam s jaderným odpadem?', *Český deník*, 22 February, 3.

Drew, K. (1994) 'Dam disasters revive tensions with Hungary', *Prague Post*, 30 March–5 April, 6.

Duma, (1994) 5 April, Sofia, 1.

Dzúrová, D. (1992) 'Relationship between environment quality and mortality in Czechoslovakia', *Acta Universitatis Carolinae: Geographica* 27 (1), Prague, 91–103.

Edeburn, M. (1993) 'Meeting the Pollution Control Challenge in Central and Eastern Europe', *Resources*, no. 113 (Resources for the Future), 1.

Eidsvik, H. and Deželič, R. (1993) 'Visit of the UNESCO mission to the Plitvice Lakes National Park', *IUCN Newsletter: East Europe* 1 (6), Prague, 5.

Elvingson, P. (1994) 'Europe's forests: still no improvement', *Acid News*, no. 3, 8–9.

Enache, L. (1994) 'The quality of environment in Romania', in P. Jordan and E. Tomasi (eds), *Zustand und Perspektiven der Umwelt im östlichen Europa*, Frankfurt am Main, P. Lang, 131–57.

European Bank for Reconstruction and Development (EBRD) (1994) *Strategy for Bulgaria*, Document BDS/BU/93-1 (Final), 26.

Fisher, D. (1992) *Paradise Deferred: Environmental Policy Making in Central and Eastern Europe*, London, RIIA.

Fisher, S. (1993) 'The Gabčikovo–Nagymáros Dam Controversy Continues', *RFE/RL Research Report* 2 (37), 7–12.

Fitzmaurice, J. (1995) *Damming the Danube: Gabčikovo/Nagymáros and Post-Communist Politics in Europe*, Boulder, Col., Westview Press.

Fontaine, P. (1994) *Europe in Ten Lessons*, (European Documentation), Luxembourg, 46pp.

Fronczak, K. (1995) 'Forest "Quality management": in every ranger lurks an auditor', *The Warsaw Voice*, 5 March, G4.

Galambos, J. (1993) 'An international environmental conflict on the Danube: the Gabčikovo–Nagymáros dams', in A. Vari and P. Tamas (eds), *Environment and Democratic Transition: Policy and Politics in Central and Eastern Europe*, Dordrecht, Kluwer Academic Publishers, 176–226.

Gastescu, P. (1993) 'The Danube delta: geographic characteristics and ecological recovery', *GeoJournal*, no. 29, 57–67.

Genov, N. (1993) 'Environmental risks in a society in transition: perceptions and reactions', in A. Vari and P. Tamas (eds), *Environment and Democratic Transition: Policy and Politics in Central and Eastern Europe*, Dordrecht, Kluwer Academic Publishers, 368–80.

Georgieva, K. (1993) 'Environmental policy in a transition economy: the Bulgarian example' in A. Vari and P. Tamas (eds), *Environment and Democratic Transition: Policy and Politics in Central and Eastern Europe*, Dordrecht, Kluwer Academic Publishers, 67–87.

Gergov, D. (1991) 'The use and protection of water resources in Bulgaria', in J. DeBardeleben (ed.), *To Breathe Free: Eastern Europe's Environmental Crisis* Baltimore, Md., Johns Hopkins University Press, 159–73.

Gergov, G. (1994) 'Classification of the river network in Bulgaria', in P. Jordan and E. Tomasi (eds), *Zustand und Perspektiven der Umwelt im östlichen Europa*, Frankfurt am Main, P. Lang, 187–200.

Gliński, P. (1992) 'Environmentalism in Poland', in P. Clancy, M. Kelly, J. Wiatr and R. Zoltaniecki (eds), *Ireland and Poland: Comparative Perspectives*, Dublin, pp. 263–74.

—— (1994) 'Environmentalism among Polish youth: a maturing social movement?', *Communist and Post-Communist Societies* 27, 145–59.

Godalovič, P. (1993) 'Zdokonalovat legislativní normy', *Zpravodaj: Ministerstva životního prostředí ČR*, no. 7, Prague, 3.

Government of the Czech Republic (1992a) *Decree*, no. 315/1992Sb, 29 April.

—— (1992b) *Decree*, no. 218/1992Sb, March.

—— (1992c) *Decree*, no. 114/1992Sb, February.

Government of Hungary (1990) *Decree*, no. 43/1990, February.

Government of Poland (1991) *Decree*, passed 16 October.

Grodzińska, K., Szarek, G., Godzik, B., Braniewski, S. and Chrzanowska, E. (1994) 'Mapping air pollution in Poland by measuring heavy metal concentration in mosses', in J. Solon, E. Roo-Zielińska and A. Bytnerowicz (eds), *Climate and Atmospheric Deposition Studies in Forests*, Conference papers, 19, Warsaw, IGSO PAS, 197–209.

Hiltunen, H. (1994) *Finland and Environmental Problems in Russia and Estonia*, Helsinki, Finnish Institute of International Affairs, 15–16.

Hlawiczka, S. (1995) 'Poland: Positive trends now appearing', *Acid News*, no. 3, 12–13.

Jancar, B. (1992) *Chaos as an Explanation of the Role of Environmental Groups in East European Politics*, Edinburgh, University of Edinburgh, 1992.

Jancar-Webster, B. (1995) 'Environmental Degradation and Regional Instability in Central Europe', in J. DeBardeleben and J. Hanningan (eds), *Environmental Security and Quality after Communism*, Boulder–San Francisco–Oxford, Westview Press, 43–68.

Janecki, J. (1995) 'Liga Ochrony Przyrody: The League for Conservation of Nature in Poland', *IUCN Newsletter: Central and Eastern Europe*, 2 (15), Prague, 6.

Jeřabek, M. (1994) 'Evolution of cultural landscape in the Northern Bohemian coal mining region on the background of socio-economic transformations', *GeoJournal* 32, 215–9.

Jordan, P. and Tomasi, E. (1994) 'Environmental management, education and policy in Romania', in P. Jordan and E. Tomasi (eds), *Zustand und Perspektiven der Umwelt im östlichen Europa*, Frankfurt am Main, P. Lang, 159–69.

Juhasz, J. (1993) 'Environmental conflict and political change: public perception on low level radioactive waste management in Hungary', in A. Vari and P. Tamas (eds), *Environment and Democratic Transition: Policy and Politics in Central and Eastern Europe*, Dordrecht, Kluwer Academic Publishers, 227–47.

Jurkiewicz, J. (1994) 'Time to repair environmental damage', *The Coastal Times* 7, (13), Sopot 5.

Kabala, S.J. (1992) 'EC helps Czechoslovakia pay debt to the environment', *RFE/RL, Research Report* 1 (20), 15 May, 56–7.

—— (1993) 'Environmental affairs and the emergence of pluralism in Poland: a case of political symbiosis', in A. Vari and P. Tamas (eds), *Environment and Democratic Transition: Policy and Politics in Central and Eastern Europe*, Dordrecht, Kluwer Academic Publishers, 40–66.

Kalvoda, J. (1992) 'Geomorphological hazards and risks of selected nuclear power plants in the ČSFR', *Acta Universitatis Carolinae: Geographica* 27 (1), Prague, 111–13.

Kiszel, V. (1993) 'Göncöl Alapitvány: (Göncöl Foundation)', *IUCN Newsletter: East Europe* 4 (9), 9.

Knight, G.C. (1995) 'The emerging water crisis in Bulgaria', *GeoJournal* 35, 415–23.

Knox, K. (1994) '"Green" minister may be ousted', *Prague Post*, 30 March–5 April, Prague, 3.

Kopačka, L. (1994) 'The Transition of the Czech Industry and its Energetic and Ecological Consequences', *Acta Universitatis Carolinae, Geographica* 29 (1), Prague, 81–98.

Kozłowski, J. and Baranowska-Janota, M. (1993) *Integrating ecological thinking into planning revisited*, FS II 93–402 (Wissenschafts zentrum Berlin für Sozialforschung), Berlin, 33.

Kramer, J.M. (1995) 'Energy and the Environment in Eastern Europe', in J. DeBardeleben and J. Hannigan (eds), *Environmental Security and Quality after Communism*, Boulder–San Francisco–Oxford, Westview Press, 89–104.

Krupnik, A.J., Harrison, K.W., Nickell, E.J. and Toman, M.A. (1993) 'Assessing the health benefits of improved air quality in Central and Eastern Europe', *Resources for the Future* 113, 7–11.

Kučera, B. (1995) 'Change in Czech nature conservation organization' *IUCN Newsletter: Central and Eastern Europe* 3 (16), 11.

Lawson, M.L. (1994) 'Russian aid offered for Slovak reactors', *Prague Post*, 11–17 May, 5.

Lakos, L. (1993) 'Ministry for Environment and Regional Policy (Hungary)', *IUCN Newsletter: East Europe* 1 (6), 8.

LeBor, A. (1995) 'Fears over nuclear plant in Slovakia', *The Times*, 27 March, London, 4.

Lehotský, M. *et al.* (1989) 'Evaluation of motorway impact on the landscape: an environmental approach', *Geografický Časopis* 41, 71–92. (In Slovak with an English summary.)

Lipietz, A. (1995) *Green Hopes: the future of political ecology*, London, Polity Press, 160.

Marsh, A. (1994) 'Slovak Dam Project remains a source of international tension', *Prague Post*, 4–10 May, 5.

Mazurski, K.R. (1993) 'Zur aktuellen Situation der Landwirtschaft Oberschlesiens', *Oberschlesisches Jahrbuch*, vol. 9, 163–79.

—— (1994) 'Warunki ekologiczne życia a umieralność w Polsce-ilościowa próba analizy współzależności', *Prace*, no. 65 (Karkonoskie Towarzystwo Naukowe), 88–97.

—— (1995a) 'Die Transformation der Wirtshafts- und Gesellschaftsordnung Polens', *Geographie und Schule*, Vol. 17, Cologne, 20–30.

—— (1995b) 'Nasilenie antropopresji na obszarach zagrożenia ekologicznego w Polsce', *Przegląd Geograficzny* 67 (1–2), 91–9.

Mering, K. (1994) 'Hazardous materials: like nailing mercury to a wall', *The Warsaw Voice*, 4 December, G2.

Merritt, G. (1991) 'Cleaning Eastern Europe's sugean stables', *Eastern Europe and the USSR: The Challenge of Freedom*, London, Routledge and Kegan Paul, 189–207.

280

Mezósi, G. and Mucsi, L. (1993) 'Critical environmental areas in Hungary (a GIS based approach)', *Acta Geographica Szegediensis*, Vol. 31, 99–107.

Mikhova, D. and Pickles, J. (1994) 'Environmental data in Bulgaria: problems and prospects', *Professional Geographer* 46, 228–35.

Mladá fronta Dnes (1995), 21 June, Prague, 3.

Morka, A. (1995) 'Nafta do naprawy', *Gazeta Wyborcza: dodatek 'Polska Nafta'*, 22 March, 1.

Mundy, J. (1993) *Land Redistribution and Nature Conservation in Selected Countries of Central and Eastern Europe*, Discussion Papers, no. 1, (East European Programme), IUCN, Cambridge.

Mussapi, R. and Thomas, S. (1994) 'Improving safety in nuclear power generation in the former Soviet Union and Eastern Europe:the effectiveness of current initiatives', *Post Soviet Business Forum Briefing*, no. 2, London RIIA (Royal Inst. Int. Affairs) IIA.

Nefedova, T. *et al.* (1994) *Atlas of Eastern and South Eastern Europe*, Vienna, Österreichisches Ost- und Südosteuropa Institut, 1992.

Nefedova, T. (1994) 'Industrial Development and the Environment in Central and Eastern Europe', *European Urban and Regional Studies* 1 (2), 168–71.

Niehörster, K. (1993) 'Umwelttechnik: Geld mitbringen Beim Aufbau einer Umweltinfrastruktur ist Polen auf westliche Hilfe angewiesen', *Wirtschaftswoche*, 29 January, Düsseldorf, 5.

Nikolova, I., Stoinev, I. and Mihov, P. (eds) (1994) *Rodopi*, Sofia, WWF 44.

Nikolova, M. and Rukova, P. (1994) 'Zdravoslovno s'stojanie na naselenieto v Burgaskija rajon', *Geoekologija*, Sofia, B'lgarsko geografsko druzhestvo. 148–53.

Nilsson, M. (1995) 'Air Quality: new standards in the offing', *Acid News*, no. 2, 1–4.

Okołów, C. (1993a) 'The new national parks in Poland', *IUCN Newsletter: East Europe* 4 (9), Prague, 4.

—— (1993b) '"Green Lungs of Poland" internationally', *IUCN Newsletter: East Europe*, 2 (7), Prague, 7.

Okólski, M. (1992) 'Anomalies in demographic transition in Poland', *Geographia Polonica* 59, 41–53.

Oszlányi, J. (1994) 'Loss of needles in relation to the biosociological position of trees in spruce ecosystems', in J. Solon, E. Roo-Zielińska and A. Byłnerowiez (eds) *Climate and Atmospheric Deposition Studies in Forests*, Conference Papers, no. 19, Warsaw, IGSO PAS, 261–4.

Pape, R. (1993a) 'The Black Triangle: Upper Silesia, not much change' *Acid News*, no. 4, 13.

—— (1993b) 'Group of 24: aid to East inventoried', *Acid News*, no. 3, 21–3.

Pape, R. (1994) 'Phare Program: Black Triangle still unaided' *Acid News*, no. 2, 12.

Paul, W. (1994) 'Die Elbe – ein europäischer Strom', *Sudetenjarhbuch der Seliger-Gemeinde*, vol. 43, 89–94.

Paulsen, C.M. (1993) 'Cost-effective control of water pollution in Central and Eastern Europe', *Resources for the Future* 113, 28–9.

Pavlínek, P., Pickles, J. and Staddon, C. (1994) 'Democratization, Economic Restructuring and the Environment in Bulgaria and the Czech Republic', in P. Jordan and E. Tomasi (eds), *Zustand und Perspektiven der Umwelt im östlichen Europa*, Frankfurt am Main, P. Lang, 57–82.

Pavlovec, R. (1994) *Alternative Ways of Meeting Power Needs in the Slovak Republic in the Light of a Withdrawal from Nuclear Energy*, Vienna, Global 200 Research Institute, 25.

Pearce, F. (1995) 'Trouble waters', *Business Europa* 12, 24–8.

Persanyi, M. (1990) 'The rural environment in a post-socialist economy: the case of Hungary', in P. Lowe *et al.* (eds), *Technological Change and the Rural Environment*, London, Fulton, 33–52.

——— (1993) 'Danube Dams of Democracy', in B. Jancar-Webster (ed.), *Environmental Action in Eastern Europe: Responses to Crisis*, New York, Armonk, 134–57.

Piños, J. (1993) 'Czech Republic: plight of Libkovice', *Acid News*, no. 4, 15.

Piotrowska, H. (1993) 'Odpady', in R. Andrzejewski and M. Baranowski (eds), *Stan środowiska w Polsce*, Warsaw, PIOŚ, 84–8.

Plamínková, J. (1995) 'North Bohemia: after five years', *Acid News*, no. 1, 1–5.

Poborski, P.S. (1993) *Air Pollution in Upper Silesia*, Katowice, Information Centre for Air Protection, 54.

Polska Agencja Prasowa, (1990) 4 September, Warsaw, 2.

Pozes, M. (1991) 'Razvoj podeželskih naselj v občini Koper' *Geographica Slovenica* 22, 1–114.

Pravda (1995), 21 June, Bratislava, 1.

Przybylak, Z. (1994) 'Organic farming: conquering the chemical craze', *The Warsaw Voice*, 6 November, G2.

Pudlis, E. (1994) 'The market can support organics', *The Warsaw Voice*, 6 November, G1–G3.

——— (1995a) 'Cleaner production programme: breath of fresh air from Norway', *The Warsaw Voice*, 8 January, G3.

——— (1995b) 'Big can also be beautiful', *The Warsaw Voice*, 5 February, G1–G2.

——— (1995c) 'Ekofundusz: debt-swapping flies', *The Warsaw Voice*, 14 May, 10.

——— (1995d) 'From opposition through professionalism to politics?', *The Warsaw Voice*, 7 May, G1–G2, 134.

——— (1995e) 'Polish Ecological Club: No longer under the table', *The Warsaw Voice*, 7 May, G3.

Quitt, E. (1989) 'Onesnaževje in onesnaženost ozračja v ČSR', *Geographical Slovenica* 20, 199–209.

Ratajczyk, A. (1994) 'Black Triangle cleanup: power plant modernization okayed', *The Warsaw Voice*, 20 November, B3.

Redclift, M. (1989) 'Turning nightmares into dreams: the Green Movement in Eastern Europe', *The Ecologist* 19, 177–83.

Reeves, A.D. (1993) 'The present state of the Hungarian environment', *Area* 25, 141–3.

Retvari, L. (1994) 'Natural resources and the environment: problems of their utilization in Hungary', *GeoJournal*, no. 32, 337–42.

Ridjanović, J. (1993) 'Neue Ergebnisse von Regionalforschungen und Umweltunter-suchungen im Jugoslawischen Adriagebiet', *Annales Universitatis Scientiarum Budapestinensis de Rolando Eötvös Nominatae: Sectio Geographica*, vol. 22–3, 85–98.

Rocks, D. (1994) 'Attempting to clean up another fine Soviet mess', *Toronto Globe and Mail*, 16 February, 4.

Romanian Press Service (1993), 28 January, Bucharest, 6.

Rubin, A. and Kaspar, M. (1993) *The Environmental Relations between the EC and the Countries of Central and Eastern Europe: An Overview of Policies, Programmes and Economic Cooperation*, Rheinbach and Bonn, Euronatur, 30.

Russell, J. (1991a) *Energy and Environmental Conflicts in East–Central Europe: The Case of Power Generation*, London, RIIA.

——— (1991b) *Environmental Issues in Eastern Europe: Setting an Agenda*, London, RIIA.

Rychtaříková, J. (1994) 'Contemporary mortality problems in Czechoslovakia', *Acta Universitatis Carolinae: Geographica*, 27 (1), Prague, 69–89.

Rykowski, K. (1993) 'Lasy', in R. Andrzejewski and M. Baranowski (eds), *Stan środowiska w Polsce*, Warsaw, PIOS, 50–2.

Šaler, A. (1993) 'The role of geosciences, especially geomorphology, in the site selection of radioactive waste disposal in S.R. Croatia', *Annales Universitatis Scientiarum Budapestinensis de Rolando Eötvös Nominatae: Sectio Geographica*, vol. 22–3, 303–10.

Schreiber, H. (1989) 'Debt-for-Nature Swap – an instrument against debt and environmental destruction', *Zeitschrift für Umweltpolitik und Umweltrecht*, Vol. 12, 331–52.

Skoberne, P. (1995) 'Slovenia: country profile', *IUCN Newsletter: Central and Eastern Europe* 3 (16), 7–9.

Sleszyński, J. (1992) 'New ideas in the Polish environmental policy; transferable permits as an incentive', *Working Group of Environmental Studies Newsletter*, no. 7, Florence, European University Institute, 11–12.

Špes, M. (1994) 'The state of the environment in the new state of Slovenia. A Geographical Survey', in P. Jordan and E. Tomasi (eds), *Zustand und Perspektiven der Umwelt im östlichen Europa*, Frankfurt am Main, P. Lang, 121–9.

Stachowiak, T. (1994) 'Furniture companies have large capacities', *Poznań Fair Magazine* 6 (125), 26.

Staniewska-Zatek, W. (1993) 'Ecological farming in the curricula of agricultural schools in Poland', *IUCN Newsletter: East Europe* 2 (7), Prague 7.

Stec, S. (1993) 'Public participation laws, regulations and practices in seven countries in Central and Eastern Europe: an analysis emphasising impacts on the development decision-making process', in A. Vari and P. Tamas (eds), *Environment and Democratic Transition: Policy and Politics in Central and Eastern Europe*, Dordrecht, Kluwer Academic Publishers, 88–119.

Steinhauserová, K. (1994) 'Boj o Labe', *Magazín Mladé fronty dnes*, 6 January, Prague, 10–11.

Šulc, P. (1994) 'Quality of surface water in the Jizera basin', *Acta Universitatis Carolinae: Geographica* 29 (2), Prague, 133–47.

Świecka, A., Dukaczewski, P. and Skwarcan, D. (1995) 'Population Trends: dampened ardor', *The Warsaw Voice*, 26 March, 18–19.

Synge, H. (1994) 'New plan for Europe's protected areas', *IUCN Newsletter: East Europe* 8 (13), Prague, 2.

Szalay-Marszó, E. (1993) 'Hungary: country profile', *IUCN Newsletter: East Europe* 2 (7), Prague, 4–6.

Szépeszi, A. (1994) 'Forest Damage Surveys in Hungary', in J. Solon, E. Roo-Zielińska and A. Bytnerowicz, (eds.), *Climate and Atmospheric Deposition Studies in Forests*, Conference Papers no. 19, Warsaw, IGSO PAS, 65–72.

Szirmai, V. (1993) 'The structural mechanisms of the organisation of ecological–social movements in Hungary', in A. Vari and P. Tamas (eds), *Environment and Democratic Transition: Policy and Politics in Central and Eastern Europe*, Dordrecht, Kluwer Academic Publishers, 146–56.

Templin, B. (1995) 'Turoszów power plant: kicking the habit', *The Warsaw Voice*, 5 February, G4.

Toman, M.A. (1993) 'Using economic incentives to reduce air pollution emissions in Central and Eastern Europe: the case of Poland', *Resources for the Future* 113, 2–7.

Trud (1995), 19 June, Sofia, 13–14.

Tuček, J. (1994a) 'Ekologické zlepšení je možné, ale drahé', *Mladá fronta Dnes*, 15 April, Prague, 2.

—— (1994b) 'Stát má odpovědnost za životní prostředí, tvrdí ministr Benda', *Mladá fronta dnes*, 27 August, Prague, 7.

Urban, F. (1993) 'Czech Republic: country profile', *IUCN Newsletter: East Europe* 1 (6), 6–7.

Varga, P. *et al.* (1990) 'Water quality of the Danube in Hungary and its major determining factors', *Water Science Technology* 22, 113–8.

Vavroušek, J. (ed.) (1992) *State of the Environment in Czechoslovakia*, Prague, Vesmir.

—— (1994) 'Environmental management in Czechoslovakia and succession states', *Environmental Impact Assessment* 14, 2–3.

Verseck, K. and Herre, S. (1994) 'Der Donau bleibt das Wasser weg', *Wirtschaft und Umwelt*, 1 February, 4.

Vološčuk, I. (1995) 'Slovakia: country profile', *IUCN Newsletter: Central and Eastern Europe* 2 (15), Prague, 4–5.

Wałdoch, A. (1992) 'Green House Effect: horror story or anomaly?', *The Warsaw Voice*, 23 August, 5.

Walls, M.A. (1993) 'Motor vehicles and pollution in Central and Eastern Europe', *Resources*, no. 113, Washington D.C., Resources for the Future, 2–7.

Wątróbski, L. (1994) 'Czysta woda to życie', *Dziennik Polski*, 3 May, London, 6.

Wedmore, L.D. (1994) 'Czech nuclear power plant controversy', *RFE/RL Research Report* 3 (15), 27–32.

Wiska, A. and Hindson, J. (1991) 'Protecting a Polish paradise', *Geographical Magazine*, 63 (6), 1–2.

Witkowski, J. (1993) 'The quality of the natural environment and demographic processes in the large towns of Poland', *Geographia Polonica* 61, 367–77.

Wysienska, M. (1993) 'The industry of the Gdańsk agglomeration and the environmental hazards of the region', in A. Duro (ed.), *Spatial Research and the Social–Political Changes*, Pecs, Centre for Regional Studies, 59–66.

Yeager, P.C. (1991) *The Limits of Law: The Public Regulation of Private Pollution*, Cambridge, Cambridge University Press.

Žák, L. (1994) 'Krkonoše: Soumrak nad lesy. Stan se naše nejvyšší hory pustinou?', *Mladý svět*, 7 January, Prague, 16–17.

Zupančić, I.Ž. (1994) 'Značilnosti mestnega podnebja', *Geografski Obzornik*, 41 (4), Ljubljana, 20–1.

INDEX

The numbers in italics refer to maps or tables, where these are separate from their textual reference.